Reading Human Nature

Reading Human Nature

Literary Darwinism in Theory and Practice

Joseph Carroll

Published by State University of New York Press, Albany

© 2011 State University of New York

All rights reserved

Printed in the United States of America

No part of this book may be used or reproduced in any manner whatsoever without written permission. No part of this book may be stored in a retrieval system or transmitted in any form or by any means including electronic, electrostatic, magnetic tape, mechanical, photocopying, recording, or otherwise without the prior permission in writing of the publisher.

For information, contact State University of New York Press, Albany, NY
www.sunypress.edu

Production by Ryan Morris

Marketing by Michael Campochiaro

Library of Congress Cataloging-in-Publication Data
Carroll, Joseph, 1949-
  Reading human nature : literary Darwinism in theory and practice / Joseph Carroll.
       p.   cm.
  Includes bibliographical references and index.
  ISBN 978-1-4384-3523-7 (hardcover : alk. paper)—
ISBN 978-1-4384-3522-0 (pbk. : alk. paper) 1. Literature and science.
2. Literature—History and criticism—Theory, etc. 3. Darwin, Charles, 1809–1882—Influence. 4. Evolution (Biology) in literature. 5. Human behavior—Philosophy. 6. Evolutionary psychology. I. Title.
  PN55.C375 2011
  809'.9336—dc22
                                                              2010031820

Contents

List of Illustrations   vii
Introduction   ix
Acknowledgments   xv

PART ONE: ADAPTATIONIST LITERARY THEORY

Chapter 1.   An Evolutionary Paradigm for Literary Study, with Two Sequels   3

Chapter 2.   An Evolutionary *Apologia pro Vita Mea*   55

Chapter 3.   A Meta-Review of *The Art Instinct*   61

Chapter 4.   Three Scenarios for Literary Darwinism   71

PART TWO: INTERPRETIVE PRACTICE

Chapter 5.   Aestheticism, Homoeroticism, and Christian Guilt in *The Picture of Dorian Gray*   91

Chapter 6.   The Cuckoo's History: Human Nature in *Wuthering Heights*   109

Chapter 7.   Intentional Meaning in *Hamlet*: An Evolutionary Perspective   123

PART THREE: EMPIRICAL LITERARY STUDY:
AN EXPERIMENT IN WEB-BASED RESEARCH

Chapter 8. Agonistic Structure in Victorian Novels: 151
            Doing the Math

Chapter 9. Quantifying Agonistic Structure in 177
            The Mayor of Casterbridge

PART FOUR: EVOLUTIONARY INTELLECTUAL HISTORY

Chapter 10. The Power of Darwin's Vision 197

Chapter 11. The Science Wars in a Long View: 259
            Putting the Human in Its Place

Chapter 12. A Darwinian Revolution in the Humanities 271

Notes   279
References 309
Index   333

Illustrations

Figure 8.1	Motive factors in protagonists and antagonists	159
Figure 8.2	Criteria for selecting long-term mates in protagonists and antagonists	161
Figure 8.3	Personality factors in protagonists and antagonists	162
Figure 8.4	Emotional responses to protagonists and antagonists	164
Figure 9.1	Motive factors in four main characters in *Mayor*	185
Figure 9.2	Personality factors in four main characters in *Mayor*	186
Figure 9.3	Criteria for selecting mates in *Mayor*	188
Figure 9.4	Emotional responses to characters in *Mayor*	190

Introduction

Ever since I began integrating literary study and evolutionary psychology in the early 1990s, people have asked what moved me in this direction. Being in graduate school in the late seventies, I came of age, intellectually, just at the time that traditional methods in the humanities were giving place to poststructuralism. Not wanting to miss out on anything good, I read some of the standard books from which my more avant-garde colleagues were learning the new idioms—by Derrida, Foucault, de Man, and their acolytes. That was frustrating, because I found little in these books that answered to my own deepest intuitions about human nature and literature. Moreover, the books seemed fundamentally incoherent, not just illogical but overtly hostile to the principles of rational order that are a common heritage for descendants of the Enlightenment. Because I was irretrievably oriented toward rational inquiry, essays in deconstruction and Foucauldian cultural critique could not stimulate my own constructive efforts. Fairly baffled, I asked fellow graduate students what it was precisely about these works that they found so compelling. They could not tell me, at least not in ways that carried conviction for me, and for the most part they did not even try to offer reasons and explanations. Instead, they suggested that one must simply immerse oneself in the rhetoric of the chief poststructuralist authorities, imitate them, and achieve conviction by way of osmosis, learning to occupy the same imaginative space as the masters. I was skeptical about the epistemic underpinnings of this strategy, and in any case it did not work for me, so going with the inexorably advancing poststructuralist tide was never really an option.

As an alternative to adopting "Theory," I tried for a few years just to pursue historical scholarship and take up difficult interpretive cases. My book on Matthew Arnold was standard intellectual history, and my book on Wallace Stevens offered close readings as evidence for an original interpretive thesis—that Stevens was essentially a religious poet. I then took up Walter Pater's *Marius the Epicurean*, an enigmatic autobiography displaced into historical fiction.[1] In the year I spent on Pater, one specific moment stands out for me as emblematic for the decisive turn my career would next take. I was at a regional MLA conference giving a talk on psychosexual symbolism in *Marius*. After I had finished, someone popped up from the audience and remarked, smugly, "Ah, but Professor Carroll hasn't considered the way Foucault has problematized the concept of individual identity." Actually, I had considered it, in my own mind, and rejected it. This moment brought into sharp relief the growing realization that it was not possible for me simply to ignore poststructuralism and go about my own business. Scholarship is a social enterprise, and if one is going to be heard, one has to engage the ideas of other scholars. Since I could not accommodate myself to poststructuralism, my only alternative was to formulate a completely different basis for literary study and to set that new basis into active opposition with the prevailing paradigm.

I started by sketching out large-scale taxonomic schemes, reworking the territory colonized by Northrop Frye, while also formulating philosophical reasons for reaffirming three core ideas in traditional humanism: individual identity, authorial intentions, and reference to a real world. Then, in the early 1990s, I finally got around to reading *On the Origin of Species* and *The Descent of Man*. As I explain more fully in chapter 2, "An Evolutionary *Apologia pro Vita Mea*," these two books had a massive impact on my imagination. They made it clear to me that all things human are contained within the scope of biological evolution. "All things human" include the products of the human imagination. Any effort to build a new framework for literary study would have to start there, with an evolved and adapted human nature, a human body and brain, giving due heed to biologically grounded motives, passions, and forms of cognition. About the same time that I was reading Darwin, I became aware that a neo-Darwinian revolution was already transforming the social sciences. Energized by the prospects thus emerging, I began the reading and writing that culminated, in 1995, in *Evolution and Literary Theory*.[2]

In a widely disseminated essay, D. T. Max identifies *Evolution and Literary Theory* as the "founding text" of literary Darwinism.[3] It was indeed an early rallying point for scholars dissatisfied with "Theory" and fascinated by the new Darwinism in the human sciences. Some of my closest friends, though, share my own reservations about the book—that it is too long, that it follows the Christmas pudding recipe (stick in everything in the kitchen and stir vigorously), that it spends at least as much time pugnaciously attacking the follies of poststructuralism as it does delineating a different and better model of literary study, and that it uses evolutionary biology chiefly to confirm traditional critical concepts rather than to establish a new conceptual model.

In the years following *Evolution and Literary Theory*, I produced an edition of *On the Origin of Species* and continued to write essays in literary theory, criticism, and historical scholarship, reviews of evolutionary books, and commentaries on neighboring movements such as ecocriticism and cognitive poetics. Those writings were collected in *Literary Darwinism: Evolution, Human Nature, and Literature*.[4] At the time it appeared, in 2004, this collection contained the most advanced thinking in the field. Nonetheless, almost as soon as it appeared, substantial parts of it were already becoming obsolete. The field as a whole is developing so fast, and my own thinking has been developing so continuously, that when people ask me for recommendations about guides to the best work in literary Darwinism, I no longer recommend any of the theoretical chapters in that earlier collection. For one thing, people who take evolutionary study seriously must constantly be assimilating new research in the evolutionary human sciences. Moreover, "evolutionary psychology" is not yet a fully formed paradigm. It is still in the process of integrating ideas about individual differences, flexible general intelligence, "group selection," the peculiarly human capacities for symbolic thinking, the nature of the imagination, and the ways basic animal impulses interact with evolved dispositions for "culture." While working on theoretical issues such as these, I've also been developing a better understanding of how to use evolutionary thinking for interpreting particular literary texts. And finally, in addition to refining theory and practice, I've added an item to the methodological tool kit used in *Literary Darwinism*: quantitative, empirical research.

The four parts of this book display the full scope of research in evolutionary literary study as it is now practiced—theory, interpretive criticism, empirical research, and intellectual history. Some of the chapters have been published in other forms but have been reworked to avoid repetition and to integrate ideas that have developed over time. For instance, I've written on the adaptive function of the arts in several essays that are not included in this volume. The thinking that went into those essays has been consolidated into a single substantial section of the first chapter. I've also published several introductory surveys to the field. Those surveys have been consolidated, condensed, and updated in another section of the first chapter. Jon Gottschall and I published an empirical article on characters in Victorian fiction that served as a pilot project for the more elaborate project we conducted in company with John Johnson and Dan Kruger. Like the first stage of a rocket, the pilot project has been left behind, superseded by the more advanced work included here.

The theoretical chapters in part 1 are arranged to give a progressive, developing overview of the whole field—past, present, and future. The first chapter, "An Evolutionary Paradigm for Literary Study, with Two Sequels," identifies the current position of literary Darwinism relative to the academic literary establishment, concisely summarizes the most significant contributions to the field, lays out the basic features of "human nature," and takes up the most important challenges that have been leveled at the Darwinists. The second chapter is an exercise in literary autobiography, locating my current perspective in relation to my literary experience over some five or six decades. The third chapter, reflecting on the enormous publicity generated by Denis Dutton's *The Art Instinct*, offers an opportunity for taking the pulse of current cultural trends—the largely sympathetic response to evolutionary aesthetics among educated general readers, and the various forms of reaction stirred into protest over Dutton's claims. In the final chapter in part 1, I evaluate three alternative scenarios for literary study: one in which literary Darwinism remains outside the mainstream of literary study; one in which literary Darwinism is incorporated as just another of many different "approaches" to literature; and a third in which the evolutionary human sciences fundamentally transform and subsume all literary study. I weigh the force and durability of established practices and thus assess the prospects for future developments.

All the chapters in part 2, "Interpretive Practice," display in action the theoretical principles articulated in part 1. The chapters are arranged in the sequence in which they were written, reversing

Introduction xiii

the chronological order of the works under discussion: *The Picture of Dorian Gray, Wuthering Heights,* and *Hamlet.* The chapter on *Dorian Gray* examines interactions among species-typical sexual dimorphism, Wilde's homosexuality, medieval Christian sentiment, and decadent aestheticism. The basic interpretive strategy bears a close affinity with the psychoanalytic idea that literary works are psychosexual symbolic constructs, but I do not, of course, embrace Freudian sexual psychology. Critics of literary Darwinism like to say that the Darwinists reduce humans and their imaginative constructs to a few basic impulses, ignoring subtleties of individual identity, literary form, and cultural context.[5] The chapter on *Dorian Gray* can stand as a refutation of such charges. The chapter on *Wuthering Heights* was for me something of a breakthrough. About the time I wrote it, I was finally developing a full systemic understanding of "human life history," that is, the way all the components of human nature form an integrated functional complex. The commentary on *Wuthering Heights* is the first interpretive effort in which I bring that systemic understanding to bear on the total structure of meaning in a specific literary work. Affectional bonding between mothers and children is central to human nature. In commenting on *Hamlet,* I connect Bowlby's theory of attachment with recent research into the neurophysiology of depression and an evolutionary understanding of individual differences in personality. All three of these commentaries engage critical traditions and take in reader responses as integral parts of literary meaning.

Part 3 contains two examples of empirical literary study. The research for these two chapters consists in a five-year-long collaborative project conducted by two literary scholars (Jon Gottschall and me) and two evolutionary psychologists (John Johnson and Dan Kruger). (For the second, a fifth member, Stelios Georgiades, joined the team.) Using the Web, we solicited numerical data on character attributes and reader responses to characters in about two hundred British novels of the nineteenth and twentieth centuries—Austen to Forster. We received about 1,600 responses—enough data for robust statistical results. The first chapter in this section gives an overview of the whole project. The second interprets the data on a single novel, Thomas Hardy's *The Mayor of Casterbridge.* Because *The Mayor of Casterbridge* is an unusually difficult novel, with a problematic tonal organization, the critical tradition displays an exceptionally low degree of consensus. We make the case that quantitative analysis can break through the impasse produced by preconceived interpretive models.

The fourth part contains three ventures into intellectual history, organized like Russian matryoshka dolls, working in chronological sequence, with the first chapter, the biggest doll, providing the context for the second and the second providing the context for the third. In the first chapter, on *On the Origin of Species*, I compare paradigm change in Darwinian evolutionary biology with Thomas Kuhn's model of scientific revolutions. I also give close attention to the sources of rhetorical power in Darwin's masterpiece. In the second chapter, on the science wars since Darwin's time, I concentrate on two specific exchanges: Arnold vs. Huxley, and Snow vs. Leavis. The misunderstandings in both these exchanges suggest that we are only just now, finally, coming to terms with the implications of Darwinism for the humanities. The last chapter, "A Darwinian Revolution in the Humanities," can stand as a conclusion for the whole volume. It takes up all the main themes in the preceding essays, evocatively and impressionistically delineating intellectual change from Darwin's time to our own, locating our current historical moment in a swirl of imaginative currents, some dying into eddies, and some gathering force as major new movements that will shape our future.

Acknowledgments

Like most scholars, I have a circle of colleagues with whom I exchange drafts. I've been particularly blessed to have some very good editors among my friends and family. Harold Fromm, Ellen Dissanayake, Jon Gottschall, Brian Boyd, Alice Andrews, Paula Carroll, Gwendolyn Carroll, and Jessica McKee have done heroic service on more than one occasion. Others have helped as well: Gad Saad with the chapter on three scenarios, Denis and Margit Dutton with the chapter "Agonistic Structure in Victorian Novels: Doing the Math," John Knapp with the chapter on *The Mayor of Casterbridge*, and Dick Cook with the chapter on Darwin's *Origin*. Michael Carroll has made a special study of *Hamlet*. While I was working on that chapter, we had long and (for me) helpful conversations about the play.

Harold deserves special mention for going through perhaps a dozen or more drafts of the chapter on *Wuthering Heights*. Alice went through many drafts of the chapter on three scenarios for literary Darwinism. Ellen, Harold, Jon, and Brian have read almost everything I've written over the past ten years, and they never fail to offer incisive and constructive editorial comments. Over a period of about a year, Jon, Brian, and I compiled and edited a collection of essays, *Evolution, Literature, and Film: A Reader*. For all three of us, the work we did on that was like taking an intensive advanced seminar in style—we all learned from each other.

It is customary in acknowledgments to say that the merits of a book belong to those who have helped edit it but that its defects belong solely to the author of the book. Clearly, this can't be true, and it is so evidently insincere that the sentiment never fails to rub me the wrong way. Accordingly, I'm going to take this opportunity to right the balance a little by declaring that whatever merits this book possesses are mine alone. For its manifold defects, I can blame only my friends and family.

Sources

In somewhat different versions, most of the material in this book has been previously published in other venues. The sources are indicated herewith. I'm grateful to the editors of the original venues for allowing me to include these essays in this collection.

Part of the material for chapter 1 appeared in "An Evolutionary Paradigm for Literary Study," *Style* 42 (2008): 103–35; "Rejoinder," *Style* 42 (2008): 309–412; and "The Adaptive Function of Literature and the Other Arts," an invited online posting (and my responses to comments from readers) in *Forum*, part of the project *On the Human* at the National Humanities Center, 22–26 June 2009: http://onthehuman.org.

Chapter 2 is excerpted from an online interview with David DiSalvo, *Neuronarrative*, http://neuronarrative.wordpress.com, posted 27 February 2009.

Chapter 3 appeared in *The Evolutionary Review: Art, Science, Culture* 1 (2010): 48–54.

A somewhat different version of chapter 4 first appeared in *New Literary History* 41.1 (2010): 53–67; copyright © *New Literary History*, The University of Virginia.

Chapter 5 first appeared in *Philosophy and Literature* 29 (2005): 286–304.

Chapter 6 first appeared in *Philosophy and Literature* 32 (2008): 241–57.

Chapter 7 first appeared in *Style* 44 (2010): 230–60.

A somewhat different version of chapter 8 first appeared as "Agonistic Structure in Victorian Novels: Doing the Math," in *Philosophy and Literature* 33 (2009): 50–72.

Chapter 9 first appeared in *Style* 44 (2010): 164–88.

Chapter 10 first appeared as the introduction to my edition of Darwin's *On the Origin of Species by Means of Natural Selection* (Peterborough, Ontario: Broadview Press, 2003).

Chapter 11 first appeared in *Interdisciplinary Essays on Darwinism in Hispanic Literature and Film: The Intersection of Science and the Humanities*, ed. Jerry Hoeg and Kevin S. Larsen (Lewiston, NY: Edwin Mellen Press, 2009), 19–33.

Chapter 12 has not previously been published.

PART ONE

Adaptationist Literary Theory

CHAPTER 1

An Evolutionary Paradigm for Literary Study, with Two Sequels

Sources and Occasions

Parts of this essay were included in a "target" article in a special double issue of the journal *Style*. Thirty-five scholars and scientists responded to the target article, and I then wrote a "rejoinder to the responses." Developing ideas from the target essay, I took up the issue of adaptive function again in an online discussion, the forum *On the Human* hosted by the National Humanities Center in North Carolina. This first chapter contains a revised and expanded version of the target article. I've appended two sequels: a condensed version of my rejoinder to the respondents in *Style*, and a condensed version of my rejoinder to the respondents in the forum *On the Human*.

The Current Institutional Position of Literary Darwinism

In the past few years, "literary Darwinism" has emerged as the most dynamic new movement in literary study. A steadily increasing mass of articles, books, edited volumes, and special journal issues has been devoted to this topic, and it has garnered wide public attention, with articles in leading newspapers and magazines all over the world. As it has gained in visibility, the movement has also attracted a good deal of criticism from diverse disciplinary perspectives—from traditional humanism, poststructuralism, cognitive poetics, and evolutionary social science. I have surveyed contributions to the field in several previous articles, aiming at bibliographic inclusiveness.[1] Here I won't replicate those bibliographic efforts. Instead, I shall briefly describe some of the more important contributions to evolutionary literary study, discuss key theoretical issues, and respond to representative critiques.

The central concept in both evolutionary social science and evolutionary literary study is "human nature": genetically mediated characteristics typical of the human species. In the concluding paragraph of a survey I published in 2003 I said that "we do not yet have a full and adequate conception of human nature. We have the elements that are necessary for the formulation of this conception, and we are on the verge of synthesizing these elements."[2] Over the past six years, that effort of synthesis has advanced appreciably. In a subsequent section, I lay out a model of human nature that incorporates the features on which most practitioners in the field would agree. One crucial element of human nature remains at least partially outside this consensus model: the disposition for producing and consuming literature and the other arts. Within the evolutionary human sciences, divergent hypotheses have been formulated about the adaptive function of the arts. Theorists disagree on whether the arts have adaptive functions, and if they do, what those functions might be. The alternative hypotheses on this topic involve alternative conceptions of human evolutionary history and human nature. They are thus vitally important to the whole larger field of evolutionary social science, and they also have important implications for the practical work of interpretive criticism. After describing and critiquing the main competing hypotheses, I make a case for one particular hypothesis. I also discuss two problems that are more particularly concerns for literary study: the challenge of generating new knowledge about literature, and the challenge of mediating between the discursive methods of the humanities and the empirical methods of the social sciences.

The most modest claim that could be made for evolutionary literary study is that it is one more "approach" or "school" that merits inclusion in casebooks and theoretical surveys. Along with Marxist, psychoanalytic, feminist, deconstructive, and New Historicist essays, one would thus have a Darwinian "reading" of this or that text, *Hamlet* or *Heart of Darkness*, say. Most casebooks of course do not yet include a Darwinian reading, and in truth the Darwinists have had a hard enough time even getting panels accepted at the MLA. My own favorite rejection note explained that the program committee felt that the Darwinian approach was too "familiar" and that what was wanted were proposals along more "innovative" lines—this in a year in which proposals with Lacanian, feminist, and Marxist themes achieved levels of production comparable to those of the American and Soviet military industries in the latter days of the Second World War. In his superbly witty parodies of literary schools in *Postmodern Pooh*, Frederick Crews includes a chapter on the evolutionary literary critics, ridiculing them in tandem with their peers in more firmly established schools, but this was perhaps merely an act of kindness.[3] By including them, Crews gave recognition to a struggling minority that—whatever their failings (as he might see them) in doctrinaire narrowness—shares his respect for reason and evidence. In a recent essay in *Style*, James Mellard speaks with evident alarm about "a growing army of enthusiasts for a new Darwinian naturalism."[4] So far as this description applies to the social sciences, it is apt enough. Darwinian social scientists hold key positions in prestigious universities, publish works in the mainstream journals in their disciplines, and win large popular audiences among the educated lay public. The literary Darwinists, in contrast, could most accurately be characterized not as an army but as a robust guerilla band. That standing could change fairly soon. If the rate of current publication in the field continues or increases, before long sheer numbers will tilt the balance toward inclusion in casebooks more conventional than *Postmodern Pooh*.

Institutionally, the literary Darwinists occupy a peculiar position. On the one hand, they are still so marginal that being included in panel sessions and casebooks would constitute an advance in institutional standing. On the other hand, their ultimate aims sweep past any such inclusion. At least among their most ambitious adherents, they aim not at being just one more "school" or "approach." They aim at fundamentally altering the paradigm within which literary study is now conducted. They want to establish a new alignment among the disciplines and ultimately to subsume all other possible approaches to literary study. They rally to Edward O. Wilson's cry for "consilience" among all the branches of learning. Like Wilson, they envision an integrated body of knowledge extending in an unbroken chain of material causation from the lowest level of subatomic particles to the highest levels of cultural imagination.[5] And like Wilson, they regard evolutionary biology as the pivotal discipline uniting the hard sciences with the social sciences and the humanities. They believe that humans have evolved in an adaptive relation to their environment. They argue that for humans, as for all other species, evolution has shaped the anatomical, physiological, and neurological characteristics of the species, and they think that human behavior, feeling, and thought are fundamentally constrained and informed by those characteristics. They make it their business to consult evolutionary biology and evolutionary social science in order to determine what those characteristics are, and they bring that information to bear on their understanding of the human imagination.

Virtually all literary Darwinists formulate "biocultural" ideas. That is, they argue that the genetically mediated dispositions of human nature interact with specific environmental conditions, including particular cultural traditions. They nonetheless characteristically distinguish themselves from "cultural constructivists" who effectively attribute exclusive shaping power to culture. The Darwinists typically focus on "human universals" or cross-cultural regularities that derive from regularities in human nature. They recognize the potent effect of specific cultural formations, but they argue that a true understanding of any given cultural formation depends on locating it in relation to the elemental, biologically based characteristics that shape all cultures.

Literary Darwinism and Cognitive Poetics

In their effort to bring about a fundamental shift in paradigm, the literary Darwinists can be distinguished from practitioners in a school that is in some respects their closest disciplinary neighbor—cognitive poetics. In her preface to a collection of essays in cognitive poetics, Ellen Spolsky explains that the cognitivists aim to "supplement rather than supplant current work in literary and cultural studies." She assures her audience that "these essays have no interest in repudiating the theoretical speculations of poststructuralist and historicist approaches to literature." She and her colleagues wish only to enter into "a constructive dialogue with the established and productive theoretical paradigms."[6] Her coeditor, Alan Richardson, takes a similar line. Emphatically distancing the cognitivists from the literary Darwinists, he describes the work of the Darwinists "as an outlier that helps define the boundaries of cognitive literary criticism proper." Describing the disciplinary alignments of individual contributors to the volume, he affirms that Spolsky seeks "not to displace but to supplement poststructuralist approaches to literature like deconstruction and New Historicism," that F. Elizabeth Hart seeks only "to supplement 'postmodern' accounts of language, subjectivity, and culture," and that Mary Crane "locates her work between cognitive and poststructuralist accounts of subjectivity, language, and culture."[7]

Efforts to segregate cognitive poetics from evolutionary literary study are doomed to failure. One thinks of early stages in the development of American cities. Enclaves outside the city core are inevitably swallowed up as the cities expand outward. Evolutionary social science seeks to be all-inclusive. Because it is grounded in evolutionary biology, it encompasses all the more particular disciplines that concern themselves with human evolution, human social organization, and human cognition. As a distinct school within evolutionary social science, "evolutionary psychology" can be described as the offspring of a coupling between sociobiology and cognitive psychology.[8] Evolutionary psychologists derive from sociobiology an emphasis on the logic of reproduction as a central shaping force in human evolution, and they seek to link that logic with complex functional structures in cognitive mechanisms. Hence the title of the seminal volume in evolutionary psychology: *The Adapted Mind*.[9] The human mind has functional cognitive mechanisms for precisely the same reason that the human organism has complex functional structures in other organ systems—because it has evolved through an adaptive process by means of natural selection. In the process of expanding outward from the logic of reproduction to the explanation of cognitive mechanisms, evolutionary social scientists have already given concentrated attention to many of the standard topics in cognitive psychology, for instance, to "folk physics," "folk biology," and "folk psychology"; perceptual mechanisms; the relation between "modularized" cognitive processes and "general intelligence"; the relation between emotions and conscious decision-making; mirror neurons, "perspective taking," "Theory of Mind," and "metarepresentation"; "mentalese" and language acquisition; metaphor and "cognitive fluidity" or conceptual blending; "scripts" and "schemata"; and narrative as an elementary conceptual schema.[10] If evolutionary psychology can give a true and comprehensive account of human nature, it can ultimately encompass, subsume, or supplant the explanatory systems that currently prevail in the humanities.

As things currently stand, the use of cognitive psychology in literary study can be located on a spectrum running from deconstruction at one end to evolutionary psychology at the other. At the deconstructive end, practitioners seek only to redescribe poststructuralist ideas in terms derived from cognitive science. Spolsky, for instance, argues that the supposedly modular character of the mind approximates to deconstructive accounts of the decentered and fragmented self.[11] Somewhere closer to the middle of this spectrum, Lisa Zunshine references evolutionary psychology to support her claims that the human mind has evolved special powers of peering into the minds of conspecifics—what psychologists call Theory of Mind (ToM).[12] Despite her appeal to selected bits of evolutionary psychology, Zunshine strongly emphasizes the "cognitive" aspect of her views, muting and minimizing their sociobiological affiliations. Beyond ToM, she declines to attribute any very specific structure to the adapted mind, and in citing other literary scholars, she prudently avoids reference to most of the published work in evolutionary literary study. She unequivocally locates herself in the community of practitioners who explicitly segregate their work from the evolutionary literary critics. Moving toward the evolutionary end of the spectrum, in film theory, David Bordwell has long identified his work as "cognitive" in orientation, but he has increasingly envisioned cognitive mechanisms as the result of an adaptive evolutionary process, and he firmly contrasts his naturalistic vision with the prevailing poststructuralist theories in film studies. Bordwell and his associates have done excellent work in linking evolved cognitive mechanisms with specific formal features of film.[13]

Because evolutionary psychology draws heavily on cognitive developmental psychology, all evolutionary literary critics are in some measure de facto cognitivists. They vary, though, in the degree to which they have incorporated information on cognitive mechanisms not just indirectly through evolutionary psychology but directly from cognitive psychology. Among the evolutionary literary critics, Brian Boyd has gone further than any other scholar in assimilating information directly from cognitive psychology, especially cognitive developmental psychology. Like Bordwell, but with more explicit and detailed reference to evolutionary thinking, Boyd demonstrates that the findings of cognitive psychology make sense ultimately because they are embedded in the findings of evolutionary psychology. He emphasizes the continuity between "play" in animals, human curiosity, and the generation of novelty in form, ideas he applies to classical works such as the *Odyssey*, modernist works such as *Lolita*, and avant-garde graphic narratives.[14]

Clearly, one central line of development for evolutionary literary study will be to link specific cognitive structures with specific literary structures and figurative modes, locating both in relation to evolved human dispositions. So far, the Darwinists have focused more on drama and fiction than on poetry, but Frederick Turner has correlated the length of poetic lines with the duration of perceptual units, and Michael Winkelman demonstrates that Zahavi's handicap principle can be effectively used to analyze the tension between convention and invention in Donne's poetic forms.[15] Boyd, Michelle Scalise-Sugiyama, and Francis Steen use goal-orientation and problem-solving to construct basic frameworks for the analysis of narrative, and Daniel Nettle uses goal-orientation for analyzing the structure of drama.[16]

A Selective Survey of Works in Evolutionary Literary Study

In 2005, Jonathan Gottschall and David Sloan Wilson published a collection of commissioned essays, *The Literary Animal: Evolution and the Nature of Narrative*, that set a new standard for cross-disciplinary research in the human sciences. Gottschall is a literary scholar who has made pioneering efforts in using empirical methods in literary study, and Wilson is an evolutionary biologist with wide-ranging cultural interests. The volume includes forewords by both a scientist (E. O. Wilson) and a literary scholar (Frederick Crews), and it contains an afterword written by a philosopher (Denis Dutton). The authorship of the essays is almost equally divided between evolutionary scientists and literary scholars. A collection edited by Robin Headlam Wells and Johnjoe McFadden, *Human Nature: Fact and Fiction*, has a similar range of contributors, with essays by Steven Pinker, Simon Baron-Cohen, Ian McEwan, me, and others. More recently, Brian Boyd, Jonathan Gottschall, and I compiled an anthology of some of the best work done in evolutionary literary study in the past fifteen years or so. As we went over the materials for this volume, sorting and evaluating them, we agreed that the level of professionalism—of expertise in assimilating information from the social sciences, of clarity in theoretical principles, and of sophistication in the use of theory for the purposes of practical criticism—has steadily improved. Like its predecessors, *Evolution*,

Literature, and Film contains essays by both literary scholars and scientists. In a new annual journal, *The Evolutionary Review: Art, Science, Culture,* Alice Andrews and I are following the lead of these other volumes in including works by both scholars and scientists. Aiming to demonstrate that "this view of life" can indeed encompass all things human,[17] we are publishing essays and reviews on film, fiction, theater, visual art, music, dance, and popular culture; essays and reviews of books, articles, and theories related to evolution and evolutionary psychology; and essays and reviews on science, society, and the environment.[18]

The first full-length books that could clearly be classed as works of literary Darwinism appeared in the mid-nineties, my own *Evolution and Literary Theory*, and Robert Storey's *Mimesis and the Human Animal: On the Biogenetic Foundations of Literary Representations*.[19] Like many of the early essays in the field, these two books presented themselves as polemical confrontations between biological naturalism and poststructuralist efforts to dispense with nature. They both also contain elements of constructive theory. Storey sketches in features of a "biogrammar"—a model of human nature—and I work out correlations between elementary biological and literary concepts. I define character, setting, and plot in terms of organism, environment, and action, and I delineate literary activity as a form of "cognitive mapping"—a subjectively charged image of the world and of human experience in the world. I identify three chief levels for the analysis of meaning in texts: (a) elemental or universal human dispositions (human nature); (b) the organization of those dispositions within some specific cultural order; and (c) the peculiarities of individual identity in represented subjects, authors, and readers. I also argue for the systematic analysis of individuality through the incorporation of modern research into personality.

More recent works of general theory have continued to define their principles in contrast to purely culturalist principles. On the whole, though, the polemical element has diminished relative to the efforts of constructive formulation. Ellen Dissanayake, an evolutionary theorist of the arts, offers an example. In *Homo Aestheticus* (1992), she set an evolutionary vision of art in contrast to poststructuralist views. In her most recent book, *Art and Intimacy: How the Arts Began* (2000), she concentrates on developing the positive aspects of her theories.[20] In *Literature, Science, and a New Humanities* (2008), Jonathan Gottschall gives evidence for a pervasive sense of a crisis of morale in the humanities.[21] He traces this crisis to a methodological failure to produce empirically valid and progressive forms of knowledge, but he is less interested in attacking a failed ethos than in offering an alternative. He argues that the humanities can benefit from incorporating scientific methods and, along with the methods, the ethos of empirical inquiry. Gottschall has published several articles in which he uses quantitative methods of "content analysis" to explore topics of sexual identity and characterization cross-culturally. *Literature, Science, and a New Humanities* includes several such studies as examples. In *On the Origin of Stories: Evolution, Cognition, and Fiction*, Boyd defines his evolutionary perspective in contrast to the culturalist models that still prevail in the humanities, but he occupies himself relatively little with criticizing poststructuralist formulations. Instead, he concentrates on incorporating evolutionary research in his own theories of writing and reading.[22] Harold Fromm is a founding figure in ecocriticism, and his intuitive naturalism has in recent years converged with "The New Darwinism in the Humanities," the title of a set of essays included in his most recent book, *The Nature of Being Human: From Environmentalism to Consciousness*. In an earlier book, *Academic Capitalism*, Fromm had actively engaged the prevailing poststructuralist orthodoxies. In his new book, collecting essays over a period of years, he occupies himself with three primary topics in separate but cumulative phases: ecocriticism, the new Darwinism in the humanities, and a naturalistic philosophy of consciousness like that associated with Daniel Dennett.[23]

The works of general theory just noted contain a fair amount of practical criticism but can be distinguished from works primarily dedicated to practical criticism. The first book-length work in practical criticism from an evolutionary angle was on Zamyatin's dystopian novel *We*—Brett Cooke's *Human Nature in Utopia: Zamyatin's We*. Cooke draws on evolutionary psychology to delineate features of human nature—communal eating, play, charismatic authority figures, sex, filial relations, and visceral responses—that are systematically violated in dystopian fantasies. He concentrates on Zamyatin's novel but locates it within the broader context of all utopian and dystopian fiction.[24] In *Shakespeare and the Nature of Love: Literature, Culture, and Evolution*, Marcus Nordlund produces an account of love, romantic and filial, in which he integrates evolutionary research with research into Renaissance ideas about love. That account serves as the context for his reading of several

Shakespeare plays. Nordlund contrasts his "biocultural" critique with purely culturalist perspectives on love and identity in the Renaissance.[25] In *Shakespeare's Humanism*, Robin Headlam Wells gives a detailed account of ideas of human nature active in the Renaissance and, like Nordlund, sets this account in contrast to current views that align the Renaissance writers with poststructuralist theories of cultural autonomy.[26] In *The Rape of Troy: Evolution, Violence, and the World of Homer*, Jonathan Gottschall integrates sociobiological theory with archeological and anthropological research in order to reconstruct the motivating forces in Homer's cultural ecology. Gottschall vividly evokes the Homeric ethos and convincingly demonstrates the value of a biological perspective for analyzing a specific cultural formation.[27] In a context seemingly far removed from that of Homer's barbarian warriors, Judith Saunders adopts a similar perspective, concentrating on the shaping force of reproductive logic, to analyze character and plot in the novels of Edith Wharton. Barash and Barash offer a set of sociobiological critiques geared toward a popular audience.[28]

Moving beyond the analysis of represented subject matter, several scholars have used evolutionary psychology to examine the interplay of perspectives among readers, authors, and characters. In our empirical study of Victorian novels, Johnson, Gottschall, Kruger, and I correlate the emotional responses of readers with motives and personalities in individual characters.[29] Robert Storey, Michelle Scalise Sugiyama, and I have all considered reader response from an evolutionary perspective.[30] Using game theory and the theory of "costly display," William Flesch identifies depictions of altruistic punishment as a chief means through which authors engage readers emotionally. Michael Austin delves into manipulative deceit and self-delusion in point of view.[31] The study of point of view shades over into the study of tone. In the critique of *Wuthering Heights* in part 2 of the present volume, I combine basic motives with "basic emotions" in a framework for analyzing complex interactions of tone in generic structures. In the critique of *Hamlet*, I develop ideas of tragedy by incorporating recent research on the neurobiology of depression, consider the kinds of emotional responses *Hamlet* has elicited in readers, and compare reader responses to *Hamlet* in various literary periods. Later in this chapter, illustrating a claim that the Darwinists can generate new literary knowledge, I shall return to some of these works and also describe others.

A Model of Human Nature

Until fairly recently in literary history, most writers and literary theorists presupposed that human nature was their subject and their central point of reference. Dryden following Horace, who follows others, offers a representative formulation. In "Of Dramatic Poesy," Dryden's spokesman Lisideius defines a play as "a just and lively image of human nature, representing its passions and humours, and the changes of fortune to which it is subject; for the delight and instruction of mankind."[32] The understanding of human nature in literature is the most articulate form of what evolutionists call "folk psychology."[33] When writers invoke human nature, or ordinary people say, "Oh, that's just human nature," what do they have in mind? They almost always have in mind the basic animal and social motives: self-preservation, sexual desire, jealousy, maternal love, favoring kin, belonging to a social group, desiring prestige. Usually, they also have in mind basic forms of social morality: resentment against wrongs, gratitude for kindness, honesty in fulfilling contracts, disgust at cheating, and the sense of justice in its simplest forms—reciprocation and revenge. All of these substantive motives are complicated by the ideas that enter into the folk understanding of ego psychology: the primacy of self-interest and the prevalence of self-serving delusion, manipulative deceit, vanity, and hypocrisy. Such notions of ego psychology have a cynical tinge, but they all imply failures in more positive aspects of human nature—honesty, fairness, and impulses of self-sacrifice for kin, friends, or the common good.

Postmodernists have put all such ideas of human nature out of play. Evolutionists, fortunately, have taken a different path. While literary theorists were immersing themselves in speculative theoretical systems such as phenomenology, psychoanalysis, deconstruction, and Marxism, the evolutionists were gradually developing an empirically based model of human nature, including childhood development, family dynamics, sexual relations, social dynamics, and cognition.

Writing from the perspective of a traditional humanist, Eugene Goodheart has devoted a book to repudiating Darwinian thinking in the humanities. Questioning the claims of evolutionary psychology to give us an adequate account of human nature, he says, "Human nature may not be a blank slate, but do we know enough to know what is inscribed upon it?"³⁴ In the manner in which it is posed, this is not a very serious question. Goodheart himself does not want an answer. Still, the question itself is well worth asking and deserves an answer. The literary Darwinists have committed themselves to the proposition that it can be answered in the affirmative. As Alan Richardson observes, the evolutionary critics differ from the cognitivists "in their high evaluation of the progress of scientific psychology."³⁵ This section is devoted to assessing the progress of scientific psychology in the one area that most concerns literary Darwinists—our modern understanding of human nature.

Natural selection operates by way of "inclusive fitness," shaping motives and emotions so as to maximize the chances that an organism will propagate its genes, or copies of its genes in its kin. Evolutionary psychologists commonly distinguish between inclusive fitness as an "ultimate" force that has shaped behavioral dispositions and the "proximal" mechanisms that mediate those dispositions.³⁶ The motives and emotions shaped by natural selection include those directed toward survival (obtaining food and shelter, avoiding predators) and those directed toward reproduction, a term that includes both mating effort and the effort aimed at nurturing offspring and other kin. Species vary in length of life, developmental trajectory, forms of mating, the number and pacing of offspring, and the kind and amount of effort expended on parental care. For any given species, the organization of these basic biological processes constitutes a distinct species-typical pattern of "life history." Like the species-typical pattern of life history for all other species, the species-typical pattern of human life history forms a reproductive cycle. In the case of humans, that cycle centers on parents, children, and the social group. Successful parental care produces children capable, when grown, of forming adult pair bonds, becoming functioning members of a community, and caring for children of their own.³⁷

Humans share with all animals a physiology organized in basic ways around reactive impulses of "approach" and "avoidance." They share with other social animals dispositions organized around affiliation and dominance.³⁸ Like all mammals, they have evolved systems of mother-infant bonding, and like chimpanzees, they have evolved dispositions for forming coalitions within large social groups. All of these characteristics are part of the species-typical repertory of dispositions that we call "human nature," but none of them is exclusive to humans. The traits that are most distinctively human constitute an integrated suite of anatomical, physiological, and behavioral features. Humans are bipedal, but proportional to body size they have much larger brains than other primates. Upright posture produces a narrowed birth canal. The problem of squeezing a large brain through a narrowed birth canal requires that human infants be born in an "altricial" or relatively helpless state. Human infants are heavily dependent on parental care for much longer than other animals, and they have, further, a greatly extended period of childhood development—the period previous to reproductive maturity. In ancestral environments (and typically still today), the dependency of human infants has required paternal investment—that is, care and resources provided by fathers. Humans share the characteristic of paternal investment with many birds and some other animals but with very few other mammals. Humans are the only animals that both have paternal investment and also live in large groups containing multiple males who form complex coalitions. Males of all species have evolved in such a way as to avoid investing in the offspring of other males, and living in multi-male groups reduces paternity certainty. Dispositions for pair bonding and sexual jealousy are thus prominent features in the evolved dispositions of human males. Human females are also distinctive in having menopause and thus a period of life that extends beyond the reproductive years. That period enables older women to raise their latest offspring to maturity and to aid in caring for grandchildren.[39]

Humans like other animals share fitness interests with their mates and their offspring, but, except in the case of monozygotic twins, the fitness interests of even the most closely related kin are not identical, and the logic of natural selection has shaped human dispositions in such a way that all intimate relations involve conflict. Females invest more than males in bearing and rearing children, and they also have certainty that their offspring are their own. Human males have evolved a reproductive strategy that includes both paternal investment and a disposition for low-investment short-term mating. Human females have evolved a need to secure the bonded attachment of a male willing to invest resources in them and their offspring, but they have also evolved dispositions for taking advantage of mating opportunities with higher-quality males than their own mates. Male and female relations are thus not only intense and passionate in their positive affects but also fraught with suspicion, jealousy, tension, and compromise. These relations often work smoothly enough for practical purposes, but they not infrequently break down in rejection, separation, abandonment, violent struggle, abuse, and even murder. Parents and children share a fitness interest in the success of the child—in the child reaching maturity and achieving successful reproduction. But the fitness interests of a child and parent are not identical. A child has one hundred percent fitness interest in itself. Each parent has only a fifty percent genetic investment in a child, and investment in any one child has to be deducted from investment in other children or potential children. Parents must often disperse resources over multiple offspring who each wish more than an even share. Parents preferentially invest in certain offspring, and they must also balance the effort they give to mating with the effort they give to parenting. Siblings form a natural social unit, allied in competition with nonrelated people, but they are also caught in intense competition with one another. Mating involves a coalition between two people who are not related by blood. They share a fitness interest in their own offspring, but they differ in the interest they have in the welfare of the kin they do not share with their mate. Even in nuclear families, fitness interests involve conflicts, and in step-families those conflicts are sharply exacerbated. The workings of inclusive fitness thus guarantee a perpetual drama in which intimacy and opposition, cooperation and conflict, are inextricably bound together.[40]

Because of their extended childhood development, humans have a long period in which to develop the social skills required by living in exceptionally complex social environments. Those social environments are structured by kin relations, flexible and multiple social coalitions, status hierarchies, and in-group/out-group relations.[41] Two features of the distinctively human suite of characteristics, both dependent on the expanded human brain, are particularly important in mediating these social relationships: (a) Theory of Mind and (b) language. Theory of Mind consists in the ability to attribute mental states to oneself and others, and it is thus the basis for self-awareness and for an awareness of others as distinct persons. The rudiments of Theory of Mind have been found in chimpanzees and some other animals, but the highly developed forms found in humans are unique. Self-awareness is a necessary precondition for the sense of personal identity—the sense that one has a distinctive set of traits, personality features, motive dispositions, social connections, and personal experiences, all extending continuously over a lifetime. Self-awareness is a necessary element of moral consciousness, and it is the precondition for self-esteem, embarrassment, shame, and guilt.[42] In its other-directed aspect, Theory of Mind is the capacity for envisioning the inner mental state of other humans, their beliefs, desires, feelings, thoughts, and perceptions. A key diagnostic characteristic for this aptitude is the ability to recognize that other people can have beliefs different from one's own, an ability that emerges in normally developing humans between the ages of three and four.[43] Language is the chief medium for conveying information in non-genetic ways. That kind of informational transmission is what we call "culture": arts, technologies, literature, myths, religions, ideologies, philosophies, and science. From the evolutionary perspective, culture does not stand apart from the genetically transmitted dispositions of human nature. It is, rather, the medium through which we organize those dispositions into systems that regulate public behavior and inform private thoughts. Culture translates human nature into social norms and shared imaginative structures.[44]

When we speak of "human nature," it is generally to this whole suite of characteristics—some common to all animals, some exclusive to mammals, some shared with other primates, and some peculiarly human—that we refer. These characteristics are so firmly grounded in the adaptive logic of the human species that they exercise a constraining influence on every known culture. Individuals can and do deviate from species-typical characteristics, but the recognition of the species-typical nonetheless forms a common frame of reference for all people. Adaptations emerge from regularities in ancestral environments, and the basic ground plan of human motives and human feelings forms one of the most important such regularities within the ancestral environments of modern humans. Because people are such intensely social animals, because their sociosexual relations are so extraordinarily complex and highly developed, and because successfully negotiating with other humans is one of the most important skills contributing to survival and to successful reproduction, having an intuitive insight into the workings of human nature can reasonably be posited as an evolved and adaptive capacity.[45] That adaptive capacity constitutes a "folk psychology," and it is in literature that folk psychology receives its most complete and adequate articulation.

The culture in which an author writes provides a proximate framework of shared understanding between the author and his or her projected audience, but every specific cultural formation consists in a particular organization of the elemental dispositions of human nature, and those dispositions form the broadest and deepest framework of shared understanding. Many authors make overt and explicit appeals to "human nature." By delineating the folk concept of human nature, we can reconstitute the shared framework of understanding within which authors interact with readers. That shared framework includes intuitions about persons as agents with goals, basic human motives, basic emotions, the features of personality, the phases of life, the relations of the sexes, filial bonding, kinship relations, the opposition between affiliation and dominance, and the organization of social relations into in-groups and out-groups.

Shifting the Frame of Interpretation

Whether traditionally humanistic or poststructuralist in orientation, literary criticism over the past century has spread itself along a continuum between two poles. At the one pole, eclectic general knowledge provides a framework for impressionistic and improvisatory commentary. At the other pole, some established school of thought, in some domain not specifically literary, provides a more systematic vocabulary for the description and analysis of literary texts. The most influential schools have been those that use Marxist social theory, Freudian psychology, Jungian psychology, phenomenological metaphysics, deconstructive linguistic philosophy, and feminist gender theory. Poststructuralist literary criticism operates through a synthetic vocabulary that integrates deconstructive epistemology, postmodern Freudian analysis (especially that of Lacan), and postmodern Marxism (especially that of Althusser, as mediated by Jameson). Outside of literary study proper, the various source theories of poststructuralism converge most comprehensively in the cultural histories of Michel Foucault, and since the 1980s, Foucauldian cultural critique has been overwhelmingly the dominant conceptual matrix of literary study. Foucault is the patron saint of New Historicism. Postcolonialist criticism is a subset of historicist criticism and employs its synthetic vocabulary chiefly for the purpose of contesting Western hegemony. Queer theory is another subset of historicist criticism and employs the poststructuralist vocabulary chiefly for the purpose of contesting the normative character of heterosexuality. Most contemporary feminist criticism is conducted within the matrix of Foucauldian cultural critique and dedicates itself to contesting patriarchy—the social and political predominance of males.

Each of the vocabulary sets that have come into prominence in literary criticism has been adopted because it gives access to some significant aspect of the human experience depicted in literature—class conflicts and the material base for imaginative superstructures; the psycho-symbolic dimensions of parent-child relations and the continuing active force of repressed impulses; universal "mythic" images derived from the ancestral experience of the human race; elemental forms in the organization of time, space, and consciousness; the irrepressible conflicts lying dormant within all partial resolutions, or social gender identity. All of these larger frameworks have enabled some insights not readily available through other means. They have nonetheless all been flawed or limited in one crucial respect. None of them has come to terms with the reality of an evolved and adapted human nature.

Humanist critics do not often overtly repudiate the idea of human nature, but they do not typically seek causal explanations in evolutionary theory, either. In the thematic reductions of humanist criticism, characters typically appear as allegorical embodiments of humanist norms—metaphysical, ethical, political, psychological, or aesthetic. In the thematic reductions of postmodern criticism, characters appear as allegorical embodiments of the terms within the source theories that produce the standard postmodern blend—most importantly, deconstruction, feminism, psychoanalysis, and Marxism. In their postmodern form, all these component theories emphasize the exclusively cultural character of symbolic constructs. "Nature" and "human nature," in this conception, are themselves cultural artifacts. Because they are supposedly contained and produced by culture, they can exercise no constraining force on culture. Hence Fredric Jameson's dictum that "postmodernism is what you have when the modernization process is complete and nature is gone for good."[46] From the postmodern perspective, any appeal to "human nature" would necessarily appear as a delusory reification of a specific cultural formation. By self-consciously distancing itself from the folk understanding of human nature, postmodern criticism loses touch both with biological reality and with the imaginative structures that authors share with their projected audience. In both the biological and folk understanding, there is a world outside the text. From an evolutionary perspective, the human senses and the human mind have access to reality because they have evolved in adaptive relation to a physical and social environment about which the organism urgently needs to acquire information.[47] An evolutionary approach shares with the humanist a respect for the common understanding, and it shares with the postmodern a drive to explicit theoretical reduction. From an evolutionary perspective, folk perceptions offer insight into important features of human nature, and evolutionary theory makes it possible to situate those features within broader biological processes that encompass humans and all other living organisms.

The Adaptive Function of Literature: A Controversy

Evolutionists insist that genes constrain and direct human behavior. Cultural constructivists counter that culture, embodied in the arts, shapes human experience. Both these claims are true, but some evolutionists and some cultural constructivists have mistakenly regarded them as mutually exclusive.[48] Some evolutionists have either ignored the arts or tried to explain them away as epiphenomenal to the basic processes of life. Many cultural constructivists, in contrast, have sought to collapse biology into culture, eliminating "human nature" and thus turning culture into a first cause or unmoved mover. In the past few years, evolutionists in both the sciences and the humanities have broken through this impasse, arguing that the imagination is a functional part of the adapted mind. These new ideas revise an earlier model of human cognitive evolution—a model most closely associated with evolutionary psychology (EP) as a specific school within the evolutionary human sciences. Revising that model makes it possible for us now fully to integrate the evolutionary human sciences and literary study.

In the early phases of EP, theorists seeking to counter the concept of the mind as a "blank slate" committed themselves to the idea of "massive modularity," the idea that the mind operates almost exclusively through dedicated bits of neural machinery adapted to solve specific practical problems in ancestral environments. Cognitive modules—the neural machinery dedicated to sight, for example—are characterized by automaticity and efficiency. The idea of massive modularity thus carried within itself a general sense of humans as adaptation-executing automata. To account for cognitive flexibility in this scheme, one could only "bundle larger numbers of specialized mechanisms together so that in aggregate, rather than individually, they address a larger range of problems."[49] The idea of massive modularity overgeneralizes from the most hardwired components of the brain. It is a massive oversimplification of human cognitive architecture, and it is already fading into the archives of intellectual history.[50] Its residual influence makes itself felt, though, in the ongoing debate over the adaptive function of the arts.[51]

In *How the Mind Works*, Steven Pinker locates the arts within an EP conception of human cognitive evolution.[52] As he sees it, natural selection shaped human motives to maximize inclusive fitness within a hunter-gatherer ecology. Sociality and language were part of the human adaptive repertoire. Imaginative culture was not. Creative imagination, whenever it appeared in human evolution, was just added on as a by-product of the cognitive/behavioral mechanisms that solved practical problems. To illustrate the by-product idea, Pinker draws parallels between art and pornography, psychoactive drugs, and rich foods like cheesecake. He acknowledges that fictional narratives might have informational content of some utility in providing game plans for practical problems that could arise. All the other features of the arts, he suggests, reflect only the human capacity to exploit evolved mechanisms for producing pleasure. This sort of pleasure, detached from all practical value with respect to survival and reproduction, would be equivalent to the pleasure derived from masturbation. (In "Does Beauty Build Adapted Minds?" Tooby and Cosmides modify their own earlier view that the arts are nonadaptive side effects, but they do not modify the underlying conception of mental architecture with which that earlier view is concordant.)

A second hypothesis from the side of evolutionary psychology, equally provocative, has been proposed by Geoffrey Miller. Miller argues that all displays of mental power, including those of the arts, might have had no adaptive value but might have served, like the peacock's tail, as costly signals indicating the general fitness of the person sending the signal. Miller's hypothesis identifies virtuosity in overcoming technical difficulty as the central defining characteristic of art.[53] Since Miller grants that the arts and other forms of mental activity, once underway, might have been co-opted or "exapted" for adaptively functional purposes, his argument reduces itself to an argument about the original function of the arts. Miller's wider argument about the origin of all higher cognitive powers has an obvious weakness: it requires us to suppose that the enlarged human brain—so costly, so complex and functionally structured, and so obviously useful for so many practical purposes in life—evolved primarily as a useless ornament for the purposes of sexual display. Virtually all commentators would acknowledge that human mental abilities can be used for sexual display, as can almost any other characteristic. We use bodily powers, clothing, and housing for sexual display, but we do not suppose that physical strength, clothing, and shelter have no primary functions subserving the needs of survival and the forms of reproduction not associated with display. Acknowledging that adaptively useful capacities can be deployed in a secondary way for the purposes of sexual display tells us nothing about any specific adaptive function those capacities might have.

Even if we overlook the weakness in Miller's broader hypothesis about the adaptive utility of the higher cognitive powers, his hypothesis about the arts says so little about the qualities and features that are specific to art that it has little explanatory value. Pinker's hypothesis is more challenging. He might be right that humanists object to his arguments at least in part because those arguments seem to diminish the dignity of the arts,[54] but I think many of these objections come from a deeper and more serious level—from a feeling that Pinker's hypothesis, like Miller's, fails to give an adequate account of his subject. Those who have sought to counter Pinker's hypothesis have a strong personal sense of what art and literature mean for them, and they have an intuitive conviction that their own experience of the arts cannot adequately be reduced to didactic lessons and pleasurable fantasy.

To solve the puzzle of adaptive function, we have to satisfy three criteria: (a) define art in a way that identifies what is peculiar and essential to it—thus isolating the behavioral disposition in question; (b) identify the adaptive problem this behavioral disposition would have solved in ancestral environments; and (c) identify design features that would efficiently have mediated this solution. Various writers have formulated propositions that collectively meet these three challenges. We can define art as the disposition for creating artifacts that are emotionally charged and aesthetically shaped in such a way that they evoke or depict subjective, qualitative sensations, images, or ideas. Literature, specifically, produces subjectively modulated images of the world and of our experience in the world. The disposition for creating such images would have solved an adaptive problem that, like art itself, is unique for the human species: organizing motivational systems disconnected from the immediate promptings of instinct. The design features that mediate this adaptive function are the capacities for producing artistic constructs such as narrative and verse and emotionally modulated musical and visual patterns.

The core element in this hypothesis—the adaptive problem art is designed to solve—is formulated most clearly by E. O. Wilson in *Consilience*. Wilson directly poses the question also posed by Pinker:

> If the arts are steered by inborn rules of mental development, they are end products not just of conventional history but also of genetic evolution. The question remains: Were the genetic guides mere byproducts—epiphenomena—of that evolution, or were they adaptations that directly improved survival and reproduction? And if adaptations, what exactly were the advantages conferred?

Wilson's answer to this question draws a decisive line between the mental powers of humans and other animals. Other animals are "instinct-driven." Humans are not. "The most distinctive qualities of the human species are extremely high intelligence, language, culture, and reliance on long-term contracts." The adaptive value of high intelligence is that it provides the means for behavioral flexibility—for generating plans based on mental representations of complex relationships, engaging in collective enterprises requiring shared mental representations, and thus producing novel solutions to adaptive problems. Behavioral flexibility has made of the human species the most successful alpha predator of all time, but achieving dominance in this way has come with a cost. Wilson speaks of the "psychological exile" of the species. To the modern human mind, alone among all minds in the animal kingdom, the world does not present itself as a series of rigidly defined stimuli releasing a narrow repertory of stereotyped behaviors. It presents itself as a vast and potentially perplexing array of percepts, inferences, causal relations, contingent possibilities, analogies, contrasts, and hierarchical conceptual structures. The human mind is free to organize the elements of cognition in an infinitely diverse array of combinatorial possibilities. And most of those potential forms of organization, like most major mutations, would be fatal. Freedom is the key to human success, and it is also an invitation to disaster. This is the insight that governs Wilson's explanation for the adaptive function of the arts. "There was not enough time for human heredity to cope with the vastness of new contingent possibilities revealed by high intelligence.... The arts filled the gap."[55] If instincts are defined as stereotyped programs of behavior released automatically by environmental stimuli, we can say that in humans the arts partially take the place of instinct. Along with religion, ideology, and other emotionally charged belief systems, the arts form an imaginative interface between complex mental structures, genetically transmitted behavioral dispositions, and behavior.

High human intelligence is part of a larger, systemic structure of species-typical adaptations that include altricial birth, extended childhood, male-female bonding coupled with male coalitions, dual parenting, postmenopausal survival, longevity, the development of skills for the extraction of high-quality resources, an enlarged neocortex that enhances powers for suppressing impulses and engaging in long-term planning, symbolic capacities enabling identification with extended social groups ("tribal instincts"), egalitarian dispositions operating in tension with conserved dispositions for individual dominance, and the power to subordinate, in some degree, impulses of survival and reproduction to the formal dictates of imagined virtual worlds.

The early EP conception of the mind supposes a sequence in which automatic cognitive processes evolved to solve adaptive problems specific to Pleistocene ecology, with the arts tacked on as side effects. The alternative vision formulated by Wilson supposes that human cognitive capacities evolved specifically for the purposes of generating adaptive flexibility.[56] In that alternative evolutionary scenario, dispositions to produce and consume works of imagination coevolved in functional interdependence with high intelligence. The affective neuroscientists Jules and Jaak Panksepp vividly evoke this vision of an integrated, systemic evolution of human cognitive powers:

What those vast cerebral expansions that emerged during the Pleistocene probably provided was a vast symbolic capacity that enabled foresight, hindsight, and the brain-power to peer into other minds and to entertain alternate courses of action, thereby allowing humans to create the cultures that dominate our modern world. . . .

What makes humans unique, perhaps more than anything else, is that we are a linguistically adept story-telling species. That is why so many different forms of mythology have captivated our cultural imaginations since the dawn of recorded history.[57]

We are a linguistically adept story-telling species because telling stories is one of the chief ways we give shape to our experience and thus ultimately direct our behavior. As Terrence Deacon puts it, "We tell stories about our real experiences and invent stories about imagined ones, and we even make use of these stories to organize our lives. In a real sense, we live our lives in this shared virtual world."[58]

Dispositions for creating and enjoying art form part of the larger evolutionary process known as "gene-culture coevolution." "Culture" includes technology and social organization as well as art, religion, and philosophy. Conceiving culture in this broader sense, evolutionary anthropologists often cite lactose tolerance as an instance of gene-culture coevolution.[59] Through natural selection, herding peoples have evolved enzymes that enable adults to digest milk. The cultural practice of keeping cattle serves as a selective force that alters the gene pool in a given population, and in turn the altered gene pool encourages the expansion of a pastoral economy. Language offers another clear instance of this kind of selective pressure. At some point in the ancestral past, humans had no power of speech. Mutations enabling rudimentary forms of "proto-language" would have given some selective advantage to those who possessed them.[60] That advantage would have increased the representation of those genes in the population at large, and the increase in those genes would have enhanced the linguistic character of the cultural environment, intensifying the selective advantage conferred by genes promoting the use of language.

A similar logic applies to imaginative culture. Developing the power of creating imaginative virtual worlds must have had adaptive value for our ancestors. Otherwise, capacities for imaginative culture would not now be human universals; artistic behavior would not spontaneously appear in all normally developing children; and humans would not display cognitive aptitudes specifically geared toward the production and reception of art—dispositions, for instance, for organizing pitched sounds in rhythmically and emotionally expressive sequences, for constructing visual designs that produce distinct moods and states of contemplative attention, and for constructing fictional narratives that generate excited, empathic responses in audiences.[61] These three factors—universality, reliable spontaneous development, and dedicated cognitive aptitudes—all suggest that dispositions for the arts were adaptive. If that is in fact the case, dispositions for producing and consuming the arts would have served as a selective force on the population, altering the gene pool, favoring those genes that facilitate producing and consuming works of art.

Somewhere between 100,000 and 30,000 years ago, there was a transformation in human culture that anthropologists designate the "Human Revolution."[62] Archeologically preserved forms of imaginative culture—art, decoration, ceremonial burial—appeared for the first time, and along with them, complex multipart tools, sewn clothing, and extended forms of trade, implying more complex forms of social organization. In *The Prehistory of the Mind*, Steven Mithen forcibly drew attention to the magnitude of this transformation and used it as evidence against the narrow-school EP conception of the massively modular mind. Countering the theory that cognitive flexibility arises from the multiplication of modules, all working automatically in response to regularities in the ancestral environment, he argued that the Human Revolution was generated by a genetically based cognitive transformation, a mutation involving language, that gave humans a vastly expanded flexibility in symbolic representation. His concept of "cognitive fluidity" is essentially a concept of metaphor: the power of linking images and ideas across diverse domains. To that power he attributes the sudden efflorescence of technological innovation and artistic production that characterizes the Human Revolution. Other theorists have argued for a more gradual evolution of human cognitive capacities.[63] I think the advocates of the Human Revolution will ultimately have the better part in this argument. In any case, at whatever pace it came about, there can be little doubt that modern symbolic culture—the culture of the past 100,000 years—differs in radical ways from the culture of the early and middle phases of hominid evolution.

The very existence of modern symbolic culture runs counter to the EP conception of human cognitive evolution—to massive modularity and the massively homogeneous character of the ancestral environment. Hence the virtual necessity, for acolytes of EP, for explaining away modern symbolic culture, treating it as merely a side effect to the adaptive structures that solved challenges supposed constant throughout the whole of the Pleistocene.

"Human nature" means that humans share species-typical dispositions: basic motives tied closely to the needs of survival, mating, parenting, and social interaction. Cognitive and behavioral flexibility are part of human nature, but they have not eliminated the underlying regularities in basic motives. In different ecologies and different forms of social organization, the elements of human nature combine in distinctive ways, but "culture" cannot build structures out of nothing. It must work with the genetically transmitted dispositions of an evolved and adapted human nature. The arts give imaginative shape to the experiences possible within any given culture, reflecting its tensions, conflicts, and satisfactions. One chief aim for evolutionary studies in the humanities is to analyze the way any given culture organizes the elements of human nature; evaluate the aesthetic, emotional, and moral qualities inherent in that organization; and probe the way it influences—by conformist pressure or antagonistic stimulus—specific works of literature.

To formulate plausible and testable hypotheses about the adaptive function of the arts, we have to satisfy three criteria: (a) define the arts in a way that identifies what is peculiar and essential to them—thus isolating the behavioral disposition in question; (b) identify the adaptive problem this behavioral disposition would have solved in ancestral environments; and (c) identify design features that would efficiently have mediated this solution.[64] We can define art as the disposition for creating artifacts that are emotionally charged and aesthetically shaped in such a way that they evoke or depict subjective, qualitative sensations, images, or ideas. Literature, specifically, produces subjectively modulated images of the world and of our experience in the world. The disposition for creating such images would have solved an adaptive problem that, like art itself, is unique for the human species: organizing motivational systems disconnected from the immediate promptings of instinct. The design features that mediate this adaptive function are the capacities for producing artistic constructs such as narrative and verse and emotionally modulated musical and visual patterns.

Consider the reality of our experience. We live in the imagination. For us, no action or event is ever just itself. It is always a component in mental representations of the natural and social order, extending over time. All our actions take place within imaginative structures that include our vision of the world and our place in the world—our internal conflicts and concerns, our relations to other people, our relations to nature, and our relations to whatever spiritual forces we imagine might exist. We live in communities that consist not just of the people with whom we come directly into contact but with memories of the dead, traditions of our ancestors, our sense of connection with generations yet unborn, and with every person, living or dead, who joins with us in imaginative structures—social, ideological, religious, or philosophical—that subordinate our individual selves to some collective body. Our sense of our selves derives from our myths and artistic traditions, from the stories we tell, the songs we sing, and the visual images that surround us.

We have all had moments in which some song, story, or play, some film, piece of music, or painting, has transfigured our vision of the world, broadened our minds, deepened our emotional understanding, or given us new insight into human experience. Working out from this common observation to a hypothesis about the adaptive function of literature requires no great speculative leap. Literature and the other arts help us live our lives. That is why the arts are human universals. In all known cultures, the arts enter profoundly into normal childhood development, connect individuals to their culture, and help people orient to the world, emotionally, morally, and conceptually.

If it is true that the arts are adaptively functional, they would be motivated as emotionally driven needs. The need to produce and consume imaginative artifacts would be as real and distinct a need as hunger, sexual desire, maternal and filial bonding, or the desire for social contact. Like all such needs, it would bear within itself, as its motivating mechanism, the pleasure and satisfaction that attend upon the fulfilling of desire. That kind of fulfillment would not be a parasitic by-product of some other form of pleasure, nor merely a means for fulfilling some other kind of need—sexual, social, or practical. Like all forms of fulfillment, the need for art could be integrated with other needs in any number of ways. It could be used for sexual display or the gratifications of sexual hunger or social vanity, and it could be used as a medium for social bonding. Nonetheless, in itself it would be a primary and irreducible human need.

Reduction and the Problem of Generating New Knowledge

In literary criticism, the word "reductive" is typically used with a pejorative connotation, the idea that a reading strips out essential features in a text or falsely conflates too many features with some one feature wrongly invested with elemental significance. To be "reductive" is to have failed in subtlety and justness. And yet, "reduction" is the ultimate aim in all efforts at producing real knowledge. We seek to reduce the multiplicity of surface phenomena to underlying regularities. We presuppose that some form of conceptual or causal hierarchy is built into the very nature of our subject, whether that subject is some aspect of the natural world or a literary text. Even the most rudimentary form of literary commentary—analytic summary or paraphrase—constitutes an exercise in reduction. As I already argued, both traditional humanists and poststructuralists have their own typical forms of reduction. In this respect, the Darwinists do not differ from critics in other schools. They differ only in the terms to which they seek to reduce texts.

Can the Darwinists produce new knowledge? That is the most serious challenge posed to the evolutionary literary critics.[65] The Darwinists would themselves add a further challenge alien to the relativist mind-set of their poststructuralist critics. Can the Darwinists produce formulations that are not only new but true? In one respect, the Darwinists start at what, from a poststructuralist perspective, might seem a disadvantage. If they believe that texts embody a folk understanding of human nature and that texts are communicative constructs, they must also suppose that most texts are understood reasonably well at the level of common language and common knowledge. They have no prefabricated sign systems, like those of Lacan or the Althusserian Marxists, into which they are prepared to translate the common-language content of a text. Are they then reduced to ordinary analytic summary, but without consoling recourse to the idealizing sentiments of the humanists?

Producing new knowledge—real knowledge, knowledge that is consilient with the broader world of empirical research—is difficult in literary study, as it is in other fields, but it is not impossible. The

Darwinists can aim at extending, refining, correcting, and contextualizing the common understanding. On the level of interpretive criticism, they can situate any given text or set of texts in relation to the pressure points in human nature. They can identify the biological forces that are invoked or repressed in any given work and can assess how those forces impinge on meaning. That interpretive effort opens a new range of aesthetic sensations and a new range of comparative analysis for the Darwinian critics. The governing terms in an evolutionary critique are not metaphysical abstractions, mid-level social and psychological concepts, or formalist principles. The governing terms are the urgent needs and driving forces in life—survival, reproduction, kinship, social affiliation, dominance, aggression, and the needs of the imagination. Physical realities and the rhythms of the life cycle shape the analytic categories through which Darwinians make sense of literary depictions. The governing terms in human life history are a matrix for other terms that are analytically neutral but resonant with elemental power and pregnant with qualitative differences. The shape of human life history is a basic reality shared by all authors and readers. Differences in the way any given author envisions that life history are essential to the imaginative qualities that distinguish the author, and those differences enter minutely into the subtlest nuances of tone, style, and formal organization. An evolutionary perspective can thus provide a comprehensive framework for comparing the perspectives of authors, the organization of meaning in texts, and the responses of readers. Moreover, like critics from other schools, literary Darwinists have a scholarly responsibility to adjudicate differences in the critical tradition, with all its local agendas and cultural conditioning. They also have an obligation to situate texts and critical histories in the broader context of evolutionary social science, connecting local critical perceptions with general principles of literary theory, and integrating those principles with principles of psychology, linguistics, and anthropology. They have an opportunity to synthesize ideas and insights from diverse fields, and those integrations can often lead to new concepts. Those new concepts, in turn, should provide the means for new critical insights into familiar texts.

To give some instances. Few if any literary works have been more discussed than those of Homer and Shakespeare, and yet the Darwinists have devoted a good deal of attention to both. By locating the *Iliad* within the context of modern sociobiological theory, Gottschall assimilates previous explanations of Homeric conflict within a single, unified causal structure. That unified explanation is intrinsically satisfying, and it also brilliantly illuminates the imaginative qualities of the poem. As Gottschall observes, concentrating on the struggle for women among Homer's barbarian warriors "helps to explain more about Homeric society than its relentless violence; it also sheds light on the origins of a tragic and pessimistic worldview, a pantheon of gods deranged by petty vanities, and a people's resignation to the pitiless dictates of fate."[66] Robert Storey and Michelle Scalise Sugiyama both discuss reader responses to *Hamlet* among a Nigerian tribal population, thus illuminating both the universal features of the text that are available across divergent cultures and the interpretive limitations in a culturally circumscribed critical perspective.[67] Daniel Nettle is both an evolutionary psychologist and a professional Shakespearean actor. Commenting on *Hamlet* and several other plays of Shakespeare, he correlates the generic structure of comedy and tragedy with elemental motives of status-seeking and mate selection.[68] This effort in strong reduction moves in a direction opposite to that of Brian Boyd, who uses *Hamlet, The Odyssey, Lolita,* and other texts to demonstrate the "expansiveness" of readings that can incorporate a diversity of cognitive mechanisms and evolutionary themes. In discussing Dr. Seuss's *Horton Hears a Who,* Boyd triangulates the story with appeals to highly particularized local cultural history (postwar Japan), developmental cognitive psychology, and universal human motives. In his commentary on the comics artist Spiegelman, Boyd brings his analytic repertory to bear on the cognitive strategies of avant-garde art.[69] One of the showpieces of evolutionary anthropology is a decisive demonstration that Freudian Oedipal theory is quite simply mistaken.[70] In her commentary on *Oedipus Rex,* Scalise-Sugiyama removes Sophocles' play from the distorting context of Freudian Oedipal theory and locates it within the more illuminating context of evolutionary findings on incest avoidance. Making use of the same body of research, Nancy Easterlin examines feminist psychoanalytic accounts of Wordsworth's autobiographical poetry. By invoking John Bowlby's ethological research into attachment between mothers and infants, she produces a more cogent and adequate account of her subject. (In an autobiographical essay, Dylan Evans, an expert on Lacan, describes his gradual disenchantment with Lacanian psychology.)[71]

Focusing on Sir Walter Scott's *Ivanhoe*, Ian Jobling explains how principles derived from evolutionary psychology can deepen our understanding of common critical observations. He gives evidence that critics have commonly observed a split between "dark" and "light" heroes in Scott's novels, and he argues that such observations have remained merely descriptive.

> All literary criticism relies on assumptions about human nature, but *these assumptions have never been clarified or justified.* Indeed, in the cultural determinist atmosphere that prevails in contemporary literary criticism, it is taboo even to make reference to "human nature." This absence of a theory of human nature in literary criticism is one of its major weaknesses.[72]

Brett Cooke's appeal to basic features of human nature proves effective as a thematic grid for organizing the whole field of utopian and dystopian literature. In her reading of Edith Wharton's novel *The Children*, Judith Saunders uses a sociobiological account of human life history to examine social configurations arising from the disruption of normal childhood development in the milieu of Jazz Age hedonism, thus usefully connecting human universals with local cultural history.[73]

In my critique of Oscar Wilde's *The Picture of Dorian Gray* (in part 2), I bring together various strands of criticism—the autobiographical and psychodramatic elements, the influence of Pater's aestheticism, the submerged homoeroticism of the text, and the Christian themes. I draw on recent studies in queer theory but locate them within a broader theoretical context derived in part from Darwin's own theory of the evolved moral dispositions of human nature and in part from Donald Symons' evolutionary researches into the psychological character of homoerotic sexual relations. I have also used evolutionary psychology for comparing levels of integration between human nature, cultural order, and individual identity in five novels; and I have articulated principles of literary evaluation from an evolutionary perspective in relation to novels depicting Paleolithic life. In a reading of *Pride and Prejudice*, I integrate elements from life history theory with an analysis of interplay in point of view among characters, authors, and readers.[74] In company with Jon Gottschall, John Johnson, Dan Kruger, and Stelios Georgiades, I coauthor an interpretive critique of Hardy's *Mayor of Casterbridge* that fundamentally revises the critical tradition on that novel (in part 3). We also use quantitative methods to delineate features of tone and gender across all of Austen's novels. We demonstrate that the common idiom of literary commentary can be effectively reduced to a structured set of categories lodged within a theoretically grounded model of human nature and of literature, and we maintain that by analyzing the relations among specific individual findings, we have discovered patterns of meaning that are not readily apparent when each finding is observed in isolation.[75]

For the past three decades, crucial elements of the common understanding have been ruled out of play in poststructuralist discourse—mimetic referentiality, human nature, and the individual human "subject" as an originating source.[76] If evolutionary literary study did nothing more than clear away these distorting theoretical impedimenta, returning criticism to a ground of common understanding from which critics would still be required to generate new knowledge through scholarly research, it would have performed a valuable service. At the present time, producing a firm empirical foundation for elementary concepts that are integral to the literary tradition has a distinct value in its own right. Any scholar can benefit from having a solid basis on which to dispute false psychology, false epistemology, and inadequate social theories. But the Darwinians need not rest content with this merely negative merit. The examples I have cited here should be sufficient to indicate that an evolutionary perspective can meet the challenge of generating literary knowledge that is both new and valuable.

The Question of Empirical Methodology

Assessing the prospects for cognitive literary study, Tony Jackson points to an asymmetry in the relations between an empirical theory and its interpretive applications in literary study: "An application of that theory to literature may well change something of our understanding of literature, but it is difficult to see how the interpretive practice can possibly change the theory." Even if the theory produces illuminating insights into texts, the absence of a reciprocal influence on the theory must be felt as a source of frustration for scholars who use the theory. As Jackson rightly observes, "a truly dialectical relationship between theory, method, and practice seems to provide a basic intellectual appeal to the majority of scholars."[77] Most literary scholars have no training in quantitative methods, little or no understanding of statistics, and no sense of how to contrive an experimental design aimed at falsifying a hypothesis. How then are they to respond to the problem of the asymmetry identified by Jackson?

As a defensive measure, it is always possible to deny that literary scholars have any need for empirical knowledge. Eugene Goodheart takes this tack. Like many humanists, Goodheart is an epistemological and metaphysical dualist. He believes that the sciences deal with regularities in the physical world, and the humanities deal with qualitatively unique individual texts. In practice, of course, no literary scholar concerns himself only with the qualitatively unique. All literary scholars make appeal, in however occasional or opportunistic a fashion, to regularities of human nature and of cultural tradition. Goodheart hedges his claims by arguing that literature is not exclusively but only *primarily* interested in the qualitatively unique, but in celebrating the spiritual power of humanistic experience, Goodheart himself makes appeal to archetypal motifs of spiritual redemption—that is, to regularities of literary figuration grounded in regularities of human psychology.[78] Much serious traditional scholarship concerns itself not with special cases but with regularities in the literary tradition. Goodheart's appeal to archetypal patterns is extended on a monumental scale in Northrop Frye's *The Anatomy of Criticism*. M. H. Abrams' *The Mirror and the Lamp* invokes heavy symbolic reductions to organize vast tracts of literary history. Ian Watt's *The Rise of the Novel* articulates major trends in the development of narrative and formulates fundamental principles of realist representation. Such a list could be extended almost indefinitely. The point should be clear. Trying to isolate literary study from psychological and historical generalizations is a sophistical maneuver that will not stand against the simplest appeal to factual evidence.[79]

Literary critics cannot do without appeal to regularities of human psychology. Need they then be wholly dependent on the productions of adjacent fields? That depends on their own initiative. A substantial number of literary scholars have made some efforts to incorporate empirical methods from the social sciences.[80] A smaller number have made efforts to adopt empirical methods and also to locate those methods within the field of Darwinian social science. As mentioned in a previous section, Jonathan Gottschall has successfully used word searches to study cross-cultural depictions in folk and fairy tales. Kruger, Fisher, and Jobling have used depictions of male characters in Byron and Scott to test whether contemporary readers respond to contrasting male mating strategies in predictable ways. Salmon and Symons have used empirical methods in examining romance and pornography. Stiller and others have used empirical methods in assessing group size and social links in plays and soap operas. Miall and Dissanayake have done an empirical prosodic analysis of mother-infant interaction.[81] Gottschall, Johnson, Kruger and I have used a model of human nature to examine characterization and reader response in numerous characters from dozens of novels (see part 3). In the conclusion to the book manuscript produced from that research, we explain the significance we think this kind of research has for both literary study and evolutionary social science:

> Research that uses a purely discursive methodology for adaptationist literary study remains passively dependent on the knowledge generated within an adjacent field, and it does not contribute in any very substantial way to that primary source of knowledge. The methodological barrier that separates discursive literary study from the adaptationist program in the social sciences limits the scope and significance both of literary study and of adaptationist social science. The production and consumption of literature is a large and vitally important part of our specifically human nature. An artificial barrier that leaves adaptationist literary scholars in the stance of passive consumers of knowledge also leaves adaptationist social scientists cut off from any primary understanding of one of the most important and revealing aspects of human nature. Literature and its oral antecedents derive from a uniquely human, species-typical disposition for producing and consuming imaginative verbal constructs. Removing the methodological barrier between humanistic expertise and the expertise of the social sciences can produce results valuable to both fields.[82]

We made specific predictions and tested specific hypotheses, but some of the most important conclusions we reached were surprising to us. These conclusions have broad general implications and constitute new knowledge—not just empirically confirmed formulations of commonly received and accepted ideas. (Our focal point was "agonistic structure" or the nature and organization of protagonists, antagonists, and minor characters in the novels.) We analyze our findings by reference to sources in evolutionary social psychology, research into personality, and the theory of basic emotions. We draw on concepts in both psychology and literary theory, and we regard our findings as contributions both to literary knowledge and to evolutionary psychology.

This whole area of research is less well developed than areas that depend on purely discursive techniques of theoretical formulation and literary interpretation. The capital costs in gaining appropriate expertise are steep—steep for literary scholars, who would have to gain some familiarity with empirical methods, and steep also for psychologists and anthropologists, who would need to assimilate concepts and modes of thinking characteristic of the humanities and appropriate to them. The costs are steep, but the benefits are potentially very large. The chief obstructions are not intrinsic but institutional. Given a few programs in which such work was encouraged, and in which the appropriate interdisciplinary training for graduate students was put into place, it is quite certain that ambitious young scholars, suitably geared in aptitude and interest, would quickly find ways to extend and develop the few pioneering efforts that have been made so far. The purely discursive study of science and literature, from largely culturalist and Foucauldian perspectives, has been a growth industry for nearly two decades. That approach to interdisciplinary study gives predominance to an ideologically charged rhetorical analysis of science. It is surely now time for both literary scholars and scientists to explore the possibilities of a different form of interdisciplinary study, a form that gives predominance to scientific method and the scientific ethos.[83]

Conclusion: The Future

Who knows? Perhaps in ten or twenty years, looking back, cultural historians will be denying that the humanities and the evolutionary social sciences were ever in any way at odds with one another. The integration of historical scholarship with a knowledge of human universals will have become standard equipment in literary study. Humanistic expertise in manipulating cultural figurations will have flowed into a smooth and harmonious stream with Darwinian findings on the elemental features of human nature. Humanistic sensitivity to the fine shades of tone and style in literary works will have blended seamlessly with a rigorous empirical analysis of cognitive mechanisms, and a facility in writing elegantly nuanced prose will mingle happily with the severe logic of a quantitative methodology. Scholars and scientists occupied with literary study will balance with easy grace between the impersonal, objective scrutiny of science and a passionate humanistic responsiveness. All of this is possible, and it is worth working toward. Any of it that we can realize will be a gain for ourselves and a contribution to the sum of human understanding.

The First Sequel

This first sequel consists in a condensed version of my rejoinder to the responses to the target article in *Style*. In this version, most of the particular references to individual respondents have been excised.[84]

The Range of Response

Responses to the target article can be roughly sorted into four groups distinguished by hypothetical ratings they might have assigned to the contentions in the target article: (1) Reject, (2) Revise and Resubmit, (3) Accept with Some Revision, and (4) Accept. In the "Reject" group, I would locate the three responses aimed at general, across-the-board repudiation. "Revise and Resubmit," containing five responses, could be defined as "cautious partial acceptance, with some substantial reservations." The third group (fifteen responses) is the largest. It contains commentaries by respondents who accept the idea of an evolutionary paradigm but concentrate on challenging one or more formulations in the target article. The fourth group (eight responses) consists of respondents who accept the idea of an evolutionary paradigm and occupy themselves chiefly with reflecting on its rationale, probing its conceptual structure, or extending its reach. The third and fourth groups overlap a good deal. Most of the respondents who challenge specific formulations in the target article also offer general reflections on evolutionary literary studies, and no respondent, presumably, agrees completely with every formulation in the target article.

Brett Cooke observes that "some readers might understandably quail at the prospect of an all-encompassing, apparently monolithic, critical perspective."[85] They might, and they do. Some respondents raise the question as to whether the evolutionary human sciences themselves display any recognizable consensus. Conversely, one respondent asks for clarification on the key controversies in human evolutionary research. Some respondents tacitly accept the findings of evolutionary social science but still question whether biological reductions can encompass all things human. The most important alternative to adaptationist views of human nature is cultural constructivism—the idea that culture exercises autonomous causal force in human thought, feeling, and behavior. Some theorists would explicitly repudiate cultural constructivism but still worry that an evolutionary approach will strip out specifically literary modes of thought. Others argue that the findings of evolutionary psychology might be valid but are often not relevant to specifically literary concerns. Some respondents suggest limitations in the range of literary works that can be effectively brought within the interpretive rubric of evolutionary psychology.

More than half of the respondents believe that evolutionary social science can provide the basis for a Grand Theory of Literature and acknowledge the necessity of incorporating concepts that are specifically literary. They also recognize that theories about the adaptive function of literature form a necessary bridge between these two domains. For many respondents, that is where agreement stops. The amount of attention the respondents devoted to the adaptive function of literature signals that this issue is both crucially important and heavily disputed. My reflections on the issue of adaptive function have been incorporated into a revised and expanded section, "The Adaptive Function of Literature: A Controversy," in the main body of this first chapter.

A Grand Theory

> There is grandeur in this view of life, with its several powers, having been originally breathed into a few forms or into one; and that, whilst this planet has gone cycling on according to the fixed law of gravity, from so simple a beginning endless forms most beautiful and most wonderful have been, and are being, evolved.[86]

The history of failed efforts at establishing a scientifically oriented paradigm for literary study must naturally give us pause. While contemplating this dreary history, though, we should locate it within a deeper intellectual history. The history of geology in the centuries before Lyell consisted in often fantastic speculative systems. Geologists became so disgusted with these fanciful speculations that in the decades immediately preceding the publication of Lyell's *Principles of Geology*, they called a moratorium on general speculative accounts of geological change. Instead of speculating, they assiduously set about the practical business of establishing the relative historical position of strata across the world.[87] If Lyell had taken the history of failed speculations as a guarantee that no general theory could ever be true, he would not have written his *Principles* and would not have established the paradigm that still governs the science of geology. Consequently, he would not have prepared the ground for Darwin's theory of natural selection. Had Darwin himself been sufficiently intimidated by the speculative inadequacies in the theories of Lamarck, Chambers, and Spencer, he would not have ventured to publish his views on natural selection. The list of such instances could be extended indefinitely. Mature theories do not typically spring full blown from nothing, nor are they always instantly recognized as basically correct. Darwin's theory of natural selection had to wait another sixty or seventy years before it finally achieved paradigmatic status in the Modern Synthesis that integrated genetics with evolutionary biology. The point of these comparisons is not that any specific theory in evolutionary literary study has yet achieved or merited paradigmatic status. The point is only that every ultimately successful theory has to start somewhere. A history of previous failures is no cause for dismay. One has to assess any given theory on its merits.

The time is ripe for evolutionary literary study. Evolutionary social science has emerged only within the past thirty or forty years, and it provides an indispensable intellectual context for the development of evolutionary literary study. In this respect, evolutionary social science is to literary study what Lyell's geology was to Darwinian evolutionary biology, and what Darwin's evolutionary biology was to evolutionary social science.

The Consensus Model in Evolutionary Social Science

The target article identifies a coordinated suite of species-typical features about which most evolutionary social scientists would now agree. Taken singly, each of these claims has interest and value. More importantly, individual findings, interesting as they might be, take their full value only in their systemic relations with one another. Life history theory—the cycle of birth, growth, reproduction, and death—provides a framework within which we can compare human nature with the nature of other species and can also make biological sense of the interdependence among the elements of human nature. The basic outlines of specifically human life history are well known. Those outlines impose constraints on plausible explanations about each of the specific features of human nature mentioned in the target article. Hence the consensus model.

Within the consensus model, as in any healthy field of research, there are unsolved problems. That is why research exists. Possible solutions to each problem consist in alternative, competing hypotheses. In the target article, I identify one central problem that is of particular importance both for evolutionary literary study and for evolutionary social science in general: the adaptive function of literature and the other arts. In the target article, I locate this issue within an account of a broad theoretical conflict over "massive modularity" and the nature of the EEA or "environment of evolutionary adaptedness." Evidence bearing on that conflict derives from multiple areas of research, each of which itself contains important unsolved problems.

A detailed survey of current problems in evolutionary psychology would include the following issues: the pace and nature of the evolution of language; the role of sexual selection in the development of higher cognitive faculties; homosexuality as adaptation, by-product, or dysfunction; the relative causal force of foraging, cooking, innovation, group size, and social interaction in the evolution of the enlarged human brain; the relative causal force of male provisioning and female coalitions in the evolution of human family structures; the adaptive significance of individual differences in personality; the relation between cognitive "modules" and "general intelligence"; the number and character of "basic emotions"; the interactions between basic emotions and Theory of Mind in the formation of more complex emotions; the limits of plasticity in the correlation between adaptively conditioned motive structures and affective responses; the exact character of the interactions between evolved cognitive mechanisms and fitness maximizing algorithms in human nature; the origin and nature of "altruism"; the existence of "tribal instincts" or social dispositions extending beyond kin but not restricted to direct social exchange; the exact nature of the interaction among multiple levels of selection (the gene, the individual, the kin group, and the larger social group); interactions between dominance, cooperation, and symbolic thinking in the evolution of elementary political dynamics; the way elementary political dynamics constrain complex social formations that contain advanced functional specialization and elaborate status hierarchies; the precise nature of gene-culture coevolution; the adaptive function of religion; and the adaptive function of literature and the other arts.[88]

Several of these research problems bear directly on a crucial historical issue: the emergence of symbolic culture and complex, multipart tools somewhere between 100,000 and 30,000 years ago. One group of scientists and scholars argues that the emergence of "modern" human behavior was gradual and cumulative, not involving any relatively sudden alteration in the organization of the brain. Another group argues that the epochal transition from the culture of the old stone age was in fact a "revolution," that it was relatively sudden, and that it was probably precipitated by some tipover mutational event in the brain, most likely an event involving language.[89]

As important as these various problems are, the alternative hypotheses put forward as possible solutions offer no serious threat to the consensus model of human life history delineated in the target article. As research progresses, we shall no doubt continue to refine our understanding of the elements in the model. It is virtually certain that the elements will be incorporated within ever deeper and more integrated levels of causal explanation.

Over the past several years, the evolutionary human sciences have been undergoing two substantial paradigmatic corrections. One is in the gradual convergence between the idea of fitness maximization, associated with sociobiology and "behavioral ecology," and the idea of "proximate mechanisms," associated with early evolutionary psychology.[90] This correction is closely connected with theoretical debates over "massive modularity" and the EEA. The other correction, concerning "levels of selection," is potentially even more important. In the eighties and early nineties, the idea that "selection" worked only at the level of genes and individual organisms, not at the level of social groups, limited and distorted our understanding of the evolution of specifically human adaptations for social life. Social adaptations for humans, like those for other primates, were understood to consist solely in dispositions for nepotism, direct exchange or reciprocation, and dominance/submission. Many of the most prominent theorists in the field have now come to recognize the importance of "multilevel selection." For humans, that includes selection at the level of social groups not closely related by kinship. Adaptations for group life above the level of the "band" solve the puzzles of "altruism" in a way that direct reciprocation never could.[91]

By gradually correcting the selectionist paradigm to include group-level processes for humans, evolutionary theorists have almost inadvertently also prepared the ground for developing more adequate evolutionary concepts of culture. Most evolutionary thinking about culture has thus far focused on mechanisms for developing technology and enforcing social norms. I anticipate that the next phase in the development of an evolutionary theory of culture will include the dissemination of a more complete and adequate understanding of the adaptive function of literature and the other arts. That phase has already begun: a collective theoretical effort to formulate an adaptive theory of the arts that includes social functions but that also identifies a psychological function at a more basic level—the level at which humans achieve cognitive and affective orientation to their total environment, physical as well as social.

An Indispensable Contextualism

Declaring that literary study is somehow exempt from the advance of empirical knowledge licenses literary scholars to restrict their professional expertise to producing rhetorical variations on theoretical formulas derived from obsolete versions of linguistics, economics, sociology, and psychology. The truth is, we cannot do without contextual knowledge. We can cordon off our own obsolete versions of that knowledge, cherry-picking the science that offers superficial verbal correlatives to our theoretical formulas. Or we can take seriously the challenges of responsible interdisciplinary research, gain the expertise we need to make reasonable judgments in the field, and accept the responsibility of producing knowledge that can meet at least minimally adequate standards of empirical validity.

Given the nature of disciplinary specialization in anthropology, archeology, and psychology, it is hardly surprising that evolutionary social science has thus far concentrated heavily on "deep history"—on history extending millions of years into the past—and on contemporary human behavior. The task of integrating this kind of knowledge with cultural history at the middle range, from the beginnings of recorded history to the present, would almost necessarily fall to the lot of scholars in the humanities. Taking up that task presents an immense challenge, and it offers an unparalleled opportunity.

Evolutionary Cultural Theory

The evolutionary anthropologist Kim Hill identifies hitherto inadequate conceptions of culture as the chief obstacle inhibiting more rapid development in evolutionary social science.[92] Assimilating theoretical work by Richerson and Boyd, among others, Hill identifies essential elements of specifically human culture that distinguish it from the kind of "culture" sometimes ascribed to nonhuman primates and a few other species. Human culture involves information that is not just socially transmitted but also cumulative, advancing through successive phases of innovation.[93] Chimpanzee bands develop distinctive styles of cracking nuts or fishing for termites, but they do not develop technological traditions in which one invention piggy-backs on another. In addition to the cumulative production of innovation, Hill identifies two other distinctive features of specifically human forms of culture: norms or rules of behavior that are reinforced by punishments and rewards; and "signaling" that perpetuates the rules and communicates group identity.[94] With respect to technology and the evolution of cooperative group behavior, the principles succinctly delineated by Hill represent the state of the art in evolutionary theories of culture. So far as they go, these principles seem sound.

To think effectively about culture from an evolutionary perspective, one has to thread one's way between two faulty extremes. One extreme consists in the belief that human cultural capabilities display no significant differences from the cultural capabilities of other species. The other extreme consists in the belief that human culture has kicked itself loose from human nature and now operates independently of genetically conditioned behavioral dispositions. Hill finds the right middle ground, the complementary claims that culture is the product of "evolved cognitive mechanisms" but that the emergence of culture has itself probably "uniquely shaped evolved human cognition and emotion." Commenting on the recent surge of research into "culture" in nonhuman animals, he observes that the difference in degree between human and animal "culture" is so large as to count as a difference in quality. Speaking of animal groups as separate cultures "undermines our ability to understand why *Homo sapiens* is a special case with special cognitive abilities."[95]

In speaking of culture as an element in natural selection, it is easy to conflate two ideas that would be better kept distinct. One idea is "gene-culture coevolution." In "An Evolutionary Paradigm for Literary Study" (in this volume), I explain that lactose tolerance is frequently used as an illustration for gene-culture coevolution. Through natural selection, herding peoples have evolved enzymes that enable adults to digest milk. The cultural practice of keeping cattle serves as a selective force that alters the gene pool in a given population, and in turn the altered gene pool encourages the expansion of a pastoral economy. The idea that needs to be kept distinct from gene-culture coevolution is "cultural evolution." This is the idea that certain cultural forms can be "selected" and "inherited" independently of their effect on natural selection. The argument for "cultural evolution," as distinct from gene-culture coevolution, is that certain cultural figurations ("memes") can work against reproductive success and nonetheless persist within limited parts of a population for a few generations.[96] Cultural evolution in this sense did not drive the evolutionary process that produced key features of an evolved human nature—big brains, linguistic skills, and ultra-sociality—that are preconditions for developing high levels of culture. All three of those features contribute directly to inclusive fitness. The evolutionary process that produced them is ordinary natural selection.

If we separate these two ideas—"gene-culture coevolution" and "cultural evolution"—we could say that the capacity for culture consists in big brains, linguistic skills, and ultra-sociality. As with lactose tolerance, the genes supporting these characteristics enter into a feedback loop with culture. Natural selection favors genes that produce big brains, language, and ultra-sociality in humans; these characteristics produce culture; and culture, in turn, acts as a selective force that alters the gene pool, further developing the capacity for culture. This kind of feedback loop constitutes gene-culture coevolution. To explain the evolution of culture, then, we need make no appeal to "cultural evolution" as a causal force that has "autonomous" power to direct the course of evolution.

Any theory of quasi-independent "cultural evolution" necessarily involves some kind of non-genetic "replicator." The most familiar version of such replicators, Dawkins' notion of "memes," is deeply problematic and has never achieved general acceptance among evolutionary thinkers. In Dawkins' conception, memes are ideas or cultural practices that supposedly function in ways parallel to but independent of genetic evolution. This supposed parallel will not stand inspection. No idea or cultural practice contains a molecular mechanism adapted by natural selection to replicate itself. Ideas and cultural practices can be disseminated and perpetuated only by activating psychological responses that affect behavior. Those psychological responses are themselves constrained by dispositions that have evolved through natural selection. Memes could thus not be "autonomous" in the way that genes are autonomous. Ideas and cultural practices are secondary and subordinate, causally, to the proximate mechanisms produced by natural selection.

Richerson and Boyd develop arguments both for gene-culture coevolution and for cultural evolution, but they identify no significant cultural structure that is not constrained by evolved dispositions. Their views on gene-culture coevolution are widely shared by other evolutionary theorists. Their arguments for "cultural evolution" remain more speculative and controversial.[97]

We should by no means neglect group-level processes. The common lexicon, to take an obvious example, contains the collective and shared intelligence of untold generations, a wealth of "knowledge" that could not possibly be generated in the life of any individual or even any succession of individuals. Culture extends the imaginative life of each of us far beyond the boundaries of individual experience, and in that crucial respect, it separates us from the other animals, even those animals that are capable of empathy for the other animals with which they come directly into contact. Even so, in taking due heed of the collective cultural mind, we need to be careful not to make the simple mistake of supposing that Dawkins' notion of memes offers the best available way of thinking about culture. Dawkins' notion is catchy in part because of the term itself, selected, for that purpose, with all the shrewd care an advertising executive gives to the selection of words in a jingle. At a deeper level, the notion of "memes" is catchy because it appeals to our natural cognitive disposition to handle difficult theoretical issues by thinking in analogies. Genes vary and are selected and inherited. So then might also "memes," we think, and in so thinking are led down many a blind alley shrouded in mist.

Memes possess no internal replicative mechanism, and no one has yet been able to identify structural principles for memes that would approximate to the sharply demarcated molecular structure of genes. Moreover, genes and organisms are tightly locked into functional interdependence. The organization of physiological processes in organisms displays a deep systemic coherence regulated by astonishingly complex genetic interactions. Specific genes are functionally organized to replicate organisms of a specific type, and specific types of organisms are functionally organized to replicate specific genes. The relation between memes and human organisms constitutes no such functionally integrated structure in a replicative process. Specific ideas and cultural practices (memes) might or might not contribute to the inclusive fitness of individual human organisms, but human organisms are not functionally organized to replicate specific memes. Throughout its life, an individual human organism retains the same genetic makeup but can undergo many profound changes in ideas and cultural practices. In these various ways, then, genes and memes display fundamental structural differences. As a result, using the analogy between genes and memes as a general heuristic for thinking about culture is almost certain to lure theorists into problematic speculation.

Conflict and Repression

In "An Evolutionary Paradigm," I argue that "the logic of natural selection has shaped human dispositions in such a way that all intimate relations involve conflict. . . . The workings of inclusive fitness . . . guarantee a perpetual drama in which intimacy and opposition, cooperation and conflict, are inextricably bound together." Conflict is integral to the whole Darwinian conception of natural relations. This observation has a wide bearing on the formal structure of dramatic depictions. All dramatic relations that end in perfect harmony necessarily repress the recognition of unresolved conflicts. Even so, life is more or less adaptively functional. Important local resolutions do in fact take place. A naturalistic perspective on human affairs would thus have good reason to be skeptical of any purely nihilistic vision of human life.

In applying the principle of inevitable conflicts, Darwinian criticism would converge with the "hermeneutics of suspicion" that characterize poststructuralist thought, but with a major difference. From the poststructuralist perspective, the norm against which all historical and actual power relations are measured is a norm of universal cooperative behavior—a world that is free of competing interests, and free of conflict. This utopian norm is a world in which "power," the differential exercise of force in social relations, no longer exists. Measured against this utopian norm, all historical and actual exercises of power are necessarily forms of gratuitous oppression. From a Darwinian perspective, in contrast, conflicting interests are an endemic and ineradicable feature of human social interaction. A Darwinian critic would thus not be disposed to evaluate historical structures of power relative to the norm of an imaginary world in which power does not exist.

I observe that Darwinians "can identify the biological forces that are invoked or repressed in any given work and can assess how those forces impinge on meaning." At least one response to the target article indicates that this formulation can be mistaken for an appeal to the Freudian concept of repression. Frederick Crews cites research that brings the specifically Freudian concept of repression into grave doubt. Yacov Rofé notes that the specifically psychoanalytic concept of repression involves "*pathogenic effects* on the individual's psychological and physical functioning, preventing both an accurate perception of reality that is necessary for adequate coping and a discharge of harmful tension."[98] Crews has now for decades been almost obsessively preoccupied with discrediting the Freudian notions to which he himself once subscribed, so it is understandable that he would have seized upon a single word with possible Freudian associations and would have located it, inappropriately, in a Freudian theoretical context. I have myself never subscribed to a Freudian conception of human mental architecture. The word "repressed," as I use it, has no specifically Freudian meaning. A glance into any standard dictionary will confirm that Freud did not invent this word and that psychoanalysis does not currently have an exclusive claim on its denotation. In the common language, "repression" means only to keep under control, to suppress, quell, or to put down by force. In the context in which I use it, the most pertinent meaning is that people commonly filter information about the world and about themselves, often eliminating or distorting evidence that does not accord with their own preconceptions, preoccupations, or desires.

As Rofé notes, repressing information need not be traumatic. Quite the contrary, avoiding unpleasant realities often serves the needs of mental health—of self-confidence and well-being. Having an accurate view of reality and of oneself also has adaptive utility, though, and the need for accuracy often conflicts with the need for distorting information in comforting ways. In a book delineating a biological conception of the unconscious parts of the mind, Timothy Wilson observes that "the conflict between the need to be accurate and the need to feel good about ourselves is one of the major battlegrounds of the self."[99] Information that is repressed—avoided, ignored, suppressed, or distorted—does not have to enter into any personal subconscious repository, where it supposedly festers as neurosis. It can simply disappear, or it can hover on the margins of consciousness, vaguely threatening, occupying a liminal sphere of latent consciousness, and thus exercising pressure against imaginative constructs from which it has been excluded. It can make itself felt not through neurotic symptoms emerging out of a personal subconscious but aesthetically, through the sense of incompleteness or falsity in an imaginative construct.

If death is absent from a given author's work, or if sexuality is envisioned only in conventional terms of domestic sentiment, those are significant critical facts. If a homosexual author feels constrained to disguise homosexual dispositions by transposing them into depictions of heterosexual relations—as Pater, Forster, Wilde, Cather, Proust, and Maugham do—that transposition crucially affects the imaginative character of the author's works. When a whole culture enforces a public norm of selfless altruism, we can be virtually certain that depictions according with that norm will produce sensations of false and forced sentiment, and we shall be alert to ironic and satiric evasions of that false norm. If an author tells us that some couple "lived happily ever after," we can reasonably suppose that in this particular depiction some significant portion of life—some frustration, antagonism, endurance, tolerance, and ennui—has been repressed in favor of a generic convention designed to produce facile satisfaction. Most generally, for literary purposes, when we examine any author in a critical, interpretive way, we probe the limits of that author's imaginative universe. We inquire closely into how the author filters information, forming plots and characters that conform to the author's own emotional needs, the norms of the culture to which the author belongs, and the generic conventions within which the author is working.

The Second Sequel

The short essay I posted on the forum *On the Human* at the National Humanities Center consisted in a reworked version of the section on the adaptive function of the arts written for the target article in *Style*. This second sequel contains a condensed version of my rejoinder to the responses elicited by the forum posting.

Testing Adaptationist Hypotheses

Bill Benzon cites Steven Pinker's review of *The Literary Animal*.[100] At the time Pinker published his review, Brian Boyd was hosting a symposium on evolutionary studies in the arts. Ellen Dissanayake, Jon Gottschall, Denis Dutton, and others were part of the symposium. I was there, too. We followed up the symposium with a lengthy online discussion of Pinker's review and the general question of the adaptive function of the arts. I'll copy here part of my contribution to this discussion.

Pinker's favored instances of adaptationist hypotheses are neither more nor less testable than the adaptationist hypotheses he disputes. He argues that sugar provides energy for a metabolism designed to process it. Ellen Dissanayake and Brian Boyd argue that the arts fix attention on adaptively salient concerns and promote social cohesion. I argue that literature provides emotionally saturated images for a psyche designed to assimilate such images and use them for evaluative, affective, and ultimately behavioral orientation. At this level of explanation, all these arguments are structurally parallel. To make further progress in understanding, we have to move from that level of conceptual parallelism into the contexts of paleoanthropology and psychological mechanism. At the present time, metabolism is better understood than cognitive processing, but a theory of psychology that limited itself to the metabolic processing of sugar would not get us very far into the workings of the human psyche. There is no prima facie reason for supposing that the workings of the human psyche are ultimately any more inaccessible to specific analysis than the workings of metabolism.

Jon Gottschall is unhappy with what he takes to be the untestability of current adaptationist hypotheses about the arts, but that theoretical discontent reflects a concept of empirical testing that is too narrow for most questions of major interest not just to the arts but to all complex human behavior. The discontent seems integral with a concept of testing that limits itself to a one-to-one relationship between question, test, and answer: one question, one test, one answer. To get a better feel for the gratuitous limitation involved in this conception of empirical testing, try this thought experiment: Read the introductory and concluding chapters of *On the Origin of Species*, specifically bearing in mind three questions: (1) What constitutes a significant problem? (2) What constitutes empirical evidence? and (3) How can complex and interrelated sets of evidence be used to test complex empirical hypotheses?

Darwin's argument is a complex, multipart argument involving interlocking pieces of evidence from multiple overlapping or contiguous fields—from embryology, anatomy, paleontology, biogeography, ecology, climatology, animal domestication, geology, and other fields. The adaptive function of literature is not so complex a question as "the origin of species," of course, but still, it is complex enough so that several component hypotheses are needed to approximate to an explanation adequate to the subject itself.

Darwin's example should give us warning that the complexity of a problem is no warrant for limiting our explanatory scope to simple, one-step solutions for simple, one-step problems. To solve complex problems, we have to put all the pieces together into a complex design. By doing that, we can, to use Jon's own terminology, shrink the space of possible explanation. Actually, that's already happening. Even Pinker has done us an inadvertent service. He has introduced useless confusions, but analyzing those confusions enables us to reject a certain range of hypotheses that lack adequate explanatory power, and they thus enable us to narrow the scope of plausible adaptationist argument. As that scope narrows, we move ever further into the range of problems that can be solved, one by one, with simple appeals to quantified evidence from empirical tests.

Any universal and reliably developing phenomenon could be casually deprecated as a by-product or "spandrel." Even complex functional structure offers no absolutely clinching proof of adaptive design. If one is determined on a radical skepticism about adaptive hypotheses—as Gould was, for instance—the by-product hypothesis is a default hypothesis. But in mainstream evolutionary thinking, the by-product hypothesis has no default status. Quite the contrary. If a behavior is universal and reliably developing; if it also has complex functional structure; and if the behavior produced by that structure can be reasonably and usefully integrated into a larger set of serious and scientifically grounded explanatory hypotheses about human evolution, the default assumption is adaptationist.[101] This is the core of the theoretical argument made by Tooby and Cosmides in "The Psychological Foundations of Culture," and they are themselves merely following Darwin, Hamilton, Williams, Dawkins, and others. Pinker's own general theory is derivative from that of Tooby and Cosmides, and in his main theoretical expositions of human adaptive designs, he himself propounds this conception of adaptive design.

The idea of a taste for cheesecake as a "by-product" of the human taste for fat and sugar is fundamentally misleading. A by-product is an adventitious aspect of some adaptive characteristic—for instance, the color of blood. The redness of blood does not contribute to its functional capacities but is rather an adventitious aspect of its functional capacities. The taste for cheesecake, in contrast, is not an adventitious aspect of the human mechanism geared toward consuming fats and sweets. The tastiness of cheesecake is a hyperstimulus for an adaptive mechanism.

The human gustatory and digestive system is not designed specifically for the consumption of cheesecake, but it is designed specifically for the consumption of fats and sugars. Cheesecake is merely a special instance of fats and sugars. In a modern ecology, the consumption of too much fat and sugar is harmful, but the human psychological and physiological mechanisms geared toward the consumption of fat and sugar are nonetheless adaptations. In a long-enough evolutionary span, those adaptations might prove maladaptive. (So it goes with most species, the vast majority of which have become extinct. All their adaptations, by definition, became maladaptive.) In given ecological contexts, even at present, the mechanisms geared toward the consumption of fat and sugar are still beneficial and thus, presumably, "adaptive." Adaptiveness can be measured only by inclusive fitness, and we can seldom assess the long-term fitness consequences of a current behavior.

If one compares art to cheesecake (the arts in general or literature specifically), the obvious parallel to draw would be between cheesecake as a "hyper-stimulus" and the oral antecedents of literature as hyper-stimuli. Even if one accepts the cogency of the parallelism, neither cheesecake nor literature could most accurately or plausibly be designated as "by-products." If literature and its oral antecedents were to be interpreted as hyper-stimuli for adaptive human cognitive dispositions, the question as to the adaptiveness of these hyper-stimuli would have to be left to long-range evolutionary history, and atomic technology will probably render all such considerations moot. Enjoy while you can. If the Bomb doesn't get us, human genetic technology will in any case probably complicate and confound all "natural" adaptive tendencies beyond all recognition.

Again, once we have cleared away basic confusions, the chief question we have to answer is this: can the oral antecedents of literature best be understood, within an evolutionary context, by segregating them into component parts, as hyper-stimuli? Or can they best be understood by regarding them as an integrated set of features that in their integrated form constituted a distinct suite of adaptively functional mechanisms—that is, whether they are themselves an "adaptation"?

When the question is posed in this way, we are back to the question of adaptive "design." With respect specifically to narrative, that question can be analyzed in three problem areas: (1) how the mind is designed to assimilate narrative, (2) the way narrative influences behavior, and (3) the way both these areas can be located within our developing knowledge of human cognitive evolution. All three areas are wide open for empirical testing.

Summing Up the Forum Discussion

Thanks to all who have participated in this discussion. I see some broad areas of consensus. Humans naturally, universally generate imaginative, artistic representations. Those representations engage evolved cognitive dispositions and fulfill deep emotional and cognitive needs. Imagination is in some form an integral part of functional human cognitive equipment.

Each of these simple propositions leaves open multiple possible formulations with different implications. We are at the point now where we all need to be thinking hard about how to bring such formulations within the range of testable propositions.

In synoptic overview, that's what looks like consensus to me. But I'm sure that if this thread were to go on indefinitely, every item in my own impression of the consensus view would be disputed with energy and conviction. Hence the need for integrating empirical methods to reduce the scope of possible plausible formulation—"narrowing possibility space," as Jon Gottschall would have it.[102] The literary people need to continue moving in the direction of empirical methodology, and the people with social science backgrounds need to recognize the crucial significance of the imagination and its works. They need to bring this subject area into their active research agenda, not leaving it to the formulations of purely speculative commentary.

The weakest aspect of "biocultural" theory so far has been the "cultural" part. Getting past this limitation is the single most important challenge facing the evolutionary human sciences. Collaboration between people with humanities expertise and people with expertise in scientific methodology will be almost indispensable in taking the next major step toward turning the evolutionary human sciences into a truly comprehensive explanatory framework for all things human.

So far, we have seen only a few instances of that kind of collaboration. The literary people are afraid of scientific methodology, and the scientists are afraid of moving into areas of culture that seem to them nebulous and mysterious. There is a lot of resistance based on prejudice and the comfort of routine practices—with the literary people harboring a distaste for the impersonal and technical character of scientific methods and the scientific people harboring a distaste for the imprecision and subjectivity of literary thinking.

I understand the resistance—have felt it all myself, from both sides. But we are now at something like a bottleneck, an impasse. Until we break through the routine of our current practices, our habitual attitudes and methods, we aren't going to get a comprehensive, integrated theory of culture. And until we get that, the evolutionary human sciences are going to be spinning their wheels just below the point at which they can begin to explain specifically human nature, and the humanists are going to be spinning their wheels in endless theoretical discussions, exercising their rhetorical ingenuity but not getting very far with positive results, just as we have been doing here. (I note that this discussion included perhaps just one evolutionary human scientist. That is bad for us, and bad for the evolutionary human sciences.)

So, the two cultures are still with us. And within at least one of those two cultures, there are subcultures that scarcely speak to one another. Reading over the discussion that followed Katherine Hayles' post, I was once again struck, rather depressingly, with the Balkanization of studies in the humanities. The folks in that other discussion speak a different language, with different references and different assumptions. I don't think there is much hope for "conversion" between these two sects. As with the evolutionary humanists and evolutionary human scientists, everybody is pretty comfortable with their routines. I do think change will come—how swiftly, I can't predict. And I think it will come by grandfathering or grandmothering out the whole population that relies fundamentally on continental speculative theory divorced in principle from the evolutionary human sciences.

Perhaps I'm being overly optimistic, but I do think, without forcing the issue in my own mind, that truth and reality will ultimately carry the day against entrenched institutional ideologies. The biggest barrier to the development of the humanities in the direction of the evolutionary human sciences is that bright young people are systematically prohibited from taking up this line of research. But a few small chinks have occurred in the armor of resistance and suppression. Purely defensive fortresses can never hold forever against the pressure of sustained friction. Cliffs wear away under the grinding of the waves.

Perhaps optimistically, then, the chief factor that will ultimately determine the future direction of the humanities is the potential for the development of knowledge. Despite routine, fear, prejudice, and entrenched interests, I am myself confident that in that one crucial factor, the "biocultural" approach is the only possible road to the future.

CHAPTER 2

An Evolutionary *Apologia pro Vita Mea*

In an interview for his online forum *Neuronarrative*, David DiSalvo asked, "What is your favorite work of literature?" The following essay is my answer.

I'm going to fudge on this question, expanding it to take in more than one genre and more than one phase of my own imaginative life—not a single "favorite," but some few favorites. The most intense and vivid imaginative experience I ever had was in reading the major poems of Wallace Stevens' culminating visionary phase, especially "The Owl in the Sarcophagus." I've also had some fine high moments with Keats, sensually rich and meditatively pure. Hardy's *Tess of the d'Urbervilles* gave me my richest, warmest, most lyrical and emotionally absorbed experience in reading a novel. When I first read George Eliot's *Middlemarch*, I had the kind of epiphanic experience—expanding my own imagination to its limits—that I had also with the late visionary poetry of Wallace Stevens, though the mode, of course, was different. I have to confess that when I first read Stevens, I was a half-witting participant in the late Romantic effort to preserve some imaginative realm for "spiritual" experience. As I was writing my book on Stevens, that belief faded and failed, and I had to finish the book in grim scholarly determination just to tell the truth about Stevens, a truth few other critics had even glimpsed—the simple observation that he is essentially a religious poet. Something similar happened in my history with *Middlemarch*, which has a divided worldview. One view is shrewdly realistic and ironic (a perspective embodied in the character Mary Garth). The other is idealistic, spiritual, moralistic, a perspective embodied in Dorothea Brooke. In the moralistic vein, Dorothea does what Stevens did in the visionary, lyrical vein—offers a secular imaginative approximation to a religious worldview. I bought into that, thus giving evidence that at that time, in my early twenties, I was still only gradually withdrawing from a religious worldview. That "melancholy, long, withdrawing roar" has been a chief trajectory of the modern imagination. My own trajectory recapitulated it in brief and in small. Nowadays, Dorothea's ardent spiritual yearnings just get on my nerves. Stevens doesn't, though. I wrote an article for a Cambridge Companion to Stevens a few years ago and revisited all his work and my own writing on it. It was like reliving the most intense love affair of one's youth. As in a museum, perfectly preserved, untarnished, lovely in memory, but no longer part of the actual world.

I lost all literal religious belief—became a confirmed atheist—when I was sixteen, but it took another fourteen years or so to drain out the last of the late Romantic imaginative spiritualism. In this gradual fading, my own experience is something like that of Darwin, who never underwent any convulsive loss of religious faith (unlike many of his contemporaries). The final paragraph of *On the Origin of Species* invokes "the Creator." After that, as Darwin explains in his autobiography, his sense of things faded into the light of common day. That kind of perspectival change radically alters one's whole repertory of imaginative response.

Reading *On the Origin of Species* and *The Descent of Man* were transformative experiences for me. When I was sixteen, I had read in a biology textbook that all features of all organisms were the product of interactions between genetically transmitted dispositions and environmental conditions. That observation had instant, axiomatic conviction for me, and it was the first step in completely altering my metaphysical perspective—leading to the loss of religious faith. (If all behavior is ultimately determined in this way, "free will" in any ultimate sense is illusory, and the idea of divine punishment and reward is outrageous.) Then a few years later I read H. G. Wells' *Outline of History*, a big two-volume work that started with the history of the earth and went on through the evolution of hominids before settling into the standard rise and fall of civilizations. Wells was T. H. Huxley's student and had an excellent grasp of the logic of adaptation by means of natural selection—hence his classic science fiction works *The Island of Dr. Moreau* and *The Time Machine*. I absorbed Darwin's theory through Wells. So, I was a Darwinian at that point without ever having read Darwin. I first read *The Origin* and *The Descent* in 1990. I had already been working for a couple of years at reconstructing literary theory from the ground up—trying to rescue it from the postmodernists, but working only with broad general categories of theme and genre. Reading Darwin made vividly apparent to me that all things human, including the products of the human imagination, simply had to be conceived within the total evolutionary development of all living things. Wells was good, but not *that* good. Darwin gave me my first real imaginative sense of deep evolutionary time. When I speak about the way imaginative works help us organize the sphere of our experience, that's the sort of thing I have in mind. It's one thing to understand a theory, be able to recite its terms, and even believe it. It's another thing to have an imaginative grasp of that theory so that you never see anything in the world in quite the same way again. Darwin had a vision of deep time, and he located all living things in that vision. As has happened with many other theorists—biologists, anthropologists, psychologists, and now literary and aesthetic philosophers—the imaginative power of Darwin's vision has fundamentally shaped my own sense of the world. That would worry me a lot if I weren't as certain as I can be that Darwin got it right, as right as it can be gotten at the present time.

One of the main ways science has fundamentally altered our imaginative experience over the past few centuries is that simply getting it right now counts for so much. The Romantics rebelled and wanted to insist that passion and aesthetic quality are themselves ultimate arbiters of imaginative vision. Beauty is truth, truth beauty. That's a mistake. The current adherents to this sort of mistake are less likely to be aesthetes than utopian ideologues. The postmodern version of Keats' dictum would go something like this: beauty is politically correct, political correctness is beautiful, and truth is a bourgeois fiction. For evolutionists, in contrast, truth comes first and is nonnegotiable.

The truth is, humans are a tiny blip in the most recent moments in the almost unimaginable trajectory of deep time. Nonetheless, in our miniscule habitation in a remote corner of the universe, we are able to look back over deep time and recognize our own place in it. That makes us special. So far as we know, within the horizon of all our discoveries, there is nothing quite like the human imagination anywhere else in the universe. If there is, we shall be most interested to find out about it. Meanwhile, we make sense of what we know. The imagination is one of the things we know, and it is the means through which we know everything else. It is worth a lot of study, and really, we have only just begun to think about it.

Stevens, Eliot, and Darwin have been among the major relationships in my imaginative life, but I have been highly promiscuous, with lots of little affairs along the way. I love movies and had rich imaginative moments, in my youth, with early Bergman, especially *Wild Strawberries*. When I first saw it, Jancsó's *The Peach Thief* was one of the finest films I had ever seen. Kronenberg's *The Fly* has a "touchstone" value for me, forming a symbolic cluster that stimulates creative thinking even to this day. I have a personal fondness for Annaud's *Quest for Fire*. Annaud succeeds in imagining what it might be like to be a scarcely articulate early human shivering in a swamp, with nothing to protect you but your own wit and courage and the few simple tools you can construct. Despite everything that can legitimately be said against it, I think Polanski's *Tess* is a cinematic masterpiece. I probably won't live long enough to see that judgment vindicated. And of course, maybe I'm wrong.

Heinrich Heine's cultural histories captivated my imagination. Friedrich Schiller's aesthetic theory in *Über Naïve und Sentimentalische Dichtung* so enthralled me that I named my first dachshund Friedrich. Guy de Maupassant's earthy sensual human warmth still seems to me one of the great good things in life. Matthew Arnold, despite his mutton chops, remains a guiding star in my sense of what a cultural, critical vision can do. When I was a child, I loved and annually reread *Huckleberry Finn*, *Little Men*, *The White Panther*, *Rifles for Watie*, *Half Magic*, and *A Wrinkle in Time*. Having kids of my own gave me a welcome opportunity to revisit those wonders and add many others, including *Across Five Aprils*, *The Phantom Tollbooth*, and *The Adventures of Stanley*

Kane. Some things you can come to early; for others you have to wait. I was middle-aged before the symphonic, orchestral magnificence of *King Lear* became imaginatively intelligible to me. In contrast, when I last reread *Little Men* and followed it up with *Jo's Boys*, a couple of years ago, Alcott's insidious strategy for undermining and suppressing specifically male motivational dispositions, as personified in Dan, irritated me. What was I thinking, at the age of ten?

Der Zauberberg, Catch-22, Le Siècle de Louis Quatorze, Tom Jones, La Jument Verte, The History of Mr. Polly, Salammbô, the *Annales* of Tacitus, *Coming Up for Air*. . . . Once one starts down memory lane, it's hard to stop. Becoming a professor of literature is something like taking a vow of poverty. There are so many of the "good things," as Trollope lovingly calls them, one must give up—money, status, security, fine houses, rich clothing, ease, and luxury. But then, one gets to spend one's life having love affairs with books.

Over the past twenty years or so, I've branched out and had passionate flings with works in personality psychology, sociobiology, and anthropology. Doing that has been fun, but it has also helped solve a very serious problem I had been having for many years—the problem of finding things to do in literary study that weren't just fun but also serious, constructive, adult. In *Middlemarch*, one of Eliot's protagonists, the medical doctor Lydgate, has serious scientific ambitions. "He was fired with the possibility that he might work out the proof of an anatomical conception and make a link in the chain of discovery." I know just how he felt. There is a passion for discovery, for constructive, creative thought. Close reading satisfied that need in literary study for just a little while, a few decades in the middle of the previous century. During that same period, people could do serious scholarly work, not the most exciting kind of thing, but solid, constructive work, producing editions, collecting letters, writing biographies. Then, in the last quarter of the century, "Theory," as it rather fatuously designated itself, gave people the exhilaration of creative, speculative thought, but the whole enterprise was shot through with sophistical fallacies, so the excitement was febrile, half delirious, corrupted. How to produce serious, real knowledge, constructive knowledge, within the field of literary study? By incorporating it within the whole broader field of the evolutionary human sciences, retaining what is peculiar and special to the nature of literary experience, making full professional use of all one's own experience, but integrating all that with the broader world of empirical, scientific knowledge about human nature.

CHAPTER 3

A Meta-Review of *The Art Instinct*

Evolutionary Aesthetics Hits the Big Time

The Art Instinct is a major publishing phenomenon—with a book tour featuring sell-out crowds at Ivy League schools, radio spots on prominent talk shows, and even a rambunctious interview on *The Colbert Report*.[1] Some of Dutton's colleagues felt that the Colbert bit lacked the dignity to which evolution is entitled. It offended their feeling that "There Is Grandeur in This View of Life." Well, no, Colbert doesn't do Grandeur. But he does do publicity, and while exchanging good-natured wisecracks with Dutton, he also held the book up to the camera, with a finger delineating the title—a gesture as close to "product placement" as any eager agent could wish. And indeed, Dutton's agent might well be looking at retirement condos in posh sections of Miami or San Diego. *The Art Instinct* has been high on the bestseller list at Amazon for months, and it has received mostly rave reviews reaching the broadest sectors of the educated lay public—*Newsweek*, *The New York Times*, *The Guardian*, *The New Yorker*, *Atlantic Monthly*, *TLS*, *The Washington Post*, *The Boston Globe*, *The Philadelphia Inquirer*, *New Scientist*, *The American Scholar*, *The Wilson Quarterly*, and others. (For links to reviews and interviews, see http://theartinstinct.com.)

For evolutionists eager to demonstrate the comprehensiveness of "this view of life," all this is very good news. Evolutionary books on human cognition and behavior have been bestsellers for many years. In 1975, E. O. Wilson's *Sociobiology* elicited howls of rage from the Marxists, but Wilson was undaunted, and in 1978, he received a Pulitzer for *On Human Nature*—a book still in print after three decades. Starting in 1994 with *The Language Instinct*—the namesake for *The Art Instinct*—Steven Pinker's books for the educated lay public have helped make evolutionary psychology a household word, at least in educated households. Frans de Waal, Matt Ridley, Richard Dawkins, Nicholas Wade, David Buss, Daniel Goleman—all these writers must also have agents driving up prices in Miami and San Diego. But no one before Dutton has had anything like this kind of popular success with an evolutionary book in the humanities. The Gottschall and Wilson collection *The Literary Animal* was a success with critics outside the postmodern establishment—in scientific journals and the lay press—but was not a blockbuster at the box office. Things seem to have changed, rather suddenly. Brian Boyd's *On the Origin of Stories* is now vying with *The Art Instinct* for sales success.[2]

Before Dutton and Boyd, the closest any studies in evolutionary aesthetics had come to major popular success were in two books that devoted just a chapter each to the arts: Pinker's *How the Mind Works* and Wilson's *Consilience: The Unity of Knowledge.* Despite Dutton's affiliation with *The Language Instinct,* Pinker came not to praise the arts but to bury them—to relegate them to the realm of idle self-indulgence largely irrelevant to the serious purposes of life. Wilson, in contrast, came not only to praise the arts but to identify them as integral, functional parts of the adapted mind.[3] The contrasting views of art offered by Pinker and Wilson involve very different concepts of human cognitive evolution. Readers seeking a trenchant analytic account of these differences will not find it in *The Art Instinct.* Dutton does not dwell on theoretical conflicts within the evolutionary camp. Like Pinker, he is essentially a synthesizer. He takes what seems most constructive from many evolutionary thinkers and integrates them with concepts from the philosophy of art—his own home field. In the most theoretically probing part of his book, Dutton offers subtle and incisive arguments against rigid distinctions between "adaptations" and "by-products" (*AI*, 90–102). Still, arguing for a "middle ground" seldom solves the deepest theoretical problems. Beneath the middle ground, there is almost always some seismic fault likely at any time to shift and topple the buildings above it. If Dutton sometimes leaves nagging theoretical issues unanswered, he offers, in compensation, a full, rich account of art in all its multifarious aspects, high and low, popular and elite, appealing to all the senses, engaging the deepest passions, absorbing cultural conventions, and fulfilling the most fundamental human needs.

Given an intellectual climate in which one or another kind of "sophistication" sets itself off strenuously against the common understanding, formulating some broad, basic home truths about the nature of our aesthetic experience is in itself a signal accomplishment. And to do it within a serious meditation on the evolutionary context—well, very few have managed that, certainly not in the full and sustained way Dutton has. Dutton's closest competitor and chief antecedent is Ellen Dissanayake, an evolutionist whose arguments on the evolutionary origins and psychosocial functions of the arts have won dedicated audiences among specialists in art and music education, art and music therapy, art and craft theory, ethnomusicology, and evolutionary aesthetics. Dissanayake's first two books adopt an ethological perspective on the way the arts are integrated into ritual and religion in premodern societies. In her third book, *Art and Intimacy*, she locates the origins of art in mother-infant interactions.[4] Dutton handsomely acknowledges Dissanayake, but he concentrates more on the high arts of advanced civilizations, and he casts new evolutionary light on issues that have often exercised theorists of modern aesthetics. Scholars and scientists can read his book with profit and pleasure, but his genial and conversational manner is pitched in the first place to the educated lay reader. Little wonder then that his book has been such a success. There must be many people who have a taste for the arts, who avidly consume popular books in evolutionary psychology, and who have been frustrated at not being able to bring those interests into conjunction. Dutton accomplishes this aim splendidly.

Though he recounts few incidents from his personal life, Dutton's book has an indirectly autobiographical aspect. He is a genuine connoisseur, especially of classical music, but also of the visual arts, and as founding editor of the journal *Philosophy and Literature*, he has a patent claim to an insider's knowledge about literature, too. Through all his multifarious references to particular works of music, painting, sculpture, and literature, he gives a strong, vivid impression of what it's like to have lived his life in absorbed, delighted responsiveness to the arts. That's something neither Pinker nor Wilson, for all their virtues, could possibly have conveyed. Nor did it come through much in the rather academic essays, most geared toward narrative theory, in *The Literary Animal*. Dutton's personal responsiveness to the arts protects him from a danger to which some evolutionists, discussing aesthetics, have succumbed—formulating general ideas that fall short on the side of common aesthetic experience.

Many evolutionists, even in the humanities, tend toward a naïve realism. They look at art as merely a source of adaptively relevant information, or they consider it, half consciously, as a direct, realistic imitation of common, average experience. Not Dutton. He intuitively understands that art creates and articulates "meaning," that it gives imaginative shape to things, and that imagination has no necessary fealty to ordinary reality. He lucidly conveys the ideas of tone or mood as the emotional continuum in a work of art, and he conveys also the central importance of "point of view," the individual perspective as a locus of experience, a little different for every artist and for every reader, listener, or viewer. As commonplace as such ideas are in conventional literary study and aesthetic philosophy, some evolutionists have been slow to grasp and integrate them with ideas about human nature.

Criticizing the Critics

Creationists don't figure very largely in debates on evolutionary aesthetics. They are out of it, below the level of serious discussion. But Dutton picked up at least one devotee of medieval religiosity among his critics. Had I religious convictions opposed to evolutionary aesthetics, Maureen Mullarkey's review in *The Weekly Standard* would make me uneasy. It would make me feel that God, in all His glory, was on Dutton's side. For one thing, the reviewer's name has unmistakably Dickensian overtones. Surely this is no coincidence, but rather the very Hand of Providence. Moreover, as if willfully supporting the hypothesis that the review is a divinely inspired, though backhanded, commendation of evolutionary aesthetics, Mullarkey says things like this: "Grace of mind—a signal to the old Scholastics of the beauty of moral harmony—is not explicable in physical terms."[5] A contemporary writer who makes appeal to Scholasticism—virtually a byword for sterile speculative abstraction—seems to be cloistering herself within an antiquarian enclave. And yet, Mullarkey has contemporaries among the most modern practitioners in the humanities. Like both traditional humanists and postmodern adherents of "Theory," she is essentially a metaphysical dualist, cordoning off the world of imagination or spirit from the world that can be explained "in physical terms."

I have sometimes envied Darwin because he had such an easy foil: the creationist view of species. His "one long argument" in the *Origin* is an argument for "descent with modification through natural selection."[6] The only real alternative to "descent with modification" is that each species was specially created. That view does not stand up to scrutiny, but it provides a wonderful frame on which to weave all the contrary evidence—geographical, anatomical, ecological, embryological, paleontological—that supports the idea of descent with modification.

I see now that there is no need to envy Darwin. Though creationism is, intellectually, dead and gone, evolutionists in the humanities and the human sciences can still find an easy foil in the metaphysical dualists. In current mainstream literary study, dualism most often takes the form of "cultural constructivism"—the idea that culture has autonomous causal force and is not constrained by innate dispositions. Cultural constructivists say things like "all identity is culturally constructed," and "biological sex has no relation to socially constructed gender roles." Such statements are close kin to declarations that "grace of mind is not explicable in physical terms," "consciousness can never be reduced to neurological events," and "there is no evidence of any connection between the brain and mind." This final statement is a direct quotation from one of my postmodern literary friends, a Jamesonian/Althusserian Marxist, who thus makes himself a bedfellow, strange enough, with the old Scholastics.

As a scientific proposition, the declaration that the mind and brain have no connection can be disconfirmed by a simple experiment conducted at home, with no special equipment, except a hammer. As soon as your postmodern friends utter such phrases, whack them in the head with the hammer, not too hard, and then ask them if they experienced any mental event concordant with the blow to the skull.

If God is generous, He is also just, and there is a kind of backhanded, evil-humored justice in His decision to hypnotize some of Dutton's reviewers so as to get them to fixate on Dutton's arguments in favor of "sexual selection."[7] Dutton has a higher opinion of Geoffrey Miller's causal hypothesis in *The Mating Mind* than most evolutionists in the humanities have had, and he mingles his own arguments for the adaptive functions of art with Miller's argument that the arts are merely forms of costly display designed to attract members of the opposite sex.[8] The equivocations that attend on this effort of synthesis are a weakness in *The Art Instinct*. I can myself forgive that weakness since I regard it as peripheral to Dutton's central concepts, merely a wrinkle in a broad exposition of all the different functions art fulfills. Forgive, but not overlook. We'll have to give a little attention to the boggle at the heart of this issue.

Insofar as "sexual selection" can be basically distinguished from natural selection, it means costly display, the selection of traits *that have no primary adaptive value*. Dutton repeatedly invokes Miller's notions, but he does not ultimately argue that the arts have no primary adaptive value. Miller does argue this, sometimes, just as he argues, sometimes, that the enlarged human brain itself has no primary adaptive value (*AI*, 17–18). In this most extreme form of his argument, Miller joins company with Steven Jay Gould, with whom, on that very issue, Dutton explicitly disagrees (*AI*, 92–94). What Dutton actually argues is that the arts have primary adaptive cognitive and social value and that they are then picked up as attractive features—attractive *because* they are adaptive—and shifted into overdrive by males and females selecting one another for those specific features.

There is more than one confusion latent in all this. The peacock's tail is an instance of costly display only because it is driven to an extreme. Tails, in their original form for the ancestors of peacocks, had adaptive utility, just as they had for other birds. They become nonadaptive—costly, dangerous, dysfunctional—through "runaway" selection. In their hypertrophic form, they are genuine instances of "costly signaling"—nonfunctional features selected precisely as a signal of underlying vigor. Does Dutton argue that the arts are nonfunctionally hypertrophic features that give evidence of underlying vigor? Ultimately, he does not. With respect specifically to narrative, this is what he argues instead:

> The features of a stable human nature revolve around human relationships of every variety: social coalitions of kinship or tribal affinity; issues of status; reciprocal exchange; the complexities of sex and child-rearing; struggles over resources; benevolence and hostility; friendship and nepotism; conformity and independence; moral obligations, altruism, and selfishness; and

so on. These themes and issues constitute the major themes and subjects of literature and its oral antecedents. Stories are universally constituted in this way because of the role *storytelling can play in helping individuals and groups develop and deepen their own grasp of human social and emotional experience*. (*AI*, 118; emphasis added)

The same basic argument applies to the other arts. They cultivate emotional intelligence and refine powers of qualitative judgment—the judgment of "values" in their broadest sense. "The arts intensify experience, enhance it, extend it in time, and make it coherent" (*AI*, 102).

There is one review that I imagine Dutton would find more disturbing than any other. Unlike the few overtly hostile reviewers, Sebastian Smee professes himself in sympathy with Dutton's evolutionary worldview, and indeed, claims that it is all so uncontroversial, so self-evident, as to be nugatory. That's the trick. And there is a trick. Behind the velvet tongue, there is a two-pronged dagger.

One prong is just the well-nigh universal disposition of all reviewers to one-up the reviewee. Smee is better at it than many people are. He understands that the most effective putdown is the putdown that gives the appearance of disappointed sympathy, as if the reviewer is saying *I have no reason at all to disparage this person; I am not threatened; I have no temperamental aversion; indeed, I rather like this person. All the greater the pity, then, that he is such a lightweight, that his arguments are trivial and insignificant. It's fine fluff. If you have nothing better to do, you might spend a pleasant hour in this chap's company, shooting the breeze; just don't expect it to amount to anything.*

The second prong of the dagger just slightly shows its steel with an early reference to Gould and Lewontin, followed by a concession that, like the duo of San Marco, Dutton rejects (supposedly) "hyper-adaptationism": "It's wrong, in other words, to see the arts as adaptations of a process of natural selection in the same way that our eyes, our spines, or our inner organs are."[9] Smee evidently knows very little about evolutionary theory, and what little he knows he got from Gould and Lewontin. He is, consequently, deeply confused about what "adaptation" means. And, be it said, Dutton feeds into this confusion. His own equivocal formulations about the supposed contrast between natural and sexual selection create just enough of an opening for Smee to insert his spandreled foot.

Smee's chief line of attack has now become familiar in the humanities, where scholars often proclaim themselves "post-theory." Deconstruction and Foucauldian cultural critique supposedly disenchanted us all with "meta-narratives": the belief that any general explanatory system could account for our experience. Post-theorists are thus aficionados of the particular, the odd bits of lore to be picked up in archives, treasured for their qualitative singularity, and offered as evidence against the validity of general ideas. In this curious move, the post-theorists presuppose the validity of the "Theory" that they think they have left behind, and despite all their professions of nontheoretical antiquarianism, they do reach general conclusions, the same conclusions their elders reached more explicitly when announcing the Death of the Author, the all-pervasiveness of "discourse," the transcendent force of "power," and the reduction of science to rhetoric and ideology.

In the move to post-theory, one grants the general validity of evolution, at least in its Gouldian forms, but also then declares that it is irrelevant, that it changes nothing, that it alters not one jot the way we would read this or that text or describe this or that historical cultural moment. Such particulars, we are to understand, even if they are in some remote way concordant with evolutionary theory, operate on such a rarefied level of "emergent" particularity that invoking evolution can have no more impact for us than affirming, without quite understanding it, Newton's theory of gravity, or Einstein's theory of general relativity. The only relevant categories, if we invoke any categories, are those of specific cultures. In reality, then, "post-theory" is just the latest incarnation of cultural constructivism. It avoids culturalist transcendentalism—the idea that culture is a first cause or unmoved mover—only by paying lip service to the reality of biology.

A hostility to general ideas is not the exclusive property of the post-theory people. It belongs as well to a certain brand of traditional humanistic thought. Among Dutton's reviewers, the most explicit defense of traditional humanism comes from William Deresiewicz. Denouncing *The Art Instinct* in company with works by Brian Boyd, Jon Gottschall, and the present writer, Deresiewicz envisions a desolate future landscape in which all aesthetic responsiveness will have been replaced by ontologically vacant generalizations emitted by humanoid statisticians wearing white lab coats. Defending humanity against that dystopian future, he argues that literary study, unlike the social sciences, "is not concerned with large classes of phenomena of which individual cases are merely interchangeable and aggregable examples. It is concerned, precisely, with individual cases, and very few of them at that: the rare works of value that stand out from the heap of dross produced in every age."[10]

The idea that commentary on the arts concerns itself only with particular instances and not with general categories and large-scale themes is of course false. Humanists both number the streaks of the tulip and also try to understand the general properties of flowers. They characterize the particular qualities of individual works and also consider genres, historical trends, common themes and forms, artistic traditions, and the relations among multiple works and historical conditions extending over decades or centuries. They register specific aesthetic qualities but at the same time try to increase the sum of valid general knowledge. The old opposition between the particular and qualitative on the one side and the general and impersonal on the other is a remnant of metaphysical dualism—the evolved cognitive bias toward dividing the world into inanimate physical things and animate agents.[11] Physical things, the bias leads us to believe, can be measured, manipulated, reduced to elements, and explained by appeal to causal laws. Animate agents, in contrast, are spiritual beings, and it is the nature of spirit that it could never be reduced to its elements or explained by causal laws. It is a mystery that manifests itself in almost magical moments of poetic inspiration. In cognitive science, such moments are called "qualia," and even in cognitive science, they are sometimes the objects of occult mystification.

Commentaries like those of Smee and Deresiewicz have no constructive intent and offer no substantive propositions that could be used for developing a full and adequate understanding of literature and the other arts. Their chief virtues are performative and rhetorical. As contributions to knowledge, they have at best a negative, symptomatic value. If they are useful at all, it is only to remind us of the challenges we face in constructing a progressive, comprehensive program of study.

To generate adequate interpretive commentary from an evolutionary perspective, we must construct continuous explanatory sequences linking the highest level of causal explanation—inclusive fitness, the ultimate regulative principle of evolution—to particular features of human nature and to particular structures and effects in specific works of art. A comprehensively adequate interpretive account of a given work of art would take in, synoptically, its phenomenal effects (tone, style, theme, formal organization), locate it in a cultural context, explain that cultural context as a particular organization of the elements of human nature within a specific set of environmental conditions (including cultural traditions), register the responses of readers, describe the sociocultural, political, and psychological functions the work fulfills, locate those functions in relation to the evolved needs of human nature, and link the work comparatively with other artistic works, using a taxonomy of themes, formal elements, affective elements, and functions derived from a comprehensive model of human nature. Dutton understands all this, and he thus joins a cadre of advanced theorists that is still rather small. *The Art Instinct* should do a good deal to increase the size of that cadre.

CHAPTER 4

Three Scenarios for Literary Darwinism

Weighing Probabilities

Thirty years ago, the idea of creating a specifically evolutionary theory of literature would scarcely have seemed imaginable and would certainly not have seemed within the range of practical possibility. Nonetheless, over the past fifteen years, "literary Darwinists" have been making rapid progress in integrating literary study with the evolutionary human sciences. What is the likely future trajectory of this movement? We can probe this question by comparing three alternative scenarios: one in which literary Darwinism remains outside the mainstream of literary study; one in which literary Darwinism is incorporated as just another of many different "approaches" to literature; and a third in which the evolutionary human sciences transform and subsume all literary study.

For the first two scenarios, we can easily enough extrapolate from past and current beliefs and practices, but we also have to factor in the continuing development of the evolutionary human sciences outside of literary study. That would have an impact on the way life would be lived within the isolated enclave of literary study. It is one thing to be a small village in a world consisting only of small villages. It is another thing to be a small village surrounded by a world empire in confident possession of the practices and beliefs through which it has achieved unification and mastery. For the third scenario, we have to envision how literary study would develop within an evolutionary perspective that encompasses all the human sciences.

Where Are We, and How Did We Get Here?

Before considering the three scenarios, I shall quickly describe the trajectory that brought us to our current state. Through the first two-thirds of the twentieth century, most literary study operated under a shared set of beliefs and values extending back to the Victorian cultural theorists, particularly Matthew Arnold. Giving up on religion, the Victorians looked for existential "meaning" in two main areas: utopian social futures, and the arts, especially poetry. They thought the arts condensed the best wisdom of our collective humanity and also gained access to whatever amorphous spirituality was left over after deducting the historical validity of the Bible, the divinity of Christ, and the immortality of the soul. One of the things left over in amorphous spirituality was the idea of a divinely ordered progression of history leading to some ultimate condition of social harmony and intellectual fulfillment. The arts, and especially poetry, would be the chief medium for recognizing and participating imaginatively in that blessed dispensation. However quaint such beliefs might now appear, until about 1980 they provided an overarching rationale for the two main kinds of study that occupied literary scholars: (1) hard-core scholarship—establishing texts, producing editions, collecting letters, writing biographies and literary histories; and (2) detailed interpretive analysis of individual texts and descriptive histories of literary traditions. Some of this work was animated by explicit invocations of Marxist, Freudian, Christian, or Jungian ideas, but most of it was eclectic, oriented to the common language and the common understanding. This whole phase can be designated the traditional humanistic paradigm.[1]

By the late seventies, signs of overproduction had become unmistakable. Most of the major projects in hard-core scholarship had been adequately completed. Critics interpreting single works were forced into ever more tenuous and improbable speculations. To publish interpretive commentary, one has to say something new, and most of what could reasonably be said at the level of common observation had already been said. The solution, of course, was to turn to European speculative philosophy, first structuralism, and then, almost immediately, "poststructuralism." The structuralists, supposedly, had demonstrated that structure is autonomous, a matrix or primary source, transcending content, and the poststructuralists demonstrated, supposedly, that structures are not only autonomous but anarchic, chaotic, impossible to pin down, and impossible to escape. "There is no outside the text," and the text itself is a house of mirrors—fun-house mirrors—signifiers generating signifiers, with no signified anywhere in sight to anchor the endless recession of distorted images.

Deconstruction swept through departments of literature like flag-waving cadres of the French Revolution, galvanizing all the inhabitants, striking terror in some, provoking others into obstinate resistance, but in most exciting rapturous enthusiasm. The inferiority complex that had long dogged literature professors vis-à-vis the scientists, who actually got things done, suddenly gave way to an extraordinary hubris in which literature professors believed they had unique access to the ultimate nature of things. The world at large, exemplified, say, by *Time* magazine, was skeptical but intimidated, uncertain at first, but willing to acknowledge any new form of glamour that could command attention. For three or four years, the deconstructors played word games, discovered their inner verbal child, fashioned exquisitely ambiguous titles for theoretical articles, and, in their more sober moments, adopted postures of cosmic nihilism. To a watching world, all this ultimately seemed rather silly, but the main force at work undermining the deconstructive regime was internal. People go into literature not just to play games with words. Literature gives access to the most intimate and powerful aspects of experience. Deconstruction offered a general stance of radical subversion to all existing values, but it offered very little in the way of positive human content.

Foucault provided the content. He absorbed deconstructive irrationalism and gladly assented to the transcendental status with which the deconstructionists had invested "Discourse," but he also had real bones to pick with the Western cultural tradition. He did not just adopt radical subversion the way a teenager adopts insolence, as a style. He went after the meat of the matter, systematically critiquing ideas of sanity, criminality, and sexuality, disdaining all social norms as arbitrary manifestations of "Power." This was a creed by which literary scholars could live, for three decades anyway, right up to the present time. It gave them a program and a stance: to reread all texts as insidious machinations of political power. Theorists and critics who have adopted this stance have a mission in life: to serve as the conscience of their race. Their constituencies are the victims of oppression in traditional power structures: especially women, ethnic minorities, homosexuals, and colonized peoples.

Three decades into the new postmodern hegemony, we are now also at least a decade into "the crisis in the humanities." The subversive metaphysical and political fervor that fuelled the post-structuralist revolution has long since subsided into tired routine. The question that generated the revolution, "What next?" is being asked again, and with increasing desperation. In a recent essay on the parlous state of the humanities, Louis Menand professes himself willing to consider almost any possible option, only just not one particular option: "consilience," that is, integrating literary study with the evolutionary human sciences. That option, he declares, would be "a bargain with the devil."[2]

Scenario One: And Never the Twain Shall Meet

In the first scenario—a continuation of the status quo—a large majority of literary scholars continue to share Menand's aversion to any connection with the evolutionary human sciences. The literary Darwinists stand wholly separate from the mainstream literary establishment, massively ignored, unable even to get panels accepted at the annual conferences of the Modern Language Association, assiduously though silently expunged from citation lists and from surveys of critical theory, not merely neglected but actively and aggressively shunned. In this scheme of things, the literary Darwinists write essays critical of mainstream practices but have no productive interaction with the mainstream.

If literary scholars reject literary Darwinism, what other kinds of work can they produce? The same kinds they have been producing for three decades: arcane theoretical systems of a purely verbal, speculative character, diverse in superficial terminology, but alike in their commitment to "cultural constructivism."[3] Along with the generation of more verbal systems, we would also have to have more readings of standard texts in terms of identity politics. This kind of thing might not seem susceptible to endless repetition. Hence the need for the constant proliferation of superficial variations in the verbal systems used for interpretation.

The poststructuralist revolution was based on no actual discoveries and no ideas more substantial than willful paradox and sophistical quibble.[4] That kind of intellectual foundation could vanish overnight, leaving nothing even for archeologists to sift through. Would it be possible then for literary study to cycle back through a traditional humanist phase? Possible, but not very likely. Traditional humanists are committed to literature itself as the deepest source of insight and wisdom. They are thus committed to the common idiom, and that idiom has already pretty much exhausted itself as a source of commentary on the standard texts. In contrast, the fundamental poststructuralist axiom is that meaning is "constructed." If that is the case, the supposedly determinate structure of meanings in a finite body of canonical texts would exercise no constraint on the proliferation of interpretive terms. Hence the greater likelihood that poststructuralism will have achieved, in a steady-state world, a permanent hegemony in literary study. It offers the hope of something always new to do, even if that novelty consists only in variations in analytic terminology.

If literary study continues indefinitely in the poststructuralist vein, it will do so under two forms of degenerative pressure: the inner inanition that is already so frequent a source of complaint among its own practitioners, and the ever growing prestige and power of the scientific understanding of human nature. Under that external pressure, "Theory" will have to become ever more elusive, avoiding all direct formulation of propositions that obviously conflict with established results of scientific research. The strategy for eluding science need consist only in refinements of a procedure already widely practiced: formulating all propositions simultaneously in two separate versions: the radical and the truistic. The radical version gives the appearance of a substantive proposition startling in its novelty, and the truistic version gives the appearance of logical invulnerability. The blending of the two versions gives the delusory appearance of propositions that are both new and true—the holy grail of all research. For instance, "There is no outside the text." Radical version: "Nothing exists outside of verbal constructs; only verbal constructs exist." Truistic version: "Everything we can talk about we can talk about only by using words; all our verbally mediated experience is verbally mediated." The radical version gives a fallacious appearance of profound novelty, suggesting a fundamental alteration in folk epistemology—that is, common sense. The truistic version, mingling indistinguishably with the radical version, invests the radical version with the self-evidence of tautology. When critics make damaging arguments against the radical version, the deconstructor can smoothly retreat into truism. "All I really meant to say was . . ." or, preemptively, "This is not to say . . ." Anyone willing to participate vicariously in the conceptual blur produced by the mingling of the two versions can enjoy the characteristic deconstructive *frisson*, the little shiver of cognitive pleasure at the manifestation of the Uncanny. To give substance to this *frisson*, one need only transpose the logic of equivocation into a slightly more concrete proposition: "All identity is socially constructed." Radical version: "The only constituents of identity are arbitrary social conventions; even something as basic as biological sex is purely and exclusively a construct of arbitrary social conventions." Truistic version: "Humans are social animals; all human experience is influenced in some way by participation in social life." In the blur between these two versions, most criticism has persisted now for decades, and could persist into the indefinite future.

We can fancy, in this scenario, that poststructuralism never dies, but we cannot fancy that it does not age. Time passes. Gollum dwindles and shrivels, becoming less human, but retaining physical vigor. Tithonus shrivels into a cricket but chirps perpetually. The Struldbrugs, in the third book of *Gulliver's Travels*, grow ever older, becoming more ill-tempered, narrow-minded, and senile, but happy, they and those who live with them, in the assurance that they will never die.

While Gollum dwindles, Tithonus chirps, and the Struldbrugs drool, what of the literary Darwinists? The first monograph in literary Darwinism, *Evolution and Literary Theory*, appeared in 1995. The number of books and articles published since 2007 and now in press—a four-year span—far exceeds the number published altogether in the twelve years from 1995 through 2006. In a steady-state scenario, this exponential growth could not continue. Otherwise, within just a few years, literary Darwinism would have come to dominate literary study, violating the premise of the scenario. So, we have to assume that the rate of growth in literary Darwinism not only levels off but actually declines—and all this while poststructuralist literary study is losing heart, on the one side, and the evolutionary human sciences are making giant strides on the other. Unlikely, but so goes the scenario. Within this scenario, we need say only that the literary Darwinists would continue to do the kind of work they have been doing all along. (For detailed information on what they have been doing all along, see the first chapter in this volume.)

The most important institutional blockage limiting further growth in literary Darwinism is that only one or two graduate programs, so far, allow students to pursue this line of work. In the steady-state scenario, then, we have to assume that older scholars continue to prohibit their students from taking up this line of investigation. Consequently, the work published in literary Darwinism would continue to be produced mostly by scholars who had already gained tenure on the strength of more conventional kinds of research.

Scenario Two: Joining the Party

In this second scenario, we can slot in the description of mainstream literary study from the previous scenario, assuming it would remain much as it now is or will be. The only thing that would change in this second scenario is that literary Darwinism would not be shunned. Nor would it become a dominant, commanding perspective, altering the whole paradigm of literary study. It would simply be recognized as yet one more "approach" to literary study. Two institutional markers would signal the realization of this scenario: evolutionists would have panels accepted at the annual conference of the Modern Language Association and its regional affiliates; and interpretive essays in literary Darwinism would regularly be included in casebooks of canonical literary texts. Most such casebooks now include essays exemplifying Marxism, Freudian psychoanalysis, deconstruction, feminism, and New Historicism (that is, Foucauldian cultural critique).

Among some of my colleagues with an evolutionist bent, this second scenario seems the most likely of the three. It takes account of the rapidly increasing visibility and prestige of literary Darwinism outside the academic literary establishment—for instance, the notices that have appeared in journals and newspapers around the world, from *Science* and *Nature* through *The New York Times, The Guardian, TLS, The Chronicle of Higher Education,* and periodicals in Sweden, Denmark, Germany, Hungary, Russia, India, Brazil, and Japan. Since the Darwinists have vindicated their claim that evolutionary ideas can be used for literary interpretation, and since they form a rapidly growing minority of literary scholars, is there any reason that this second scenario might not almost inevitably take place sometime within the next few years? I think there is. Marxism, Freudianism, and deconstruction are all totalizing in their own ways, but they can also all be converted into forms that make them parts of the standard postmodern blend. Althusserian Marxism and Lacanian psychoanalysis are essentially compatible with Foucauldian discourse theory. And indeed, "poststructuralism" as a school can be most concisely defined as the subordination of Marxist social theory and Freudian psychoanalytic theory to deconstructive semiotics. That is the message in Foucault's definition of "discursive practices."[5]

Can Darwinism be subordinated in this way to the transcendent power of the sign? Efforts along this line have not been wanting. In *Darwin's Plots*, Gillian Beer takes Darwinian themes as precursors for Derridean indeterminacy. George Levine takes a similar line in *Darwin and the Novelists*. Ellen Spolsky adopts the idea of "cognitive domains" from evolutionary psychology and uses this idea as evidence for the Derridean claim that cognition is necessarily incoherent.[6] Still, no specifically Darwinist form of poststructuralist interpretation has emerged from these efforts. Poststructuralism yields causal primacy to language. To think in evolutionary terms, in contrast, is almost automatically to adopt a perspective of deep time, a perspective in which "life," self-replicating DNA, precedes thought, to say nothing of language. One can speak of DNA itself as a form of "language," but this is just a metaphor, and it does not take one very far into the formation of personal and social identity. "Constructivist" and biological notions of personal and social identity seem inherently incompatible. Biology is too deep, broad, and basic to be easily or convincingly depicted as just another semiotic gambit.

The powerful disciplinary motives behind literary academics' resistance to biology form a natural bond with ideological motives. If human nature were "socially constructed," it could easily be changed to fit more neatly into whatever moral and political forms one might favor.[7] Causal force would reside primarily not in underlying biological realities but rather in the formulation of social ideals. One would need merely to think an ideal, using it to guide one's commentary on literature and life, in order to bring about desirable social change. This idealist approach is a particular manifestation of a pervasive and perhaps universal human cognitive disposition: the disposition for wishful thinking. Wishful thinking offers the solace of comforting illusion and could possibly even have adaptive, therapeutic value, easing stress and making it easier to endure insoluble problems. Nonetheless, pleasurable fantasy necessarily operates in tension with adaptive dispositions for finding out how things actually work. Literary academics at the present time are perhaps particularly susceptible to wishing away real social problems, rather than understanding them, because they have painted themselves into a disciplinary corner. Having abjured the prospect of gaining real knowledge, they have inevitably placed a heavy emphasis on moral and political judgment as the chief justification for what they do. If they cannot offer objective knowledge about their subject, the rationale for their professional existence must be that they occupy a superior ideological perspective. This professional *raison d'être* is a politicized, poststructuralist version of the humanist idea that a literary education makes one a better person. Poststructuralist ideologues envision a world in which conflicting interests and differential distributions of power no longer exist. Accordingly, they look with disapproval on all actual forms of social and political organization. They thus guarantee for themselves a perpetual stance of ideological superiority. Darwinism is by no means incompatible with an informed and humane moral creed,[8] but it is most definitely incompatible with the utopian ideal of a world order in which conflicting interests and differential distributions of power do not exist.

Despite the inherent incompatibility between Darwinism and Foucauldian cultural critique, for the purposes of the scenario, let us imagine that the Darwinists are brought into the casebooks. Would they consider themselves just one more approach among many? Some no doubt would. "Pluralism" is a chronic symptom of theoretical confusion in the humanities. The idea is that the world is divided into two main parts: a physical part that can be understood by science—reduced to components, quantified, and unified—and an imaginative, cultural, spiritual, or personal part—qualitative, consisting of unique irreducible moments of experience and unique irreducible effects, aesthetic and imaginative. By its very nature, this second world could never be reduced to a unified set of underlying regularities. It could only be described and evoked. Its essence is not reductive law but phenomenal particularity. The best way to deal with it is to bring as many perspectives as possible to bear on a subject and thus to illuminate as many diverse aspects of the subject as possible. The diversity of aspects would never add up to a single, unified phenomenon, and explanations of those aspects would never add up to a single, unified explanation. Though denied the ultimate satisfaction of unified causal explanation, adherents of this worldview can look forward to an endless succession of incomplete and incompatible interpretive responses to the same finite body of novels, poems, and plays. This, more or less, is the pluralist metaphysic. However diverse their overt professions of theoretical allegiance, this metaphysic defines the deepest convictions in most practitioners in the humanities today.

What, then, would a Darwinist contribution to a casebook look like? To qualify as Darwinist, a reading would have to bring all its particular observations into line with basic evolutionary principles: survival, reproduction, kinship (inclusive fitness), basic social dynamics, and the reproductive cycle that gives shape to human life and organizes the most intimate relations of family. While retaining a sense of the constraining force of underlying biological realities, literary Darwinism would also have to emulate the chief merit of Foucauldian cultural critique—its understanding that the forms of cultural representation are highly variable, that these variations subserve social and political interests, and that every variation has its own specific imaginative quality. As it is currently practiced, cultural critique usually arrives at its conclusions in a theoretically illegitimate way, by assuming the causal primacy of representation. This is what it means to say that reality and social identity are "constructed." Despite the obvious fallacies in this idea, Foucauldian critique often has rich descriptive power. The Foucauldians have achieved dominance in literary study partly because they recognize that the chief purpose of literary study is to examine the forms of cultural imagination. To compete for space in casebooks, then, the Darwinists would almost necessarily have to eschew their own tendencies toward literalist representationalism—the idea that literary texts merely depict a preexisting reality in a true and faithful way.

Vulgarity accompanies theoretical movements the way camp followers—hawkers, prostitutes, and idlers—accompany an army in the field. Just as there is a "vulgar Marxism," there is also a "vulgar

Darwinism." Yet further, there is a vulgar form of literary Darwinism. In its most naïve form, literary Darwinism consists in merely pointing to the existence of Darwinian themes in various works of literature. Madame Bovary wants a mate with more status than her husband. Anna Karenina is bored with her respectable husband and gets charmed into an illicit relation with a Byronic type better suited for short-term mating. No wonder she ends up throwing herself beneath a train. Tom Jones just can't resist a roll in the hay with Molly Seagram, and that gets him into hot water with Sophia Western, but he is only doing what comes naturally to males, so she forgives him in the end. Had Sophia herself been found dallying with Molly's brother, the outcome could not have been so favorable. The sexual double-standard is just part of human nature.

In its short history, vulgar literary Darwinism has already become established as a convenient target for critics eager to dismiss the possibility of evolutionary criticism in its more sophisticated forms.[9] Practitioners of the more sophisticated forms recognize that literature does not simply represent typical or average human behavior. Human nature is a set of basic building blocks that combine in different ways in different cultures to produce different kinds of social organization, different belief systems, and different qualities of experience.[10] Moreover, every individual human being (and every artist) constitutes another level of "emergent" complexity, a level at which universal or elemental features of human nature interact with cultural norms and with the conditions of life that vary in some degree for every individual. Individual artists negotiate with cultural traditions, drawing off of them but also working in tension with them. The tension derives from differences in individual identity, the pull of universal forms of human nature, and the capacity for creative innovation in the artist. Individual works of art give voice to universal human experience, to the shared experience of a given cultural community, and to the particular needs of an individual human personality. Literary meaning consists not just in what is represented—characters, setting, and plot—but in how that represented subject is organized and envisioned by the individual human artist. Moreover, literary meaning is a social transaction. Literary meaning is only latent until it is actualized in the minds of readers, who bring their own perspectives to bear on the author's vision of life. A thorough interpretive effort would subsume represented subjects and formal organization into an overarching concept of literary meaning, and it would expand the concept of meaning to include its transmission and interpretation. Still further, instead of looking only at intentional meanings and the responses of readers, a thorough evolutionary critique would look at the kinds of psychological and cultural work specific literary texts actually accomplish—the functions they fulfill—and it would locate those functions in relation to broader ideas of adaptive function, thus bringing the interpretation of individual works to bear as evidence on the larger, still controverted question of adaptive function.

The more any Darwinian critique succeeded in achieving this kind of total reading, the less compatible it would be with the pluralism implicit in casebooks. If Darwinism becomes just another approach included in casebooks, it will probably do so by carving out its own distinctive niche in a way parallel to that of the deconstructionists, Freudians, Marxists, and feminists. Like their fellow practitioners in other schools, Darwinists would need to make their interpretive essays distinctive by making them crude and sensationalistic. Casebook essays typically earn their keep by riding hobby horses into the ground. They sacrifice justice and sensitivity in favor of programmatically rehearsing terms that distort the actual structure of meaning in a literary text. If the Darwinists wish badly enough to be included in casebooks, they should be able to meet these requirements with no more difficulty than that encountered by practitioners of the other critical schools.

Scenario Three: Back to the Future

If literary Darwinism were to be dominated by its vulgar form, the evolutionists would have some chance of getting into the casebooks but no chance of ultimately transforming literary studies. Transformation involves renovation from the ground up, eliminating the endemic confusion of "pluralism" and carrying through on the implications of a Darwinian vision. It is not the case that there is nothing outside the text. It is not even the case that there is nothing outside of life. Before life evolved, there was a physical universe in which it could evolve. It *is* the case, though, that there is nothing in life outside of evolution. That means both less and more than it might seem to mean. It does not mean that the forms of literary development—genres and traditions—exactly parallel the macro-structures of evolutionary development. It does not mean that all human experience is driven in a simple and direct way by the biblical injunction *go forth and multiply*. It does not mean that all literary characters exemplify average or species-typical forms of behavior. It certainly does not mean that all authors, even ancient, medieval, Renaissance, and neoclassical authors, are crypto-Darwinists. What it does mean is that all humans past and present have evolved under the massively constraining force of adaptation by means of natural selection. It thus means that the species as a whole has a characteristic structure of "life history." That life history entails a species-typical set of motive dispositions and emotional responses, and along with them a species-typical range of personality characteristics. Individuals can and often do vary from the species typical, but the species typical provides a common frame of reference. Individual persons vary from that base line in ways that have systemic effects on the motivational and emotional characteristics of the whole system. Individuals can mate with members of their own families, prefer sexual partners of their own sex, murder their parents or children, live celibate lives in religious orders, consign themselves to perpetual hermitage in deserts, starve themselves to death, throw themselves on hand grenades, blow themselves up in crowded market squares, devote their lives to charitable purposes, sacrifice worldly ambition for the sake of art, or write books declaring that reality is purely a social construct. All of these forms of behavior can be traced to the only possible source of all behavior: the interaction between genetically transmitted dispositions and specific environmental conditions. Consequently, none of these behaviors is "unnatural," and indeed, there is no such thing as an unnatural form of behavior. Every form of behavior consists in some discernible combination of the elements of human nature interacting with specific environmental conditions. Every form of behavior has its own distinct set of affects; everything comes with a cost; every form of satisfaction sacrifices some other possible form of satisfaction; every fulfilled impulse works in tension with some other impulse left unfulfilled; and every act shapes the total organization of feeling and perception in the whole organism and in the larger social groups in which virtually all individual humans are embedded. The motives and passions that have derived from an adaptive evolutionary process constitute what we call "human nature." Intuitive perceptions of these motives and passions are products of "folk psychology"—the common, shared basis for the understanding of intentional meaning in other human beings. Folk psychology is the *lingua franca* of social life and of literature.

The Darwinian literary study that, in this scenario, will ultimately absorb and supplant every other form of literary study will assimilate all the existing concepts in literary study—traditional concepts of style, genre, tone, point of view, and formal organization, substantive concepts of depth psychology, social conflict, gender roles, family organization, and interaction with the natural world. It will not just take those concepts ready-made and tack them together like a shack made of flattened cans and scraps of cardboard on the edge of a third-world city. It will use them as heuristic guides to the emergent structures that are most relevant to literary study as a subject matter with its own peculiar features and concerns, but it will rebuild each of those concepts *de novo*—reshaping, breaking down, consolidating, and adding—by direct and explicit reference to the rapidly expanding research in all the contiguous disciplines of the human sciences.

Most of the literary Darwinists now at work have been trained in the old schools and have been teaching themselves new concepts and methods, striving and sometimes struggling to gain an assured perspective on disciplines in which they have no specialist expertise—evolutionary biology, genetics, psychology, anthropology, linguistics, personality theory, and cognitive and affective neuroscience. At the same time, they have been integrating these concepts with traditional concepts in literary study, building theoretical principles that could explain and direct their efforts, and seeking to vindicate these theoretical constructs through Darwinian readings of specific texts. All this is necessary, but it is not enough.

There are no real ontological or epistemological barriers separating the humanities and the evolutionary human sciences. We do not occupy parallel universes, stepping comfortably out of one when we drive a car or visit the dentist and into another when we read a novel, look at a painting, or listen to a piece of music. It is all the same world, intelligible by the same instruments.[11] The barriers separating these two worlds are the barriers merely of convention based on ignorance. "Pluralism" elevates those conventions to the dignity of a theoretical position, and that position provides a rationalization for maintaining the habitual limitations in the scope of our subjects and the methods by which we investigate those subjects. In this third scenario, the pace of production in Darwinist publication will continue or increase; the institutional resistance of the postmodern establishment will crumble from within, almost silently, softly metamorphosing into dust, like the Soviet empire, as a result of intellectual dry rot. A few hammer blows no doubt will be needed to knock down actual obstructions, like the Berlin Wall, but these blows are more symbolic than substantive. The real barriers are in the minds of men and women. As these changes occur, the Darwinists will not be elevated into comfortable hegemony, simply taking possession of the seats of power vacated by the erstwhile commissars of the postmodern politburo. They will be in something like the same position as the former states of the Eastern Bloc, running hard just to catch up with their more prosperous neighbors to the West, working day to day to maintain life while simultaneously rebuilding their whole institutional infrastructure.

In this third scenario, high school students will all take introductory courses in statistics, which are, after all, less demanding mathematically than the more advanced forms of math in the standard high school curriculum. Undergraduates, as part of their general education, will take more advanced courses in statistics and will also take courses in empirical methodology. This will not be so much an added burden as it might seem, since the whole undergraduate curriculum will be much more unified than it now is. Courses in the "social sciences" will themselves all be integrated from an evolutionary perspective—the perspective that prevails now, for instance, in journals such as *Behavioral and Brain Sciences*. The evolutionary human sciences will be closely integrated with required courses in evolutionary biology, molecular biology, and the sciences of the brain. Students in the humanities will develop basic proficiency in these disciplines in the same way virtually all European students, in all disciplines, now develop a good working knowledge of the English language.

When undergraduate English majors write papers on Shakespeare or Virginia Woolf, Chaucer or Charlotte Brontë, they will in some ways do what they have always done—talk about characterization, personal and social identity in the characters and in the author, style, point of view, tone, the organization of narrative, and cultural contexts and literary traditions. But in other ways, all this will be different. In writing of personal and social identity, they will not have recourse to obsolete and misleading ideas from Freud, Marx, and their degenerate progeny. They will have recourse instead to empirically grounded findings in the evolutionary human sciences. In speaking of tone and point of view, they will make use of cognitive and affective neuroscience. They will consider local affects in relation to the actual brain structures and neurochemical circuits that regulate emotions, to "mirror neurons," Theory of Mind, and "perspective taking." In assessing style and the formal organization of narrative or verse, they will take account of underlying cognitive structures that derive from folk physics, folk biology, and folk psychology. They will still bring all their intuitive sensitivity to bear, registering the affective qualities that distinguish one work from another, communing in spirit with the author, or holding off skeptically from authors with whom intimacy for them is repugnant. They will not regard their own subjective responses as wholly arbitrary nor as somehow incommensurate with the brain structures that regulate behavior, thought, and feeling in ordinary life. When they locate literary works in relation to cultural context, they will have recourse to new forms of history, both forms that use brain science to create an ecological and psychopharmacological profile of a given era,[12] and also forms that delineate large-scale laws of social organization deriving from elementary processes of inter-group conflict and intra-group organization.[13] They will draw on knowledge both of the actual social and political situation and of the deep evolutionary background for that situation. We already see works of literary scholarship that answer to this description.[14]

When they come to graduate study, aspiring literary scholars will have open before them a wide spectrum of methodological choices, ranging from the purely discursive, essayistic forms of commentary that now dominate the humanities to the rigorously quantitative, empirical methods that now prevail in the sciences.[15] Some no doubt will tend more in one direction than in another, but none will think that quantitative and discursive forms of study occupy separate and incommensurate universes. They will not cast about desperately for novelty, taking recourse in superficial verbal variations ensconced in sophistical theoretical ambiguities. They will, rather, wake up like kids at Christmas, delighted with the endless opportunities for real, legitimate discovery that are open to them.

Conclusion: Belief in Things Unseen

In one way, the third scenario is the hardest about which to make concrete predictions. To predict a continuation of the status quo, one need only extrapolate from what one can actually see and factor in the consequences of degenerative pressure, internal and external. The process at work is something like that in which profilers for police agencies take a photograph of a person missing for years, apply known principles for the way people's faces change over time, and come up with a reasonable approximation to what the missing person would look like now. So also, with the second scenario, one holds the mainstream practices steady while adding to them the current practices of literary Darwinism. To make literary Darwinism fit comfortably into the culture of casebooks, one need only standardize its current tendencies toward vulgarity. The third scenario allows us to stipulate the conditions for rebuilding literary knowledge from the ground up, but by its very nature as a progressive, empirical discipline, it exceeds prediction. It promises discovery, things not yet dreamed of, lying latent in the bosom of reality, at levels of causal structure we have not yet penetrated, and at levels of complexity we do not yet, perhaps, have the skills even to envision. If one were able to travel back in time, visit some far-seeing investigator in the Renaissance, an astronomer, say, or an anatomist, take him by the elbow and give him a tour of the modern world, would it not all seem to him truly alien, strange, wonderful beyond all imagining? And yet, all these wonders were lying latent in the world, and he would himself have been taking the first steps toward their discovery.

PART TWO

Interpretive Practice

CHAPTER 5

Aestheticism, Homoeroticism, and Christian Guilt in *The Picture of Dorian Gray*

Oscar Wilde's *The Picture of Dorian Gray* offers two special challenges to Darwinian criticism. First, the novel is saturated with homoerotic sexual feeling, and it thus defies any simple reading in terms of behavior oriented to reproductive success. Second, the central conflicts in the novel involve two competing visions of human nature, and in their conceptual structure neither of those visions corresponds very closely to the quasi-Darwinian conceptual structure implicit in most realist and naturalist fiction. One vision derives from the aestheticist doctrines of Walter Pater, and the other from a traditional Christian conception of the soul. Pater's ideas about human motives and the human moral character are at variance both with Christianity and with Darwinism. Christianity and Darwinism share certain concepts of the human moral and social character, but they couch those concepts in different idioms, and they would invoke wholly different causal explanations for how human nature came to be the way it is. Wilde does not develop his themes in Darwinian terms, but the novel can still be read and understood from a Darwinian perspective. If Darwinian psychology gives a true account of human nature, including its homoerotic variations and the affective and ethical dimensions of religious beliefs, it can explain the meaning structure of *Dorian Gray*.

In weighing the effects of Wilde's homosexuality on the meaning of the novel, I shall use the incisive Darwinian analysis of homosexual behavior provided by Donald Symons in *The Evolution of Human Sexuality*. I shall not concern myself with the still controverted—

and for my purposes irrelevant—question as to whether homosexuality is or is not an adaptive form of behavior. I shall instead compare the psychological character of homosexual and heterosexual relationships. In analyzing the conflict between homoeroticism and the Christian ethos, as Wilde conceives it, I invoke a Darwinian conception of species-typical evolved sex differences, and I correlate homoeroticism with male sexual psychology and the Christian ethos with the maternal female character. I argue that Wilde associates aestheticism with homoeroticism and that he sets both in opposition to the idea of lasting affectional bonds and self-sacrificing love. As an aesthete devoted solely to sensual pleasures, Wilde's protagonist repudiates the idea of affectional bonds, and it is that repudiation which produces the mood of guilt and horror in which the novel culminates. Wilde partially identifies with his own protagonist, and he is himself riven by the conflict between homoerotic aestheticism and Christian pathos. The unresolved conflicts in the plot of the novel reflect deep divisions in his own personal identity.

In recent years a number of studies have discussed the specifically homosexual character of *Dorian Gray*. By making this issue into an explicit theme these studies have taken a crucial new step toward a true understanding of the deep symbolic structure in Wilde's novel. But most of these studies have been written from a liberationist standpoint; most have been written from within a Foucauldian framework of sexual theory, treating of homosexuality as a discursive construct or a literary trope; and none has made use of evolutionary psychology.[1] Both liberationist commitments and poststructuralist ideas lead critics away from the central artistic purposes and the basic structures of meaning in Wilde's novel. A commitment to a liberationist standpoint typically involves a determination to envision all homosexual experience in a positive light. As a result, most of the recent gender criticism of Wilde's novel has avoided registering the elements of guilt and self-loathing in Wilde's self-image, and those elements are central to the meaning of the story—to its characterization, plot, theme, style, and tone. (Three studies of *Dorian Gray* have acknowledged negative elements in Wilde's depiction of homosexual experience.)[2] Poststructuralism repudiates, in Jonathan Dollimore's phrase, "the model of deep human subjectivity."[3] If the fundamental artistic motive in *Dorian Gray* is to articulate the conflicts in the depths of Wilde's own identity, the poststructuralist affiliations of current gender theory would necessarily join with its liberationist commitments in casting a veil over the meaning of his novel. In order to gain a true understanding of the deep symbolic structure of Wilde's novel, we must combine a recognition of deep human subjectivity with a recognition of Wilde's own conflicted feelings about his homosexuality. If we deploy this combination, we are in a position, for the first time, fully to grasp Wilde's meaning.

Deep human subjectivity depends fundamentally on the physical, biological reality of human life—on physical sensations that put people in touch with a real physical world; on animal needs, impulses, and reactions such as hunger, thirst, visceral terror, and sexual desire; on basic emotions lodged deep in human nature: maternal and filial love, sympathy, hatred, and the craving for companionship; and on forms of cognition that emerge spontaneously in all normally developing minds. All of these sources of subjectivity precede the evolution of modern *Homo sapiens*. They thus necessarily precede the evolution of language and of complex forms of culture. And not just precede in time. Evolved, genetically transmitted dispositions inform and constrain language and culture. Language and culture articulate and formalize innate dispositions. The sense of deep human subjectivity is a biological reality.

A Darwinian critique of *Dorian Gray* would acknowledge the way in which all its symbolic figurations—sexual, religious, and philosophical—are culturally and historically conditioned, but it would also identify the way in which those culturally conditioned figurations organize the elemental, biologically grounded dispositions of human nature. The symbolic figurations in Wilde's story cannot be limited to the socially encoded values and conventional literary meanings available within a specific cultural context. Wilde, like all artists, assimilates the cultural configurations available to him, but he penetrates to their sources in human nature, and he uses these configurations as a medium through which to articulate his own individual identity—his own sexual, social, moral, and intellectual character.

Dorian Gray is a wealthy young man of exceptional beauty. His friend Basil Hallward paints a portrait of him that captures that beauty. While he is posing for the painting, Basil's friend Lord Henry Wotton tells Dorian that youth and beauty are the only things worth having in life and admonishes him to live fully, since his own youth must soon fade. Dorian exclaims that he wishes he could change places with the painting so that the painting would grow old but that he would remain young. His wish is granted, though he does not realize it until some time later. He becomes engaged to a young actress, Sybil Vane, whose talent as an actress he admires. When she falls in love with him, her acting deteriorates, he rejects her, and she kills herself in despair. Under Lord Henry's tutelage, Dorian finds that he can regard her death coldly, as an aesthetic event, and he then notices that the painting has changed; it has acquired a look of cruelty about the mouth. Dorian hides the painting in his old school room, and as he ages, pursuing a life divided between aesthetic cultivation and debauchery, he never changes in appearance. He remains young and beautiful, while the portrait grows steadily older and more hideously ugly, manifesting in its deformity the moral corruption of Dorian's "soul." Years later, Basil hears rumors that Dorian is secretly leading a depraved life and asks Dorian to tell him the truth about his behavior. In response, Dorian shows him the painting. Basil is horrified and calls on Dorian to repent and reform. Instead, Dorian stabs Basil and kills him, and the hand in the painting becomes stained with blood. Some time after, in a thematically irrelevant episode that bulks out a slender narrative, Sybil Vane's brother discovers Dorian's identity. He blames Dorian for Sybil's death and plans to murder him but is himself killed in a hunting accident. Having escaped destruction, Dorian makes an effort to behave generously to a girl by not seducing her. He hopes his generosity will be reflected in the painting, but the face in the painting only takes on a new expression of cunning hypocrisy. In loathing and revulsion, Dorian stabs the painting in the heart. The knife stroke kills Dorian himself, and he and the painting once again change places. The image in the painting becomes young and beautiful; and Dorian Gray, as a corpse, is old and loathsome.

The three chief male figures in the novel all embody aspects of Wilde's own identity, and that identity is fundamentally divided against itself. The novel is thus a "psychodrama." Writing in a period before poststructuralism had cordoned off "deep human subjectivity," Barbara Charlesworth gives a succinct formulation to this view of the novel. "Wilde, even more consciously than most writers, split himself into various characters and saw in all of them some portion of his actual or potential self. . . . [his] was a nature of contradictions from which he could find no escape. . . . With the intelligence to understand all the conflicts of his age, yet without the ability or the will to resolve them, Wilde was finally broken by them."[4] In a letter to a friend, Wilde himself suggests an autobiographical dimension for the characters in the novel, but his own commentary tacitly smoothes over both the sinister aspects of the three characters and the conflicts among them. He says that *Dorian Gray* "contains much of me in it. Basil Hallward is what I think I am: Lord Henry, what the world thinks me: Dorian what I would like to be—in other ages perhaps."[5] What he does not say in his letter is that Dorian is beautiful but selfish, sensual, and cruel; Lord Henry is a worldly cynic incapable of registering the moral horror that leads Dorian to murder and suicide; and that Basil is enthralled by Dorian's beauty but appalled at the moral quality of his life. The conflicts emerging out of these values and dispositions constitute the central structures of meaning in the story. The idea that the characters embody aspects of Wilde's own conflicted identity stands in sharp contrast with the poststructuralist idea that the characters embody various aporias, gaps, and paradoxes inherent in "textuality."[6]

For Wilde, identity consists of two main elements, sensual pleasure and moral pathos, and in his moral universe these two elements are usually set in opposition to one another. Sensual pleasure associates itself with egoism, worldly vanity, and cruelty. Moral pathos is sometimes associated with devoted love, but it manifests itself primarily as pity for the poor and as tenderness toward children. Erotic passion allies itself with sensual pleasure. The morally negative side of Wilde's identity is distinctly male and predatory, and the positive side distinctly female and maternal. In Wilde, the moral sense couches itself explicitly and imaginatively in Christian terms—in terms of self-sacrificing love, sin, remorse, redemption, and the soul.

The most overt and explicit manifestations of Wilde's polar thematic structure appear in his fairy tales—stories that have medieval characters and settings and that are saturated with the spirit and mood of medieval religious experience. In the fairy tales, Christian pathos usually triumphs over egoistic cruelty and sensual pleasure. The Happy Prince and the swallow who serves as his messenger sacrifice themselves for love and pity, and God sanctifies their sacrifice. The Selfish Giant repents of his selfishness, embraces the Christ Child, and is taken to heaven. The Young King renounces wealth and pomp that feeds off the suffering of the poor, and when his subjects revolt, God himself intervenes and crowns him with glory. The Star Child is arrogant and cruel, but he is sore afflicted, repents, humbles himself in self-sacrificing penance, and as a reward is crowned king. The Nightingale impales her heart on a thorn, sacrificing her life for love. The young man and woman for whom she makes the sacrifice are not worthy of it, but Wilde's own sublime lyricism implicitly affirms its intrinsic beauty:

> So the nightingale pressed closer against the thorn, and the thorn touched her heart, and a fierce pang of pain shot through her. Bitter, bitter was the pain, and wilder and wilder grew her song, for she sang of the Love that is perfected by Death, of the Love that dies not in the tomb.[7]

When Basil invokes the specter of guilt at living solely for selfish pleasure, Lord Henry tells him that "'mediaeval art is charming, but mediaeval emotions are out of date.'"[8] Clearly Lord Henry has not been reading Wilde's fairy stories, and his flippant dismissal of guilt underscores his inadequacy as an interpreter of Dorian's experience. It is nonetheless the case that in *Dorian Gray* the Christian ethos manifests itself only negatively, as guilt and anguish. There is no moment of transfiguring redemption at the end. It is not a fairy tale but a horror story, and in that respect, it is perhaps more true to Wilde's own life than the stories that depict redemptive transfigurations.

Dorian is not all of Wilde, but he is part of him, and the qualities exemplified in Dorian's career have two main sources in Wilde's own experience, one an intellectual source, and the other a personal, sexual source. The chief intellectual source is the philosophy of aestheticism propounded by Walter Pater. The personal, sexual source is the homoerotic sensibility that places a maximal value on youth, beauty, and transient sensual pleasure. Pater was himself homosexual, though possibly celibate, and in Wilde's own mind aestheticism and homoeroticism converge into a distinct complex of feeling and value. Dorian's life turns out to be something like an experimental test case for the validity of Pater's aestheticist philosophy, and the experiment falsifies the philosophy. Dorian lives badly and dies badly, but the retributional structure does not simply eliminate the Paterian component from Wilde's sensibility. That component is inextricably linked with Wilde's temperament and his sexual identity. (Several of the scholars who have commented on Wilde's use of Pater in *Dorian Gray* have recognized Wilde's ambivalence toward Pater but have emphasized the negative, satiric aspects of Wilde's treatment.)[9]

The key tenets of Pater's philosophy are divulged in one highly condensed and vastly influential passage in the conclusion to *Studies in the Renaissance*. Pater treats of humans as egoistic isolates for whom reality consists only of transient sensory impressions:

> Every one of those impressions is the impression of an individual in his isolation, each mind keeping as a solitary prisoner its own dream of a world.... To such a tremulous wisp constantly re-forming itself on the stream, to a single sharp impression, with a sense in it, a relic more or less fleeting, of such moments gone by, what is real in our life fines itself down. It is with this movement, with the passage and dissolution of impressions, images, sensations, that analysis leaves off—that continual vanishing away, that strange, perpetual, weaving and unweaving of ourselves.[10]

Throughout *Dorian Gray*, Wilde echoes the explicit ethical doctrines of this brief essay. Pater declares that "not the fruit of experience, but experience itself is the end" (*Renaissance*, 188). And Lord Henry inducts Dorian into the philosophy of a "new Hedonism" the aim of which "was to be experience itself, and not the fruits of experience, sweet or bitter as they might be.... It was to teach man to concentrate himself upon the moments of a life that is itself but a moment" (*DG*, 101). In answer to Pater's evocation of the amorphous and unstable character of the individual ego, Dorian "used to wonder at the shallow psychology of those who conceive the Ego in man as a thing simple, permanent, reliable, and of one essence. To him, man was a being with myriad lives and myriad sensations, a complex, multiform creature" (*DG*, 111). Pater suggests that "our failure is to form habits" (*Renaissance*, 189). And Lord Henry proclaims, "'The people who love only once in their lives are really the shallow people. What they call their loyalty, and their fidelity, I call either the lethargy of custom or their lack of imagination. Faithfulness is to the emotional life what consistency is to the life of the intellect—simply a confession of failure'" (*DG*, 43).

By emphasizing the single moment of sensation and the isolated but amorphous ego, Pater eliminates the two central components of moral life—the bonds we have with other lives, and the continuity of identity through time. Following Pater, Lord Henry tells Dorian that "'the aim of life is self-development. To realize one's nature perfectly—that is what each of us is here to do'" (*DG*, 19). What Pater and Lord Henry fail to understand is that the "self" cannot be cultivated or "developed" in isolation from its relations with others. Nor can it be developed with an emphasis on isolated moments of sensation; it bears within it the burden of all its past acts. As Darwin understood, those two forms of extension—of the self in relation to others, and of the self extending over time—are the very basis and substance of moral life:

> A moral being is one who is capable of comparing his past and future actions or motives, and of approving or disapproving of them. . . .
>
> Man, from the activity of his mental faculties, cannot avoid reflection: past impressions and images are incessantly passing through his mind with distinctness. Now with those animals which live permanently in a body, the social instincts are ever present and persistent. . . . They feel at all times, without the stimulus of any special passion or desire, some degree of love and sympathy. . . . A man who possessed no trace of such feeling would be an unnatural monster. . . . Conscience looks backwards and judges past actions, inducing that kind of dissatisfaction, which if weak we call regret, and if severe remorse.[11]

Darwin's analysis refers to "man" in general, that is, to human nature. Generalizing from his own temperament as an isolated, introverted aesthete, Pater developed a philosophy of the goals and purposes of life that are not congruent with human nature and that are thus not functional and adequate for most people. Wilde partially accepted Pater's vision; Dorian Gray embodies Wilde's own disposition to live in absorbed and egoistic delight at pure aesthetic sensation. But in Wilde's personality that disposition is set in active and even violent tension with the sense of social bonding and the continuity of the individual identity.

In Wilde's own imagination, the egoistic sensualism of Pater's decadent aestheticism correlates with the emphasis on promiscuous and impersonal sex that is a distinguishing feature of a homoerotic sensibility, and Wilde's intuition in this regard gains confirmation in the research of Donald Symons. In *The Evolution of*

Human Sexuality, Symons collates and analyzes multiple studies of homosexual behavior. On the basis of these studies, he concludes that male homosexual behavior is characterized by promiscuous, impersonal sex. He explains this pattern of behavior by invoking the Darwinian logic of differences in the reproductive interests of males and females, and the corresponding differences in male and female sexual psychology. Males and females have coevolved, but their sexual character is partially complementary and partially conflicting. Males can benefit reproductively by promiscuous sexual encounters, and male sexual psychology is more prone to casual sex. Females benefit most by enlisting the sustained support of a male who possesses material resources and is willing to invest them in the woman and in her offspring. Men tend toward promiscuous desire; women seek lasting relationships. Men are on average adapted preferentially to value youth and beauty in a mate, and women are on average adapted preferentially to value status and resources in a mate. Because men and women have coevolved in adaptive interdependency, men are adapted to seek the status and resources women value, and women are adapted to be attentive to those aspects of beauty that motivate men. In strongly hierarchical, polygynous societies, men of high rank and wealth have multiple wives or concubines (and the lowest ranking males are consequently excluded from sexual relations altogether). In monogamous societies, males partially suppress their desire for multiple sexual partners, though pornography, prostitution, adultery, and "serial monogamy" still cater to evolved male proclivities for diffuse sexual experience. In homosexual communities, Symons explains, the male desire for promiscuous sexual encounters is not constrained to compromise with female dispositions toward long-term pair bonding. The result is that male homosexual communities produce a culture of promiscuous sexual encounters. (Lesbians, in contrast, maximize female proclivities for stable, long-term pair bonds.)[12]

Dorian Gray has an overt heterosexual plot, and there is no explicit homosexuality in the story—it could hardly have been published had there been—but the putatively heterosexual liaison with Sybil is of a purely aesthetic character, and the atmosphere of the story is saturated with homoerotic feeling and style. That feeling and style make themselves felt from the opening lines of the novel, and the first several scenes establish its sexual orientation by interweaving four chief elements: images of luxuriant sensuality, an overriding preoccupation with male beauty, the depiction of effeminate mannerisms among the characters, and a perpetual patter of snide remarks that are hostile to women, to marriage, and to sexual fidelity. None of these four elements would by itself decisively signal a homoerotic orientation, but in the combination Wilde produces, the effect is unmistakable and strongly evocative. Luxuriant sensuality is not exclusively homoerotic, but when it is closely associated with a fixation on male beauty, it invests that fixation with an erotic charge. Antagonism to heterosexual bonding is not in itself an unequivocal marker of homoeroticism. Heterosexual males can also express dislike for being tied down, but when coupled with homoerotic sensuality and with effeminacy of manners, antagonism to female desires for "fidelity" assumes a specifically homoerotic character. Recent historians of gender roles have argued that until Wilde's trials for homosexual practices, in 1895, effeminacy of manners was not unequivocally associated in the public mind with a specifically homosexual persona; they argue also that Wilde's own persona and the public response to his trials were pivotal in fixing the modern public image of the homosexual.[13] But even before Wilde's trials, effeminacy would by definition already have signaled a disruption or crossing of gender boundaries, and that disruption, since it is associated with an erotically charged fixation on male beauty, gives a sufficiently distinct signal of the sexual orientation that animates Wilde's characters. Among heterosexuals, feminine characteristics act as a stimulus or trigger for male sexual desire. One chief reason effeminacy can be so easily integrated with a homoerotic persona is that effeminacy indirectly suggests that the effeminate male could himself be an object of male sexual desire. (The original serialized version of *Dorian Gray* contains a few more overtly homoerotic gestures and expressions than the book version.)

Evoking a homoerotic atmosphere is central to Wilde's artistic purposes. From the very first lines of the novel, he uses all the resources of his style to orient the reader to his own distinctively homoerotic sensibility, and he makes a point of locating that sensibility in relation to the themes of Pater's aesthetic philosophy:

> The studio was filled with the rich odour of roses, and when the light summer wind stirred amidst the trees of the garden there came through the open door the heavy scent of the lilac, or the more delicate perfume of the pink-flowering thorn.

> From the corner of the divan of Persian saddle-bags on which he was lying, smoking, as was his custom, innumerable cigarettes, Lord Henry Wotton could just catch the gleam of the honey-sweet and honey-coloured blossoms of the laburnum, whose tremulous branches seemed hardly able to bear the burden of a beauty so flame-like as theirs; and now and then the fantastic shadows of birds in flight flitted across the long tussore-silk curtains that were stretched in front of the huge window, producing a kind of momentary Japanese effect. . . . (*DG*, 7)

Pater had proclaimed that "this, at least of flame-like our life has, that it is but the concurrence, renewed from moment to moment, of forces parting sooner or later on their ways," and he had characterized the ultimate constituent of experience, the impression, as "a tremulous wisp" (*Renaissance*, 187, 188). By importing Pater's distinctive idiom ("flame-like," "tremulous") into Basil's studio, Wilde gives Pater's abstract doctrines not just a concrete habitation and a name but also a sexual orientation. In its delicate and luxurious sensuality and its emphasis on art-like effects, the evocation of this scene strikes a new note in English fiction. It registers a distinct sensibility, and one defining aspect of that sensibility is an overwhelming preoccupation with male beauty. Dorian is first introduced, through his portrait, as "a young man of extraordinary personal beauty" (*DG*, 7). Lord Henry expands expressively on this flat denotation—"'this young Adonis, who looks as if he was made out of ivory and rose-leaves. Why, my dear Basil, he is a Narcissus'" (*DG*, 9). When he meets Dorian in person, Lord Henry reflects, "'Yes, he was certainly handsome, with his finely-curved scarlet lips, his frank blue eyes, his crisp gold hair'" (*DG*, 18). Basil does not merely register Dorian's youthful beauty; he identifies it as the central value in his own ethos: "'You have the most marvellous youth, and youth is the one thing worth having'" (*DG*, 22). After Lord Henry has told Dorian that youth is the only thing worth having, Dorian cries out that he is jealous of the portrait whose beauty will not die, while he will only get older and uglier. "'Oh, if it were only the other way! If the picture could change, and I could be always what I am now! Why did you paint it? It will mock me some day—mock me horribly!' The hot tears welled into his eyes; he tore his hand away, and, flinging himself on the divan, he buried his face in the cushions" (*DG*, 26). Scenes of women lying prone and weeping are common enough in Victorian fiction; scenes depicting males in that posture are vanishingly rare. By flinging himself on a divan, weeping over the prospect of his own lost beauty, Dorian crosses a gender boundary in two distinct ways: he displays a passionate preoccupation with his own personal appearance, and he indulges in histrionic emotional expressiveness.

In these opening scenes, delicate and luxurious sensualism, a preoccupation with male beauty, and effeminate manners combine to produce a distinctly homoerotic atmosphere. As a polemical accompaniment to this atmosphere, Lord Henry keeps up a drum beat of denigrating comments against heterosexual bonding. "'You seem to forget that I am married,'" he tells Basil, "'and the one charm of marriage is that it makes a life of deception absolutely necessary for both parties'" (*DG*, 10). In response to Basil's confession that for so long as he lives "'the personality of Dorian Gray will dominate me,'" Lord Henry responds, "'Those who are faithful know only the trivial side of love; it is the faithless who know love's tragedies'" (*DG*, 16). Faithfulness is, as Lord Henry says in a passage already quoted, "'simply a confession of failure.'" Commitment or bonded attachment is a "trivial" form of personal interaction; promiscuous and opportunistic liaisons animated by transient appetites are the "serious" and substantial forms of interpersonal relation. These contentions are not abstract, universal, and gender neutral. Lord Henry is quite clear about the sexual orientations implicit in the conflict of values he propounds. "'Always! That is a dreadful word. It makes me shudder when I hear it. Women are so fond of using it. They spoil every romance by trying to make it last forever'" (*DG*, 24). (Elaine Showalter comments on the misogyny in the novel but does not register the antagonism to long-term bonding as the focal point of Lord Henry's polemic.)[14] The folly of fidelity is one of Lord Henry's favorite themes—a complement to his themes of sensual indulgence and self-cultivation as the ultimate aims in life. "'What a fuss people make about fidelity!' exclaimed Lord Henry. 'Why, even in love it is purely a question for physiology. It has nothing to do with our own will. Young men want to be faithful, and are not; old men want to be faithless, and cannot'" (*DG*, 28). These are universalizing claims about the nature of human intimacy, but what they universalize is not human nature in its heterosexual form; it is a specifically homosexual ethos produced by isolating and totalizing male dispositions toward promiscuity.

Lord Henry's assault on normative heterosexuality is subversive and revolutionary on a grand scale. "'The longer I live, Dorian, the more keenly I feel that whatever was good enough for our fathers is not good enough for us. In art, as in politics, *les grandpères ont toujours tort*'" (*DG*, 43–44). This claim could not be more boldly sweeping. In art and politics, *the grandfathers are always wrong*. Not wrong on this or that principle or point of taste or value—wrong generally, wrong fundamentally, wrong simply by virtue of being who and what they are, wrong precisely because, as heterosexuals, they became grandfathers.

Wilde invests part of his identity in each of the three characters, and the relations among them reveal the divisions within that identity. Lord Henry is what the world thinks Wilde is because in his own essayistic writings Wilde actually says many of the same things that Lord Henry says. Lord Henry often sounds like Wilde, but unlike Wilde, Lord Henry is not himself an artist. His creativity limits itself to the formulation of epigrams. Basil is a moralist, not a wit, but he is also a true artist. For Wilde, the central enigma of personal identity is that the creative spirit, as it is embodied in Basil, is fundamentally divided against itself. Basil is devoted to Dorian as the embodiment of purely sensual beauty, but he also believes in the "soul"; he believes, that is, in the continuity of moral identity—in the bonds we have with others that form part of our own inner selves. He argues that one would have to pay "'a terrible price'" for "'living merely for one's self,'" a price in "'remorse, in suffering, in . . . well, in the consciousness of degradation'" (*DG*, 64). The plot tacitly affirms these suppositions, and Dorian himself thinks of the painting in the same terms Basil uses to explain the logic of moral consequences. The painting becomes "the visible symbol of the degradation of sin," an "ever-present sign of the ruin men brought upon their souls." As such, it would "be a guide to him through life, would be what holiness is to some and conscience to others, and the fear of God to us all" (*DG*, 77).

Basil acknowledges the reality of conscience, but as an artist he is also hopelessly dependent on Dorian. Basil works most successfully as an artist when he is most fully under the sway of Dorian's "personality," and when Dorian distances himself from Basil, Basil's art goes into decline (*DG*, 163). In his early days, Basil's conscience is blind and his art successful. In his final encounter with Dorian, his conscience awakes to the moral horror of a purely aestheticist orientation, and he calls on Dorian to repent and reform. As Dorian unveils the portrait and Basil sees it for the first time in two decades, "An exclamation of horror broke from the painter's lips as he saw in the dim light the hideous face on the canvas grinning at him. There was something in its expression that filled him with disgust and loathing. . . . It was some foul parody, some infamous, ignoble satire. He had never done that. Still, it was his own picture. He knew it, and he felt as if his blood had changed in a moment from fire to sluggish ice. His own picture! What did it mean?"

> "'Can't you see your ideal in it?'" said Dorian, bitterly. . . .
> "'There was nothing evil in it, nothing shameful. . . . '"
> "'It is the face of my soul.'"
> "'Christ! What a thing I must have worshipped! It has the eyes of a devil.'" (*DG*, 122)

Lord Henry's discourse dominates the earlier portions of the story. For the final portions, Lord Henry reveals himself as wholly inadequate to interpret the meaning of the events in which he has participated. As Marlow says of Kurtz's fiancée in *Heart of Darkness*, Lord Henry is "'out of it.'"[15] He does not know that Dorian has murdered Basil, and he does not know that Dorian's portrait—his inner self—bears the marks of corruption and degradation. Despite the appearance of his rhetorical dominance in the exchanges with Basil and Dorian, Lord Henry is less capable of registering the full meaning of the story than either of them. Dorian most fully lives out the doctrine of egoistic hedonism, but he also feels the countervailing force of conscience. Basil feels the horror of moral corruption, but he also feels the haunting pull of beauty. Both of these characters are divided against themselves, but they do at least have depths of personal identity. Lord Henry is simple and whole, but he is also flat, two-dimensional. He professes a philosophy of surfaces, and his observations on the course of Dorian's career remain wholly on the surface. He mockingly quotes a street preacher's question—what does it profit a man if he gain the whole world and lose his soul? In response, Dorian assures him, with unwonted fervor and sincerity, that "'the soul is a terrible reality. It can be bought, and sold, and bartered away. It can be poisoned or made perfect'" (*DG*, 164). Lord Henry disclaims the very existence of the soul, and Dorian's soul thus remains a closed book to him.

The plot of *Dorian Gray* is retributional, but the meaning of the novel is not exhausted by any simple moral message. Defending himself against critics who accused the novel of promoting immoral behavior, Wilde asserts that "the real moral of the story is that all excess, as well as all renunciation, brings its punishment, and this moral is so far artistically and deliberately suppressed that it does not enunciate its laws as a general principle, but realises itself purely in the lives of individuals, and so becomes simply a dramatic element in a work of art, and not the object of the work of art itself" (*Letters*, 263). This is a rather trite and bland account of the didactic message conveyed by Dorian's disastrous career. Dorian's problem is not merely that he indulges in "excess." His problem is that he fails to create or sustain affectional bonds. He betrays all the people who are closest to him; he destroys them or leads them to ruin. But a more important point, in qualification of this appeal to didactic structure, is that didacticism is a form of resolution; it is an affirmation of an assured set of normative values, and the novel affirms no such set of normative values. There is no resolution of conflict in the story itself, and Wilde as narrator occupies no position above and apart from the story. There is at no point in the novel a single dominant perspective, standing apart from all three characters and encompassing them, that provides a normative, authoritative vision of the whole. The vision of the whole is nothing more, or less, than the enactment of the conflicted, unresolved relations among the three chief characters.

The most likely candidate for the role of internal moral guide would be Basil, but Basil is fundamentally compromised by his subjection to Dorian's "personality." He is himself guilty of an unconscious complicity with the values that animate Dorian's behavior. That complicity is revealed in the crucial episode during which the supernatural transformation in the painting takes place. While Basil is finishing the painting, Dorian is listening to Lord Henry propound the doctrine of hedonistic aestheticism. Lord Henry's talk is enchanting to Dorian, and it is the immediate prelude to the supernatural interchange that takes place between himself and the painting. At the end of the sitting, Basil apologizes for fatiguing Dorian. "'When I am painting, I can't think of anything else. But you never sat better. You were perfectly still. And I have caught the effect I wanted—the half-parted lips, and the bright look in the eyes. I don't know what Harry has been saying to you, but he has certainly made you have the most wonderful expression'" (*DG*, 21). Basil has explicitly warned Dorian not to listen to Lord Henry and has told him that Lord Henry "'has a very bad influence over all his friends'" (*DG*, 19). He would not himself like what Lord Henry says to Dorian, but he very much likes the effect Lord Henry's words have on Dorian, and capturing that effect brings him to the highest point of his own artistic achievement.

In his devotion to Dorian, Basil tacitly associates himself with the aestheticist ethos, but aestheticism is not the whole of art for either Basil or Wilde. In speaking of Sybil's artistic purpose as an actress, Basil articulates a moral conception of art like that which informs Wilde's fairy tales. "'To spiritualize one's age—that is something worth doing. If this girl can give a soul to those who have lived without one, if she can create the sense of beauty in people whose lives have been sordid and ugly, if she can strip them of their selfishness and lend them tears for sorrows that are not their own, she is worthy of all your adoration, worthy of all the adoration in the world'" (*DG*, 66). In this conception, the artist does not only celebrate sensuous beauty; he also creates empathy—suppressing selfishness and making people feel for the sorrows of others. It is this more complete and adequate conception of art that dominates the fairy tales, but within *Dorian Gray* it can neither achieve dominance nor be wholly suppressed.

In his conversation with Dorian about Sybil's suicide, Basil attempts to assert his own moral perspective but fails to sway Dorian and ultimately yields to him, thus tacitly acknowledging his own dependence on Dorian's identity. In speaking of Sybil's death, Dorian's speech has been more coldly and heartlessly selfish than at any previous time; it has all of Lord Henry's cynicism with none of his whimsical humor. He describes her death as "'one of the great romantic tragedies of the age'" and contrasts it with the "'tedious'" "'middle-class virtues'" of "'commonplace lives'" (*DG*, 86). He dispenses with sorrow and seeks to see the whole episode only "'from a proper artistic point of view'" (*DG*, 86). Given Basil's temperament and values, one would anticipate that he would be profoundly shocked and alienated by this speech, but Dorian appeals to his friendship, and "the painter felt strangely moved. The lad was infinitely dear to him, and his personality had been the great turning point in his art. He could not bear the idea of reproaching him any more. After all, his indifference was probably a mood that would pass away. There was so much in him that was good, so much that was noble" (*DG*, 87). Basil's fascination with

Dorian compromises his moral judgment. He cannot distinguish between the charm of Dorian's "personality" and his own sense of the "good" and "noble." His language, recorded in free indirect discourse, is that of someone rationalizing the bad behavior of a friend or lover, and there is no evidence from the text, no verbal clues, that at this point the narrator has any ironic detachment from Basil's perspective. His confusion about Dorian's value as a person seems to reflect Wilde's own perplexity, and that perplexity is at the very heart of the story.

Darwin tells us that humans have an evolved moral sense that consists in empathic human bonds extending over time and generating a sense of personal responsibility. When that sense of human connection is violated, he explains, we feel guilt and remorse. Basil confirms these contentions, and the plot of the story gives them symbolic form. Wilde does not invoke Darwin's psychological theory. He speaks instead of "'the soul'" and the "'sense of degradation,'" but the moral and psychological content of Wilde's Christian imagery is interchangeable with Darwin's naturalistic analysis. Wilde is intoxicated by Pater's aestheticism, but his own intuitions tell him that Pater's concept of human nature is profoundly false. It is adequate to sustain a two-dimensional character like Lord Henry, who scarcely seems to exist outside the medium of his epigrams. It is not adequate to sustain either Basil or Dorian. Like Kurtz in *Heart of Darkness*, Dorian has a glimpse, before his death, of the horror of his own soul. Unlike Conrad's Marlow, though, Wilde does not try to invest that moment of vision with redemptive power. Dorian loathes himself, but, except by killing himself, he never stops being himself. Suicide is not a form of resolution. It is a capitulation to ultimate failure.

Wilde's conception of an unresolvable conflict between the aesthetic and moral sides of his own identity is not a criterion of artistic success or artistic failure. It is merely the subject and animating spirit of his novel. One central measure of the novel's success as a work of art is the degree to which its figurative structure, its stylistic devices, and its tonal qualities are adequate to articulate that subject. The sustained psychodramatic interactions of the characters and the virtuoso interplay of cynical wit, voluptuous aestheticism, and morbid horror fulfill Wilde's artistic purposes. The novel is in many ways painful and unpleasant, but it is nonetheless a small masterpiece. In order to appreciate Wilde's artistic achievement in this novel, we have to recognize that despite all its sensuous luxuriance and provocative wittiness, its culminating dramatic moment depicts a loathsome self-image stabbed to the heart. The central artistic purpose in *Dorian Gray* is to articulate the anguish in the depths of Wilde's own identity.

Dorian compromises his moral judgment. He cannot distinguish between the charm of Dorian's "personality" and his own sense of the "good" and "noble." His language, recorded in free indirect discourse, is that of someone rationalizing the bad behavior of a friend or lover, and there is no evidence from the text, no verbal clues, that at this point the narrator has any ironic detachment from Basil's perspective. His confusion about Dorian's value as a person seems to reflect Wilde's own perplexity, and that perplexity is at the very heart of the story.

Darwin tells us that humans have an evolved moral sense that consists in empathic human bonds extending over time and generating a sense of personal responsibility. When that sense of human connection is violated, he explains, we feel guilt and remorse. Basil confirms these contentions, and the plot of the story gives them symbolic form. Wilde does not invoke Darwin's psychological theory. He speaks instead of "'the soul'" and the "'sense of degradation,'" but the moral and psychological content of Wilde's Christian imagery is interchangeable with Darwin's naturalistic analysis. Wilde is intoxicated by Pater's aestheticism, but his own intuitions tell him that Pater's concept of human nature is profoundly false. It is adequate to sustain a two-dimensional character like Lord Henry, who scarcely seems to exist outside the medium of his epigrams. It is not adequate to sustain either Basil or Dorian. Like Kurtz in *Heart of Darkness*, Dorian has a glimpse, before his death, of the horror of his own soul. Unlike Conrad's Marlow, though, Wilde does not try to invest that moment of vision with redemptive power. Dorian loathes himself, but, except by killing himself, he never stops being himself. Suicide is not a form of resolution. It is a capitulation to ultimate failure.

Wilde's conception of an unresolvable conflict between the aesthetic and moral sides of his own identity is not a criterion of artistic success or artistic failure. It is merely the subject and animating spirit of his novel. One central measure of the novel's success as a work of art is the degree to which its figurative structure, its stylistic devices, and its tonal qualities are adequate to articulate that subject. The sustained psychodramatic interactions of the characters and the virtuoso interplay of cynical wit, voluptuous aestheticism, and morbid horror fulfill Wilde's artistic purposes. The novel is in many ways painful and unpleasant, but it is nonetheless a small masterpiece. In order to appreciate Wilde's artistic achievement in this novel, we have to recognize that despite all its sensuous luxuriance and provocative wittiness, its culminating dramatic moment depicts a loathsome self-image stabbed to the heart. The central artistic purpose in *Dorian Gray* is to articulate the anguish in the depths of Wilde's own identity.

CHAPTER 6

The Cuckoo's History

Human Nature in *Wuthering Heights*

Wuthering Heights occupies a singular position in the canon of English fiction. It is widely regarded as a masterpiece of an imaginative order superior to that of most novels—more powerful, more in touch with elemental forces of nature and society, and deeper in symbolic value. Nonetheless, it has proved exceptionally elusive to interpretation. There are two generations of protagonists, and the different phases of the story take divergent generic forms that subserve radically incompatible emotional impulses. Humanist readings from the middle decades of the previous century tended to resolve such conflicts by subordinating the novel's themes and affects to some superordinate set of norms, but the norms varied from critic to critic, and each new interpretive solution left out so much of Brontë's story that subsequent criticism could gather up the surplus and announce it as the basis for yet another solution. Postmodern critics have been more receptive to the idea of unresolved conflicts, but they have tended to translate elemental passions into semiotic abstractions or have subordinated the concerns of the novel to current political and social preoccupations. As a result, they have lost touch with the aesthetic qualities of the novel. Moreover, the interpretive solutions offered by the postmodern critics have varied with the idioms of the various schools. Surveying the criticism written up through the 1960s, Miriam Allott speaks of "the riddle of *Wuthering Heights*." Taking account both of humanist criticism and of seminal postmodern readings, Harold Fromm declares that *Wuthering Heights* is "one of the most inscrutable works in the standard repertoire."[1]

Brontë's novel need not be relegated permanently to the category of impenetrable mysteries. The critical tradition has produced a good deal of consensus on the affects and themes in *Wuthering Heights*. Most of the variation in critical response occurs at the level at which affects and themes are organized into a total structure of meaning. In the efforts to conceptualize a total structure, one chief element has been missing—the idea of "human nature." By foregrounding the idea of human nature, Darwinian literary theory provides a framework within which we can assimilate previous insights about *Wuthering Heights*, delineate the norms Brontë shares with her projected audience, analyze her divided impulses, and explain the generic forms in which those impulses manifest themselves. Brontë herself presupposes a folk understanding of human nature in her audience. Evolutionary psychology converges with that folk understanding but provides explanations that are broader and deeper. In addition to its explanatory power, a Darwinian approach has a naturalistic aesthetic dimension that is particularly important for interpreting *Wuthering Heights*. Brontë's emphasis on the primacy of physical bodies in a physical world—what I am calling her naturalism—is a chief source of her imaginative power. By uniting naturalism with supernatural fantasy, she invests her symbolic figurations with strangeness and mystery. From the perspective of evolutionary psychology, the supernaturalism can itself be traced to natural sources in Brontë's imagination.

An evolutionary account of human nature locates itself within the wider biological concept of "life history." Species vary in gestation and speed of growth, length of life, forms of mating, number and pacing of offspring, and kind and amount of effort expended on parental care. For any given species, the relations among these basic biological characteristics form an integrated structure that biologists designate the "life history" of that species. Human life history, as described by evolutionary biologists, includes mammalian bonding between mothers and offspring, dual-parenting and the concordant pair-bonding between sexually differentiated adults, and extended childhood development. Like their closest primate cousins, humans are highly social and display strong dispositions for building coalitions and organizing social groups hierarchically. All these characteristics are part of "human nature." Humans have also evolved unique representational powers that enable them to create consciously held values. Novelists select and organize their material for the purpose of generating emotionally charged evaluative responses, and readers become emotionally involved in stories, participate vicariously in the experiences depicted, and form personal opinions about the characters.

Beneath all variation in the details of organization, the life history of every species forms a reproductive cycle. In the case of *Homo sapiens*, successful parental care produces children capable, when grown, of forming adult pair bonds, becoming functional members of a community, and caring for children of their own. With respect to its adaptively functional character, human life history has a normative structure. In this context, the word "normative" signifies successful development in becoming a socially and sexually healthy adult. The plot of *Wuthering Heights* indicates that Brontë shares a normative model of human life history with her projected audience, but most readers have felt that the resolution of the plot does not wholly contain the emotional force of the story. Brontë is evidently attracted to the values vested in the normative model, but her figurations also embody impulses of emotional violence that reflect disturbed forms of social and sexual development.

The elements of conflict in *Wuthering Heights* localize themselves in the contrast between two houses: on the one side Thrushcross Grange, situated in a pleasant, sheltered valley and inhabited by the Lintons, who are civilized and cultivated but also weak and soft; and on the other side Wuthering Heights, rough and bleak, exposed to violent winds, and inhabited by the Earnshaws, who are harsh and crude but also strong and passionate. Conflict and resolution extend across two generations of marriages between these houses. In the first generation, childhoods are disrupted, families are dysfunctional, and marriages fail. The destructive forces are embodied chiefly in Catherine Earnshaw and Heathcliff, and Brontë depicts their passions with extraordinary empathic power. In the second generation, the surviving children, Catherine Linton and Hareton Earnshaw, bridge the divisions between the two families, and the reader can reasonably anticipate that they will form a successful marital bond. Through this movement toward resolution, Brontë implicitly appeals to a model of human life history in which children develop into socially and sexually healthy adults. Nevertheless, the majority of readers have always been much more strongly impressed by Catherine and Heathcliff than by the younger protagonists.

The differences between the two generations can be formulated in terms of genre, and genre, in turn, can be analyzed in terms of human life history. The species-typical needs of an evolved and adapted human nature center on sexual and familial bonds within a community—bonds that constitute the core elements of romantic comedy and tragedy. Romantic comedy typically concludes in a marriage and thus affirms and celebrates the social organization of reproductive interests within a given culture. In tragedy, sexual and familial bonds become pathological, and social bonds disintegrate. (On the structure of romantic comedy and tragedy, Frye, after more than half a century, remains the most authoritative source.)[2] *Wuthering Heights* contains the seeds of tragedy in the first generation, and the second generation concludes in a romantic comedy, but the potential for tragedy takes an unusual turn. In most romantic comedies, threats to family and community are contained or suppressed within the resolution. In *Wuthering Heights*, the conflicts activated in the first generation are not fully contained within the second. Instead, the passions of Catherine and Heathcliff form themselves into an independent system of emotional fulfillment, and the novel concludes with two separate spheres of existence: the merely human and the mythic. The human sphere, inhabited by Hareton Earnshaw and the younger Cathy, is that of romantic comedy. In the mythic sphere, emotional violence fuses with the elemental forces of nature and transmutes itself into supernatural agency. Romantic comedy and pathological supernaturalism are, however, incompatible forms of emotional organization, and that incompatibility reflects itself in the history of divided and ambivalent responses to the novel.

Brontë would of course have had no access to the concept of adaptation by means of natural selection, but she did have access to a folk concept of human nature.[3] To register this concept's importance as a central point of reference in the story, consider three specific invocations of the term "human nature." The older Catherine reacts with irritated surprise when her commendation of Heathcliff upsets her husband. Nelly Dean explains that enemies do not enjoy hearing one another praised. "'It's human nature.'"[4] Reflecting on the malevolent mood that prevails under Heathcliff's ascendancy at Wuthering Heights, Isabella observes how difficult it is in such an environment "'to preserve the common sympathies of human nature'" (*WH*, 106). The younger Cathy is sheltered and nurtured at the Grange, and when she first learns of Heathcliff's monomaniacal passion for revenge, she is "deeply impressed and shocked at this new view of human nature—excluded from all her studies and all her ideas till now" (*WH*, 172). In Heathcliff, human nature has been stunted and deformed. Apart from his passional bond with Catherine, his relations with other characters are almost exclusively antagonistic. The capacity for hatred is part of human nature, but so is positive sociality. No other character in the novel accepts antagonism as a legitimately predominating principle of social life. Brontë shares with her projected audience a need to affirm the common sympathies that propel the novel toward a resolution in romantic comedy.

Identifying human nature as a central point of reference does not require the critic to postulate any ultimate resolution of conflict in a novel. Quite the contrary. Darwinians regard conflicting interests as an endemic and ineradicable feature of human social interaction. Male and female sexual relations have compelling positive affects, but they are also fraught with suspicion and jealousy. Even when they work reasonably well, these relations inevitably involve compromise, and all compromise is inherently unstable. Parents have a reproductive investment in their children, but children have still more of an investment in themselves, and siblings must compete for parental attention and resources. Each human organism is driven by its own particular needs, with the result that all affiliative behavior consists in temporary arrangements of interdependent interests. Nelly Dean understands this principle. Reflecting on the ending of the brief period of happiness in the marriage between Catherine Earnshaw and Edward Linton, she explains, "Well, we *must* be for ourselves in the long run; the mild and generous are only more justly selfish than the domineering—and it ended when circumstances caused each to feel that the one's interest was not the chief consideration in the other's thoughts" (*WH*, 72). The prospective marriage of Hareton and Cathy invokes a romantic comedy norm in which individual interests fuse into a cooperative and reciprocally advantageous bond, but no such bond is perfect or permanent, and many are radically faulty. The conclusion of *Wuthering Heights* juxtaposes images of domestic harmony with images of emotional violence that reflect deep disruptions in the phases of human life history.

In modern evolutionary theory, the ultimate regulative principle that has shaped all life on earth is the principle of "inclusive fitness"—that is, of kinship, the sharing of genes among reproductively related individuals. Kinship takes different forms in different cultures, but the perception of kinship is not merely an artifact of culture. Kinship is a physical, biological reality that makes itself visible in human bodies. The species-typical human cognitive system contains mechanisms for recognizing and favoring kin, and perceptions of kin relations loom large in folk psychology. As one might anticipate, then, kinship forms a major theme in the literature of all cultures and all periods. In *Wuthering Heights,* that common theme articulates itself with exceptional force and specificity. Kinship among the characters manifests itself in genetically transmitted features of anatomy, nervous systems, and temperament. The interweaving of those heritable characteristics across the generations forms the main structure in the thematic organization of the plot.

Heathcliff and Catherine are physically strong and robust, active, aggressive, domineering. Edgar Linton is physically weak, pallid and languid, tender but emotionally dependent and lacking in personal force. Even Nelly Dean, fond of him as she is, remarks that "he wanted spirit in general" (*WH,* 52). Isabella Linton, in contrast, is vigorous and active. She defends herself physically against Heathcliff, and when she escapes from him, she runs four miles over rough ground through deep snow to make her way to the Grange. Her son Linton, weak in both body and character, represents an extreme version of the debility that afflicts his uncle Edgar. Linton Heathcliff is "a pale, delicate, effeminate boy, who might have been taken for my master's younger brother, so strong was the resemblance; but there was a sickly peevishness in his aspect that Edgar Linton never had" (*WH,* 155). Isabella's son has "large, languid eyes—his mother's eyes, save that, unless a morbid touchiness kindled them a moment, they had not a vestige of her sparkling spirit" (*WH,* 159). Despite his inanition, Linton Heathcliff can be kindled to an impotent rage that recalls his father's viciousness of temper. Witnessing an episode of the boy's "frantic, powerless fury," the old servant Joseph cries in malicious glee, "'Thear, that's t' father! . . . That's father! We've allas summut uh orther side in us'" (*WH,* 192). With respect to Linton Heathcliff, Nelly Dean participates in the brutal physical naturalism of her creator's vision. She observes that Linton is "'the worst-tempered bit of a sickly slip that ever struggled into its teens! Happily, as Mr. Heathcliff conjectured, he'll not win twenty!'" (*WH*, 186). He lives into his mid teens but in manner remains infantile—self-absorbed and querulous. The younger Cathy is as physically robust and active as her mother and her aunt Isabella. She also has her mother's dark eyes and her vivacity, but she has her father's blond hair, delicate features, and tenderness of feeling. "'Her spirit was high, though not rough, and qualified by a heart sensitive and lively to excess in its affections. That capacity for intense attachments reminded me of her mother; still she did not resemble her, for she could be soft and mild as a dove, and she had a gentle voice, and pensive expression: her anger was never furious; her love never fierce; it was deep and tender'" (*WH*, 146). The younger Cathy has not inherited her mother's emotional instability. Nor does she display her mother's antagonistic delight in teasing and tormenting others. Her cousin Hareton Earnshaw is athletically built, has fine, handsome features, and his mind, though untutored, is strong and clear. He has evidently not inherited the fatal addictive weakness in his father's character. The inscription over the door at Wuthering Heights bears his own name, Hareton Earnshaw, and the date 1500. In his person, the finest innate qualities in the Earnshaw lineage come into flower.

The interweaving of heritable characteristics across the generations progresses with an almost mechanical regularity, but the meaning invested in that progression ultimately resolves itself into no single dominant perspective. The narrative is all delivered in the first person; it is spoken by participant narrators who say "I saw" and "I said" and "I felt." Through these first-person narrators, Brontë positions her prospective audience in relation to the story while she herself remains at one remove. The story is told chiefly by two narrators—Lockwood and Nelly Dean. (To avoid the typographical clumsiness of multiply-embedded quotations, I quote both Lockwood and Nelly as primary narrators.) Lockwood, a cultivated but vain and affected young man, holds the place of a conventional common reader who is shocked at the brutal manners of the world depicted in the novel. Nelly is closer to the scene, sympathetic to the inhabitants, and tolerant of the manners of the place—characteristics that enable her to mediate between Lockwood and the primary actors in the story. She provides a perspective from which the local cultural peculiarities can be seen as particular manifestations of human universals. When Lockwood exclaims that people in Yorkshire "*do* live more in earnest," she responds, "Oh! here we are the same as anywhere else, when you get to know us" (*WH*, 49). Both Nelly and Lockwood express opinions and make judgments, but neither achieves an authoritative command over the meaning of the story. Nelly is lucid, sensible, humane, and moderate, and she sees more deeply than Lockwood, but her perspective is still partial and limited. She has her likes and dislikes—she particularly dislikes Catherine Earnshaw and Linton Heathcliff—and some of her most comprehensive interpretive reflections fade into conventional Christian pieties that are patently inadequate to the forces unleashed in the story she tells. She holds the place of a reader for whom the impending romantic comedy conclusion offers the most complete satisfaction.

Behind the first-person narrators, the implied author, Emily Brontë herself, remains suspended over the divergent forces at play in the two generations of protagonists. The resolution devised for the plot is presented in an intentionally equivocal way. In its moment of resolution, the novel functions like an ordinary romantic comedy, but the pathological passions in the earlier generation are too powerful to be set at rest within a romantic comedy resolution. The narrative offers evidence that the earth containing the bodies of Catherine and Heathcliff is still troubled, and always will be, by their demonic spirits. In the last sentence of the novel, Lockwood wonders "how any one could ever imagine unquiet slumbers for the sleepers in that quiet earth." But it is Lockwood himself, during the night he spends at Wuthering Heights, to whom the ghost of Cathy appears in a dream, crying at the window. Nelly Dean deprecates the rumor that the spirits of Catherine and Heathcliff walk the moors, but she also reports that the sheep will not pass where the boy saw the ghosts of Heathcliff and Catherine, and she herself is afraid to walk abroad at night.

Nelly introduces her story of Heathcliff by saying it is a "cuckoo's" history (*WH*, 28). It is, in other words, a story about a parasitic appropriation of resources that belong to the offspring of another organism. That appropriation is the central source of conflict in the novel. The biological metaphor incisively identifies a fundamental disruption in the reproductive cycle based on the family. Heathcliff is an ethnically alien child plucked off the streets of Liverpool by the father of Catherine and Hindley, and then, almost unaccountably, cherished and favored over Hindley. When the father dies, Hindley takes his revenge by degrading and abusing Heathcliff.

When Heathcliff returns from his travels, he gains possession of Hindley's property by gambling, and after Isabella's death, he uses torture and terror to acquire possession of the Grange. He abducts the younger Cathy, physically abuses her, and compels her to marry his terrorized son Linton Heathcliff. From the normative perspective implied in the romantic comedy conclusion of the novel, Heathcliff is an alien force who has entered into a domestic world of family and property, disrupted it with criminal violence, usurped its authority, and destroyed its civil comity. In the romantic comedy resolution, historical continuity is restored, property reverts to inherited ownership, and family is reestablished as the main organizing principle of social life. The inheritance of landed property is a specific form of socioeconomic organization, but that specific form is only the local cultural currency that mediates a biologically grounded relationship between parents and children. The preferential distribution of resources to one's own offspring is not a local cultural phenomenon. It is not even an exclusively human phenomenon. It is a condition of life that humans share with all other species in which parents invest heavily in offspring.[5] The cuckoo's history is a history in which a fundamental biological relationship has been radically disrupted.

Brontë assigns to the second generation the thematic task of restoring the genealogical and social order that has been disrupted in the previous generation. The younger Cathy serves as the chief protagonist for this phase of the story. During the brief period of her relations with Linton Heathcliff and Hareton Earnshaw, she meets moral challenges through which she symbolically redeems the failures of her elders. Despite the ill usage she has received from Heathcliff's son, she nurses and comforts him in his final illness, and after his death, she establishes a wholesome bond with Hareton Earnshaw. Linton Heathcliff is wretched and repugnant in his self-absorbed physical misery. By nursing him in his final illness, the younger Cathy introduces a new element into the emotional economy of the novel—an element of redemptive charity. Her attitude to Hareton, at first, reflects the class snobbery that had distorted her mother's marital history. Cathy feels degraded by her cousinage with Hareton, and she mocks him as a lout and a boor. By rising above that snobbery and forming a beneficent bond with him, she resolves the conflict between social ambition and personal attachment that had riven the previous generation. Linton Heathcliff had embodied the worst personal qualities of the older generation—the viciousness of Heathcliff and the weakness of the Lintons—and Hareton and the younger Cathy together embody the best qualities: generosity and strength combined with fineness and delicacy. Even Heathcliff participates in the romantic comedy resolution, though in a merely negative way. The eyes of both Hareton and the younger Cathy closely resemble the eyes of Catherine Earnshaw. Seeing them suddenly look up from a book, side by side, Heathcliff is startled by this visible sign of their kinship with Catherine, and his perception of that kinship dissipates his lust for revenge. By dying when he does, he leaves the young people free to achieve their own resolution.

It seems likely that one of the strongest feelings most readers have when Heathcliff dies is a feeling of sheer relief. In this respect, both Lockwood and Nelly Dean serve a perspectival function as common readers. Lockwood leaves the Grange just before the final crisis in the story. All around him, he sees nothing except boorish behavior, sneering brutality, and vindictive spite. The mood is pervasively sullen, angry, bitter, contemptuous, and resentful. The physical condition of the house at Wuthering Heights, where Heathcliff, Hareton, and the younger Cathy are living, is sordid and neglected. When Lockwood returns, Heathcliff is dead; there are flowers growing in the yard; the two attractive young people are happy and in love; and Nelly Dean is contented. Very few readers can feel that all of this is a change for the worse. It is something like the clearing of weather after a storm, but it is even more like returning to a prison for the criminally insane and finding it transformed into a pleasant home.[6]

Readers have often expressed feelings of pity for Catherine and Heathcliff, but few readers have liked them or found them morally attractive. The history of readers' responses to the two characters nonetheless gives incontrovertible evidence that they exert a fascination peculiar to themselves. In the mode of commonplace realism, they are characters animated by the ordinary motives of romantic attraction and social ambition, and in the mode of supernatural fantasy, they are demonic spirits, but neither of these designations fully captures their symbolic force. At the core of their relationship, a Romantic identification with the elemental forces of nature serves as the medium for an intense and abnormal psychological bond between two children. Describing her connection with the earth,

Catherine tells Nelly that she once dreamed she was in heaven, but "'heaven did not seem to be my home; and I broke my heart with weeping to come back to earth; and the angels were so angry that they flung me out, into the middle of the heath on the top of Wuthering Heights; where I woke sobbing for joy'" (*WH*, 63). Catherine plans to marry Edgar Linton because he is of a higher class than Heathcliff, but she herself recognizes that class is for her a relatively superficial distinction of personal identity. "'My love for Linton is like the foliage in the woods. Time will change it, I'm well aware, as winter changes the trees—my love for Heathcliff resembles the eternal rocks beneath—a source of little visible delight, but necessary. Nelly, I *am* Heathcliff'" (*WH*, 64). As children, Heathcliff and Catherine have entered into a passional identification in which each is a visible manifestation of the personal identity of the other. Each identifies the other as his or her own "soul." Each is a living embodiment of the sense of the other's self. This is a very peculiar kind of bond—a bond that paradoxically combines attachment to another with the narcissistic love of one's self. Self-love and affiliative sociality have fused into a single motive that transforms the unique integrity of the individual identity into a dyadic relation. Dorothy Van Ghent astutely characterizes the sexually dysfunctional character of this dyadic bond. The relationship is not one of "sexual love, naturalistically considered," for "one does not 'mate' with one's self."[7] In normally developing human organisms, a true fusion between two individual human identities occurs not at the level of the separate organisms but only at the genetic level, in the fertilized egg and the consequent creation of a new organism that shares the genes of both its parents.

 The unique integrity of the individual identity is a psychological phenomenon grounded in biological reality. Individual human beings are bodies wrapped in skin with nervous systems sending signals to brains that are soaked in blood and encased in bone. Individual bodies engage in perpetual chemical interchange with the substances of the environment—air, water, and food—but they nonetheless constitute self-perpetuating physiological systems that can be radically disrupted only by death. Brontë's imagination dwells insistently on the reality and primacy of bodies, and that naturalistic physicality extends into the depiction of the peculiar psychological fusion of individual identity in Heathcliff and Catherine. At the time of Edgar Linton's funeral, eighteen years after that of Catherine, Heathcliff describes to Nelly his necrophiliac excursion to Catherine's grave. He has the sexton uncover her coffin, knocks out the side next to his own anticipated grave, and bribes the sexton to remove that side of his own coffin when he is buried. Catherine and Heathcliff achieve consummation not in a reproductively successful sexual union but in the commingling of rotted flesh. If necrophilia can reasonably be characterized as a pathological disposition, the empathetic emotional force that Brontë invests in the relationship between Catherine and Heathcliff can also reasonably be characterized as pathological.

The pathology that culminates in necrophilia disrupts the reproductive cycle, and it arises from disruptions in an earlier phase in that cycle—in childhood development. After Catherine has had the first attack of the hysterical passion that ultimately leads to her death, she tells Nelly that she came out of a trance-like condition, and "'most strangely, the whole last seven years of my life grew a blank! I did not recall that they had been at all. I was a child'" (*WH*, 98). She feels she is "'the wife of a stranger,'" and she yearns passionately to return to her childhood. "'I wish I were a girl again, half savage, and hardy, and free.'" The fixation on childhood has a seductive romantic appeal, but the passion behind the romance derives much of its psychological force from the traumatic disruptions in the family relations of the two children. Heathcliff is an orphan or an abandoned child. Catherine's mother—like Emily Brontë's own mother—dies when she is a child, and her father is emotionally estranged from her. Both children display a hypertrophic need for personal dominance, and their capacity for affectional bonding channels itself exclusively into their relation with one another. Neither Heathcliff nor Catherine ever becomes a socially and sexually healthy adult. Heathcliff's social relations, including his relation to his own son, are all destructive, and he finally also destroys himself. Catherine is torn apart by the unresolvable conflict between her childhood fixation and her adult marital relation. She remains estranged from her husband until her death, and she dies two hours after giving birth. The violence of feeling through which she destroys herself is a symptom of a psychological stress sufficiently strong to shatter a robust physical constitution.[8]

In the folk understanding of human nature, the needs for self-preservation and for preserving one's kin have a primal urgency. From a Darwinian perspective, those needs are basic adaptive constraints through which inclusive fitness has shaped the species-typical human motivational system. In *Wuthering Heights*, the movement of the plot toward the resolution of the second generation demonstrates that Brontë herself feels the powerful gravitational force of that system. Her empathic evocation of the feelings of Heathcliff and Catherine nonetheless indicates that her own emotional energies, like theirs, seek a release from the constraints of human life history. Some of the most intense moments of imaginative realization in the novel are those in which violent emotions assert themselves as autonomous and transcendent forms of force—moments like that in which Catherine's ghost cries to be let in at the window and like that in which she haunts Heathcliff and lures him into the other world. For both Catherine and Heathcliff, dying is a form of spiritual triumph. The transmutation of violent passion into supernatural agency enables them to escape from the world of social interaction and sexual reproduction. In the sphere occupied by Hareton and the younger Cathy, males and females successfully negotiate their competing interests, form a dyadic sexual bond, and take their place within the reproductive cycle. In the separate sphere occupied by Heathcliff and Catherine, the difference of sex dissolves into a single individual identity, and that individual identity is absorbed into an animistic natural world.[9]

The fascination Heathcliff and Catherine exercise over readers has multiple sources: a nostalgia for childhood, sympathy with the anguish of childhood griefs, a heightened sensation of the bonding specific to siblings, the attraction of an exclusive passional bond that doubles as a narcissistic fixation on the self, an appetite for violent self-assertion, the lust of domination, the gratification of impulses of vindictive hatred and revenge, the sense of release from conventional social constraints, the pleasure of naturalistic physicality, the animistic excitement of an identification with nature, and the appeal of supernatural fantasies of survival after death. All these elements combined produce sensations of passional force and personal power. In the prospective marriage of the second generation, those sensations subside into the ordinary satisfactions of romantic comedy, but Brontë's own emotional investments are not fully contained in that resolution. The ghosts that walk the heath are manifestations of impulses that have never been fully subdued. In becoming absorbed in the figurations of Catherine and Heathcliff, readers follow Brontë in the seductions of an emotional intensity that derives much of its force from deep disturbances in sexual and social development. They thus follow her also into a restless discontent with the common satisfactions available to ordinary human life.

Wuthering Heights operates at a high level of tension between the motives that organize human life into an adaptively functional system and impulses of revolt against that system. In Brontë's imagination, revolt flames out with the greater intensity and leaves the more vivid impression. Even so, by allowing the norms of romantic comedy to shape her plot, she tacitly acknowledges her own dependence on the structure of human life history. She envisions her characters in the trajectory of their whole lives. The characters are passionate and highly individualized, but life passes quickly, death is frequent, and individuals are rapidly reabsorbed within the reproductive cycle. Catherine and Heathcliff seem to break out of that cycle, but in the end, they are only ghosts—elegiac shadows cast by pain and grief. Investing those shadows with autonomous life enables Brontë to gratify the impulse of revolt while also satisfying a need to sacralize the objects of elegy. That improvised resolution points toward no ultimate metaphysical reconciliation, no ethical norm, no transcendent aesthetic integration, and no utopian ideal. Brontë's figurations resonate with readers because she so powerfully evokes unresolved discords within the adaptively functional system in which we live.

CHAPTER 7

Intentional Meaning in *Hamlet*

An Evolutionary Perspective

Common Wisdom and New Knowledge

Can an evolutionary perspective advance on the common wisdom of the critical tradition? One way to approach this question is to look at an actual example. *Hamlet* is convenient for this purpose, partly because it is so important and so well known, and partly because it has already attracted considerable attention from evolutionary critics. Robert Storey, Michelle Scalise-Sugiyama, Daniel Nettle, John Knapp, Brian Boyd, and John Tooby and Leda Cosmides have all used *Hamlet* to illustrate theoretical principles about literature, and Boyd and Knapp have made more detailed interpretive comments on it. After outlining a model of interpretive criticism from an evolutionary perspective, I shall summarize their efforts, compare them with traditional humanist readings, and offer my own interpretive commentary on *Hamlet*.

Offer my own interpretive commentary on *Hamlet*? Adding to the thousands or tens of thousands already produced? The heart grows faint; the native hue of resolution is sicklied o'er with the pale cast of thought, almost. What can be said about *Hamlet* within the common idiom, having no systematic recourse to extraneous theories, has most assuredly already been said. So far, the efforts to devise new readings by invoking extraneous theories—Freudian, deconstructive, Marxist, Foucauldian, and feminist, among others—have on the whole done less to illuminate the play than to elaborate their own preconceptions. Hamlet's erotic passion for Gertrude and secret complicity with Claudius in getting the castrating Hamlet senior safely underground (Jones); Hamlet as the Phallus (Lacan); the ghost as the transcendental Signified (Adelman; Garber); Hamlet's revolt against Claudius as a nascent impulse of proto-proletarian class consciousness (Bristol); Polonius as the embodiment of the Panopticon, peeping on everyone (Neill); Gertrude as the embodiment of anarchic feminine sexuality demonized by the Patriarchy (Adelman)—all such fancies have served as Procrustean beds, distorting the common understanding of the play.[1] If there is a "deep structure" to *Hamlet*, we will not get to it by violating the folk psychology implicit in the common idiom. We will get to it only by developing analytic concepts congruent with the common idiom but encompassing the common understanding within a more systematic and integrated body of causal explanations. Shakespeare holds a mirror up to nature.[2] So must we. By repudiating the very concept of "nature," postmodern theory has moved off in a direction that could not possibly advance on the common understanding.

Is it possible to formulate a set of theoretical principles distinct enough to offer real explanatory leverage but broad and flexible enough to give a just rendering of the thematic and tonal structure of the play? I think it is. We can integrate evolutionary concepts of human nature with the common understanding embodied in the best of traditional humanistic criticism. Using that conceptual structure as our interpretive framework, we can ask basic questions about the meaning of the play and provide reasoned answers. Those answers can of course have no claim to absolute validity; they are speculative, discursive, and rhetorical, not empirical and quantitative. They are not here tested and decisively falsified or confirmed by controlled experiment. They can nonetheless make claims to cogency based on common experience and the empirical validity of the concepts to which they appeal.

Previous Evolutionary Commentaries on *Hamlet*

Storey, Scalise-Sugiyama, and Boyd all comment on Laura Bohannan's essay on *Hamlet*.[3] (Oddly, though writing several years after Storey, making many of the same points, and using sometimes nearly identical phrasing, Scalise-Sugiyama does not cite Storey or include his book in her bibliography. Boyd cites both Storey and Scalise-Sugiyama.) Bohannan is an ethnographer who in the sixties lived among the Tiv, a nonliterate Nigerian tribal people, and recorded their ways. She told them the story of *Hamlet*, and they responded volubly, commenting on the play, criticizing the actions of the characters, and interpreting the events in accordance with their own customs and beliefs. For instance, the Tiv do not believe in ghosts, so they assumed that Hamlet's vision of the ghost was the result of witchcraft, in which they do believe. They felt it was wrong for Hamlet to seek revenge himself instead of asking for help from older relatives. Marrying a deceased brother's wife is obligatory for them, so they see no reason for Hamlet to be upset by Gertrude's remarriage. Bohannan emphasizes all the little ways in which the Tiv read the actions of the play in the light of their own ethos, which they mistakenly regard as universal. She concludes that literary meanings are not universally available.

The three evolutionists who respond to Bohannan all counter this conclusion by emphasizing the many quite basic ways in which the Tiv understand the play much as we do or as Shakespeare's contemporaries did, and they all three formulate "biocultural" propositions reconciling the idea of human universals with the idea of local cultural variations. The Tiv, like everyone else, understand narratives with protagonists pursuing goals such as seeking revenge, making alliances with friends, evading or fighting enemies, uncovering deceit, tricking others, feeling passions such as anger, grief, and contempt, negotiating the rules of ethical codes, avoiding incestuous relations, and either succeeding or failing in their efforts. All of this is part of "folk psychology" and communicable in the common idiom, even in translation. Moreover, relatively superficial differences of cultural ethos are not unintelligible to any people cosmopolitan enough to have registered that their local customs and beliefs are not necessarily universal.

Boyd includes consideration of Bohannan as only one of several topics in his essay. After giving a general exposition of "biocultural" theory, he takes up *Hamlet* as a particular case to illustrate how various features of the theory could bear on a reading of a specific literary work. Boyd formulates no comprehensive interpretive thesis for the play. Instead, he provides a catalogue of possible topics of analysis that could be applied to any work and illustrates them with application to *Hamlet*. He discusses the predominance of negative emotions in Ekman's list of seven "basic" emotions, comments on revenge and justice as evolved dispositions, gives an exposition of cost-benefit analysis and applies it to the problem of catching and holding the attention of an audience, uses cost-benefit analysis to frame a consideration of using familiar dramatic materials and providing novel twists, uses the idea of minimal ontological violations—violating realism—to explain the interest in supernatural phenomena such as ghosts, argues for the evolutionary basis of a preoccupation with individual differences in persons, points to Theory of Mind as a category relevant to the dramatic interest in reading the motives and beliefs of others, taking *Hamlet* as an especially intense instance of such interest, discusses the way emotion guides decision-making (referencing Damasio), and concludes with revisiting the question of the Tiv and the tension between local cultural practices and universal forms of behavior and cognition. All these analytic categories are no doubt relevant and useful, but until they are put to work as part of a whole interpretive argument, they are like the materials and tools assembled at a building site before the actual construction begins.

Boyd argues that an evolutionary reading need not be "reductive" but can be, in contrast, "expansive."[4] To call a reading "reductive" is to say that it is crude and narrow, that it leaves out too much of what is really important. And yet, all theory and all interpretation aim at legitimate "reduction." We try to reduce the multifariousness of phenomenal surfaces to underlying structures. We identify key causal principles in complex phenomena such as wars and economic developments. In commenting on literary works, we identify central themes and dominant tonal qualities. Without such efforts at reduction, all commentary would be lost in diffuse detail, like the waters of a flash flood sinking without trace into the sands of a desert.

In "What Happens in *Hamlet*?" Daniel Nettle makes a bold effort to produce a framework for adequate interpretive reduction. Despite the title of the essay (alluding to J. Dover Wilson's book),[5] Nettle actually says next to nothing about *Hamlet*, specifically. Like a substantial portion of the essays produced thus far in evolutionary literary studies, his essay is a theoretical prolegomenon to interpretation. He works through the basic theoretical problem of reconciling universals and specific cultural configurations, invokes Aristotle on the principle of goal-oriented action as the heart of drama, and then identifies four elements of analysis for cataloguing plays: two motives (mating and status), and two outcomes (success and failure). Comedies are successful mating games, tragedies unsuccessful status games.[6]

I am highly sympathetic to the ambition behind Nettle's effort—the desire to discover the elements of "deep structure" in literary texts. The effort itself, though, I think a failure, for two reasons. First, there are too few elements invoked to account for the range of possible human concerns. And second, Nettle considers only the motives of the characters, leaving out point of view, and thus leaving out the meaning that both characters and authors invest in actions. Nettle's only interpretive comment on *Hamlet* suggests the kind of "reductiveness"—almost comical—that can result from such premature theoretical reductions. "Status games—negative outcome represents the quintessential tragedy ('all tragedies end with a death'). Hamlet not only loses his kingdom to his uncle but is killed too."[7] Losing a kingdom and getting killed happen also to Richard III and to King Lear. And are we then to see these three plays as just variants on a simple theme of seeking status? That description comes closest to *Richard III*. It leaves most of *King Lear* unaccounted for, and seems altogether peripheral to the protagonist of *Hamlet*. Thwarted political ambitions are the least of Hamlet's concerns. They are scarcely mentioned until nearly the end of the play ("He that hath kill'd my king and whor'd my mother, / *Popp'd in between th' election and my hopes*, / Thrown out his angle for my proper life," V. ii. 64–66, emphasis added). In his first soliloquy ("O, that this too too [solid] flesh would melt," I. ii. 129), Hamlet concentrates on his mother's disloyalty to her dead husband and on the contrast in quality between his uncle and his father. In the scene before he leaves for England, after watching Fortinbras' troops pass by, Hamlet berates himself, again, for failing to act. "How stand I then, / That have a father kill'd, a mother stain'd, / Excitements of my reason and my blood" (IV. iv. 56–58). No mention of thwarted ambition.

Reducing all human concerns to sex and status leaves out survival itself as a motive (smuggled into Nettle's one comment on Hamlet but not part of the analytic scheme). It also leaves out all positive sociality, eliminating the interplay between impulses of dominance and impulses of affiliative, cooperative sociality. It thus leaves out reciprocity, the sense of justice, and the revenge that flows from violated reciprocity. It leaves out all kin-related motives, filial bonding, parental love (thus leaving out the heart of *King Lear* and everything in *Hamlet* that flows from outrage at a murdered father and corrupted mother). And finally, it leaves out the imagination itself, the need, so clearly dominant in Hamlet, to achieve an adequate interpretive understanding of the events in which he is embroiled. Nettle himself evidently has some sense of how much his effort at reduction has left out. He observes that "the human mind is structured in such a way that domain-specific schemata about kinship, love, competition, and cooperation are easily evoked." Yes, indeed. Why not include them then in the effort at schematic reduction to basic principles? Rather than answering this question, Nettle formulates an open-ended escape clause: "There is no desire here to reduce the complexity and shifting nature of dramatic meaning."[8] Well, yes, there is such a desire, and the desire is wholly legitimate. It just fails to achieve its purpose.

Among all the extraneous theories that critics have used to interpret *Hamlet* over the past century, Freudian Oedipal theory has been overwhelmingly the most influential, embedding itself not just in the tradition of written interpretations but also in performance. Always on the lookout for novelty, Laurence Olivier dramatized Hamlet's supposed Oedipal impulses in the closet scene between Hamlet and Gertrude, and that theatrical device then took on a life of its own, replicated in numerous productions for stage and screen.[9] One of the chief early triumphs of evolutionary psychology was the revelation that Freudian Oedipal theory is quite simply mistaken. Humans, like all other mammals, have evolved mechanisms for avoiding incest. Particular cultural conventions codify those impulses in ways that admit of some variation. For instance, some cultural codes, like that of the Tiv, allow or even require a man to marry his deceased brother's wife. In other cultural codes—like that to which Hamlet and his father subscribe—this particular bond is felt to be incestuous. Some variation, but within very limited bounds. No cultural code allows sexual relations between parents and children. In all known cultures, when such relations occur (almost always fathers abusing female children), they are condemned as immoral and criminal. Hamlet himself gives no evidence, in any remark he makes, that he himself has any sexual desire for his mother. One could impute such desire only on the strength of an extraneous theory that presupposed its universality. Since this particular extraneous theory is false, imputing the desire to Hamlet is utterly arbitrary. It goes beyond the play, and beyond human nature. The whole Freudian tradition—with all its derivative postmodern forms—holds a distorting mirror up to the play.

John Knapp is among the first of the new psychological literary theorists to recognize just how centrally important the modern findings on incest are for literary study.[10] For a century now, psychological literary criticism has been in thrall to the false ideas of Freudian psychology, and to the Oedipal theory at the very center of those ideas. In seeking to provide an alternative to the Oedipal scheme, Knapp invokes "family systems therapy" (FST). This is clinical theory, practical in purpose, oriented to the dynamics among family members. A guiding idea in the theory is that individuals should not be looked at alone but in relation to other family members. In clinical practice, this idea can of course be useful. As a concept in literary criticism, it can also be useful, but like all preconceived analytic ideas it must be used with care, letting the explicit evidence of the text give the necessary prompts as to which concepts are most relevant. In his interpretive critique of *Hamlet*, Knapp seems to me to go beyond the evidence of the text. Operating on the basis of assumptions derived from FST, he supposes that the relationship between Gertrude and Hamlet senior was in reality deeply flawed, and in pursuit of this thesis, he casts substantial doubt on the image of Hamlet's father that we derive from Hamlet himself.

If there were serious hidden conflicts in the marriage of Hamlet's parents, Gertrude's disloyalty would not be so shocking as it is. In his first soliloquy, Hamlet dwells on his parents' evidently reciprocal devotion, and the ghost of Hamlet's father affirms that he was devoted to Gertrude. In the closet scene, Hamlet upbraids his mother for her shallowness and sensuality, and she affirms the justice of his rebuke. Nor is she merely swayed temporarily by the force of Hamlet's rhetoric. Later, speaking only to herself, she gives passionate voice to her feeling of shame and guilt.

> To my sick soul, as sin's true nature is,
> Each toy seems prologue to some great amiss,
> So full of artless jealousy is guilt,
> It spills itself in fearing to be spilt.
> (IV. v. 17–20)

To palliate the guilt, to adopt an evenhanded, clinical stance, in which Claudius, Hamlet senior, and Gertrude stand all on a moral par, is to diminish the tragic scope of the conflict, reducing it to a messy family squabble. Hamlet, Hamlet senior, and Gertrude, all three register the moral significance of her disloyalty. (Even Horatio murmurs at the unseemly haste of the remarriage.) Hamlet speaks clearly and explicitly about the moral implications of Gertrude's o'er hasty marriage to Claudius:

> Such an act
> That blurs the grace and blush of modesty,
> Calls virtue hypocrite, takes off the rose
> From the fair forehead of an innocent love,
> And sets a blister there, makes marriage vows
> As false as dicers' oaths, O, such a deed
> As from the body of contraction plucks
> The very soul, and sweet religion makes
> A rhapsody of words.
> (III. iv. 40–48)

A chief theme in the play is the nature of the human, the difference between humans and animals of a lower order. Hamlet's mother has hasted with bestial lust to incestuous sheets. A beast that wants discourse of reason would have mourned longer. As Boyd observes, humans alone "can focus our minds altogether on particular events of the past. . . . Most animals cannot afford *not* to attend to their immediate environment and cannot easily reason beyond it."[11] Humans have a unique capacity "to think beyond the immediate." Hamlet concurs:

> What is a man,
> If his chief good and market of his time
> Be but to sleep and feed? A beast, no more.
> Sure, He that made us with such large discourse,
> Looking before and after, gave us not
> That capability and godlike reason
> To fust in us unus'd.
> (IV. iv. 33–39)

From Hamlet's perspective, Gertrude's behavior is literally inhuman. Overlooking her wanton self-degradation—a degradation that she ultimately confesses even to herself—takes us outside the structure of intentional meaning in the play.

How do we get to that intentional meaning? Daniel Nettle invokes Aristotle's belief that "the aim or purpose of the protagonist is the most important aspect of a tragedy," and he cites Brunetière's claim that "'what we ask of the theatre is the spectacle of a *will* striving towards a goal.'"[12] These phenomena are clearly central to human experience and to social monitoring, but *Hamlet* gives evidence of how limited they are in characterizing the total structure of meaning in specific literary works. If Hamlet's only "goal" were to kill Claudius, there need have been no play. After the ghost speaks to him, he could simply have walked directly to the chamber in which Claudius, taking his "rouse," deep in his cups, suspecting no harm, would have been easy prey to a swift thrust of the rapier. Achieving a specific practical goal is clearly not adequate to account for Hamlet's motivation. All the less, then, will it account for what motivates the play as a whole. What was Shakespeare getting at?

One of the best critics of the nineteenth century, William Hazlitt, observes that Hamlet is more inclined "to indulge his imagination in reflecting upon the enormity of the crime and refining on his schemes of vengeance" than on putting them "into practice." Hamlet's "ruling passion is to think, not to act."[13] Thinking, not acting, is what mostly happens in *Hamlet*. Hamlet sometimes thinks about why he does not act, and berates himself for not acting, but more often, he is just thinking—about the fickleness of women and the perfidy of men, about the purpose and techniques of drama, about mortality, the transience of life, eternity, and the human condition. Generalizing from this feature of Hamlet's character, the evolutionary psychologists John Tooby and Leda Cosmides ascribe a symbolic value to the play as a whole. They declare that both *Alice in Wonderland* and *Hamlet* "are focused on an evolutionarily ancient but quintessentially human problem, the struggle for coherence and sanity amidst radical uncertainty."[14] The problem is quintessentially human because only humans are massively detached from the narrowly channeled behavioral promptings of instinct. They thus gain immense flexibility in the conscious regulation of their actions, but they also thus become vulnerable to confusion, uncertainty, doubt, perplexity. The cognitive flexibility that is a peculiarly human attribute and that has so much adaptive power—does so much to increase inclusive fitness—also has dangers and costs that are peculiar to the human condition. Hamlet exemplifies both the mind's power and its vulnerability.

The interpretive formulation put forward by Tooby and Cosmides bypasses the important but limited concern, *What is the protagonist's goal?* In place of this question, they tacitly pose a larger, more important question, *What is this play about?* That is, what are its chief themes and motivating concerns? To what pressing human issues does it give imaginative form? What is the full scope of its meaning and effect? Their answer to such questions is right, I think, as far as it goes, and not just right but powerful, astute, incisive. Still, it does not distinguish between *Alice in Wonderland* and *Hamlet*. Clearly, then, it must be heavily qualified. Tooby and Cosmides describe the symbolic implication of the play at a level so high that it leaves out almost everything specific about the characters, circumstances, and emotional qualities in *Hamlet*. The circumstances Hamlet must confront in Denmark are not the same as those Alice must confront in Wonderland. A murdered father and a salacious mother are not part of Alice's situation. Moreover, Alice's personality is very different from Hamlet's, and not nearly so well developed. As Tooby and Cosmides perceptively suggest, Hamlet and Alice share a certain giddy sense of struggling for coherence and sanity, but otherwise the emotional qualities of the two works are very different. The challenge, then, is this: to connect Hamlet's struggle for coherence and sanity with an argument about the organization of the features that distinguish *Hamlet* as a particular work of art.

Hazlitt, Bradley, and Darwin: The Great Ideal Movement or the Indelible Stamp?

The best of traditional humanist criticism—literary criticism before the postmodern era—can be conceived as the finest articulation of the common idiom. Singling out Hazlitt and A. C. Bradley as representative figures in the humanist tradition should raise few skeptical eyebrows among Shakespearean scholars. Hazlitt was writing early in the nineteenth century and Bradley early in the twentieth. Both assimilate the insights of their most astute predecessors and add something particular and valuable of their own. Both their books on Shakespeare have been steadily in print, and Bradley's book *Shakespearean Tragedy* has sold over half a million copies. Hazlitt and Bradley get to much that is true and important about the play. Consequently, by qualifying, elaborating, or correcting their ideas, a critic can offer some genuine advance on the humanist heritage of insight and wisdom about the play. To this group of humanist critics, I shall add Darwin. His insight into *Hamlet*—registered indirectly but unmistakably in an allusion at a climactic rhetorical moment in his own writing (the conclusion to *The Descent of Man*)—provides a clue to the limitations in Bradley's interpretive thesis about the play.

Hazlitt and Bradley both adopt what Northrop Frye would call the "low mimetic" approach. That is, they view Hamlet as if he were a real person. But they also take him, and the play he is in, as symbols of a general human condition. As Hazlitt puts it, "the distresses of Hamlet are transferred, by the turn of his mind, to the general account of humanity. Whatever happens to him we apply to ourselves, because he applies it to himself as a means of general reasoning." Echoing Aristotle on the superiority of poetry to history, Hazlitt declares that "this play has a prophetic truth, which is above that of history." Hamlet, then, as Hazlitt sees him, is a *representative man*. Representative of what? The characterization, though long, merits full citation:

> Whoever has become thoughtful and melancholy through his own mishaps or those of others; whoever has borne about with him the clouded brow of reflection, and thought himself "too much i' the sun"; whoever has seen the golden lamp of day dimmed by envious mists rising in his own breast, and could find in the world before him only a dull blank with nothing left remarkable in it; whoever has known "the pangs of despised love, the insolence of office, or the spurns which patient merit of the unworthy takes"; he who has felt his mind sink within him, and sadness cling to his heart like a malady, who has had his hopes blighted and his youth staggered by the apparition of strange things; who cannot be well at ease, while he sees evil hovering near him like a specter; whose powers of action have been eaten up by thought; he to whom the universe seems infinite, and himself nothing; whose bitterness of soul makes him careless of consequences, and who goes to a play as his best resource to shove off, to a second remove, the evils of life by a mock representation of them—this is the true Hamlet.[15]

Everything about this description seems correct, but still it falls short in both generality and particularity. Tooby and Cosmides characterize Hamlet as symbolizing an evolutionarily ancient condition, something permanent and universal. Hazlitt of course does not generalize that far. His evocative description of Hamlet's personality and condition is far more detailed than the interpretive account given by Tooby and Cosmides, but it is not quite so particular as it might be. Hamlet has become thoughtful and melancholy, Hazlitt suggests, "through his own mishaps or those of others." Ah, but this case is common, and if common, why seems it so particular to Hamlet? Why does he feel an inner torment that passes show? The word "mishaps"—there's the rub. Murder and incestuous levity in the nuclear family are not mishaps; they are crimes and sins; sources of psychological trauma very different from mere accident. They engage guilt, shame, and outrage; they disturb the very foundations of emotional organization in their victims.

This is where Bradley comes in. He assimilates the Romantic tradition that includes Hazlitt, but he adds to it two important elements: an acute emphasis on the trauma of Hamlet's mother's self-degradation, and a brilliant clinical analysis of Hamlet's depression. Previous critics had of course acknowledged that Hamlet was distressed at his mother's behavior, and previous critics had used the vocabulary of "melancholia" to describe his mental state. To my knowledge, no critic before or after Bradley has gotten either of these topics so clearly into focus as central features in the psychological organization of the play, and no critic has delineated them with the lucid precision and fullness Bradley brings to them. Citing the first soliloquy and deducing from it "a sickness of life" and "a longing for death," Bradley asks why. "It was not his father's death." That was a matter of common grief. Nor was it "the loss of the crown," which is not even mentioned in the soliloquy. "It was," rather, "the moral shock of the sudden ghastly disclosure of his mother's true nature." Hamlet is "forced to see in her action not only an astounding shallowness of feeling but an eruption of coarse sensuality, 'rank and gross,' speeding post-haste to its horrible delight." The experience is "devastating," producing "bewildered horror, then loathing, then despair of human nature. His whole mind is poisoned."[16]

On the level of the common idiom, Bradley's description of Hamlet's state of mind, and the cause for that state, could not, I think, be bettered. Bradley takes Hamlet's own statements at face value, and Hamlet is, after all, overwhelmingly the dominant voice in the play, the voice that most commands attention and respect. Hamlet sees into the heart of his mother and uncle, quickly pins Rosencrantz and Guildenstern to his display case of duplicitous courtiers, fools Polonius to the top of his bent, and wins the admiration of Ophelia, the intimate regard of Horatio, and the respect of Fortinbras. More, in his soliloquies, Hamlet displays a power of meditative intelligence that remains a touchstone for most literate people. We can guess around Hamlet, supposing that we know better than any intelligence embodied in the play, but the play itself offers us no good alternative to his perspective, and efforts to guess around Hamlet—in the various modern theoretical schools—have on the whole made a poor showing. Part of Bradley's wisdom and skill as a critic derives from his good sense in knowing when to accept intentional meaning for what it is worth. In the case of Hamlet, as the canonical status of the play attests, it has a worth on which academic inventiveness is not likely to improve.

In revising Bradley's interpretive thesis, then, I shall not be disputing his diagnosis of Hamlet's malady. I shall only be locating this diagnosis within a more modern and more adequate explanatory context. Where I take issue with Bradley, I take issue precisely because, in formulating his interpretive thesis, he disconnects the symbolic meaning of *Hamlet* from the specific character of the "pathological condition" he himself so astutely describes. Like Tooby and Cosmides, Bradley thinks the play exemplifies "a tragic mystery inherent in human nature," but he does not locate that mystery in the context of human evolution. Instead, he locates it in the context of idealist metaphysics—the "Schlegel-Coleridge type of theory":

> Wherever this mystery touches us, wherever we are forced to feel the wonder and awe of man's godlike "apprehension" and his "thoughts that wander through eternity," and at the same time are forced to see him powerless (it would appear) from the very divinity of his thought, we remember Hamlet. And this is the reason why, in the great ideal movement which began towards the close of the eighteenth century, this tragedy acquired a position unique among Shakespeare's dramas, and shared only by Goethe's *Faust*. . . . *Hamlet* most brings home to

> us at once the sense of the soul's infinity, and the sense of the doom which not only circumscribes that infinity but appears to be its offspring.[17]

The essential character of "the great ideal movement" is that it ascribes a transcendent power and significance to thought. The Absolute is *Nous*, transcendental Mind, detached from all biological constraint, a universal presence, first cause and unmoved mover. Accordingly, in this climactic formulation of his interpretive thesis, Bradley forgets all about truant mothers and clinical depression and instead becomes fixated on the "divinity of thought." In some vague, mystical way, thought is infinite but also, since it is the cause of all things, the cause of "doom." Perhaps Bradley means that because we can conceive infinity we are also aware of death, but then, consciousness of death is not the chief source of Hamlet's distress. Indeed, he looks to death as a release from suffering. In any case, Bradley seems to have in mind more than an awareness of death. He has disputed the Schlegel-Coleridge argument that Hamlet is hampered from acting because he overthinks his possible options, but he still attributes Hamlet's powerlessness to "the divinity of his thought." Dressed in Bradley's skillful rhetoric, the juxtaposition of divinity, helplessness, infinity, and doom is all mildly impressive, in an abstract, idealist sort of way, but it would be hard to say what it means, and it fails to connect in any concrete way to the particular circumstances of the play. Like the extraneous theories of the postmodern era, it does less to illuminate the play than to articulate its own preconceptions.

Bradley's idealist interpretive thesis is out of harmony with his own best insights. As he himself says, Hamlet's problem is not just that he thinks too much. His problem, first and most importantly, is "the moral shock of the sudden ghastly disclosure of his mother's true nature" (117). In echoing Hamlet at the end of *The Descent of Man*, Darwin gets the right relation between man's god-like intellect and his too, too solid flesh:

> We must acknowledge, as it seems to me, that man with all his noble qualities, with sympathy which feels for the most debased, with benevolence which extends not only to other men but to the humblest living creature, with his god-like intellect which has penetrated into the movements and constitution of the solar

system—with all these exalted powers—Man still bears in his bodily frame the indelible stamp of his lowly origin.[18]

Darwin echoes Hamlet's diction and captures the very cadence of Hamlet's speech in his first conversation with Rosencrantz and Guildenstern:

> What a piece of work is a man, how noble in reason, how infinite in faculties, in form and moving, how express and admirable in action, how like an angel in apprehension, how like a god! the beauty of the world, the paragon of animals; and yet to me, what is this quintessence of dust? (II. ii. 303–08)

For both Hamlet and Darwin, the enigma here is not the self-defeating character of an involuted, transcendental Reason, but rather the tension between the mind, able to soar free in its inquiries, and the pull of the flesh. That pull makes itself felt not just in mortality, the common doom, but in the thousand shocks flesh is heir to. The one shock that does Hamlet the most damage is delivered by his mother, but conflict is built into the very nature of life. Natural selection is a struggle. More are born than can survive—that is an integral piece in the logic of selection. In sexually reproducing species, males and females share fitness interests but also have conflicting individual interests. Parents must make choices between effort devoted to survival and to mating and effort devoted to parenting. Parents and offspring share some fitness interests but in other interests diverge. The same principle applies even to siblings; and it applies to all individuals who form parts of social groups. We need not look to hermetic processes of thought to uncover tragic mysteries in the human condition. Man's lowly origin provides more than sufficient material for conflict that can lead to tragic outcomes.

Attachment and Loss: An Evolutionary Perspective on the Psycho-Symbolic Significance of Mothers

Hamlet could have been a romantic comedy—we see the spoiled remnants of a love story in Hamlet's relation to Ophelia. Or it could have been a heroic tale of princely valor, as in *Henry V*. Fortinbras, in his extemporaneous eulogy, says that Hamlet "was likely, had he been put on, / To have prov'd most royal" (V. ii. 382–83). Of that too, we see only the spoiled remnants. From the first moment we overhear Hamlet in his solitude, all such worldly concerns have faded into nothing for him. All normal motives and pursuits seem to him "weary, stale, flat, and unprofitable" (I. ii. 133). His father's murder, when he learns of it, enrages him, but before he knows his father was murdered, he is already deeply disturbed, so disturbed that he yearns for death. This is the psychological core of his condition. If the play as a whole has large symbolic significance—and most assuredly it does—the symbolic meaning must in some fashion spring from Hamlet's relation with his mother. That much the Freudian critics get right. Where they have gone wrong is in following Freud's false lead in supposing that all relations between mothers and sons are neurotic.[19] Hamlet wishing for death in his first soliloquy is not Everyman articulating a universal human condition—a condition of illicit longing and repressed impulses for incest. He is any man for whom the springs of feeling have been fouled at their source.

In *The Descent of Man*, Darwin speculates that all positive social feelings originate, phylogenetically and ontogenetically, in the bonding between mothers and infants.[20] That insight lay dormant for a century until John Bowlby made it the cornerstone in the modern evolutionary understanding of human emotional development.[21] Bowlby adopts an ethological, evolutionary perspective on mother-infant bonding and associates it with a crucial insight from psychoanalytic theory: the formative influence of childhood experience on adult life. The mother-infant relation is distinct from the sexual, but it can have a major impact on the quality of sexual relations later in life.[22] If mothers are absent, abusive, or emotionally detached, their children can have severe difficulty in forming healthy affectional bonds in other relations, sexual or social, and in performing effectively as parents when they have children of their own. Freudian psychoanalysis has been attractive to literary critics in part because it gives access, in however distorted a manner, to the continuity of emotional experience in individual identity.

The evolutionary understanding of attachment has fundamentally altered the false Freudian idea that there is no natural, healthy human condition. Healthy bonding between mothers and infants is essential to emotional well-being. Failed bonding or traumatic separation leads to emotional dysfunction and, in its most severe forms, to psychiatric illness, especially to clinical depression.[23] Illness is defined precisely as a deviation from a healthy, "normal" state. Hamlet says his wit is diseased, but even more, his heart is diseased. One of the most important motifs in *Hamlet* is a motif of disease: pestilence, contagion, infection, decay, filth, rot, sores, ulcers, cancers, foul odors, and rank fluids.[24] If the play has symbolic import beyond the literal plot—if it taps into deep forms of experience not limited to the peculiar circumstances of a fratricidal uncle and a mother making a hasty and degrading remarriage—that symbolic import consists largely in a condensed representation of corruption in the emotional nucleus formed by the relation between mother and child.

Hamlet's Depression

Bradley's description of Hamlet's diseased mental state gives evidence that even a hundred years ago depression was fairly well understood on the phenomenal level. Bradley, at least, understands a good part of it, and he makes use of his insight to give a cogent explanation for the one chief feature in Hamlet that has puzzled critics for centuries—why Hamlet delays in killing Claudius:

> [Melancholy] accounts for the main fact, Hamlet's inaction. For the *immediate* cause of that is simply that his habitual feeling is one of disgust at life and everything in it, himself included,—a disgust which varies in intensity, rising at times into a longing for death, sinking often into weary apathy, but is never dispelled for more than brief intervals. Such a state of feeling is inevitably adverse to *any* kind of decided action; the body is inert, the mind indifferent or worse; its response is, "it does not matter;" "it is not worth while," "it is no good."[25]

This is half the story of depression—the "anhedonia" or absence of positive affect. The other half of the story is the predominance of negative affect—anguish, horror, and despair. Bradley absorbs that part of the story into his description of anhedonia. In the modern neurobiological understanding of depression, this is a mistake. We now know that emotion is not a unitary polarized phenomenon extending from "bliss" at the positive pole to "despair" at the negative pole. The emotional circuits regulating positive and negative emotions are in fact separate and distinct. In milder forms of depression, one or the other circuit can be activated much more strongly than the other. The "blahs," a general sense of indifference and distaste, can occur without any active sensation of anguish. Conversely, anxiety, guilt, and emotional pain can occur independently of apathy. In the most severe forms of depression, including the worst phases through which Hamlet suffers, the negative and positive emotional circuits are both compromised.

By uncovering the causal mechanisms of depression, modern research has confirmed one of Bradley's chief insights—that depression is not a normal, healthy reaction to adverse circumstances. It is "pathological," a malfunction or breakdown in an adaptive system, like diabetes, heart disease, or stroke.[26] The brain's positive and negative emotional circuits function as a homeostatic system. This system (the "limbic" system) is designed to respond to good things (elation) and to bad things (alarm, flight or fight), and then to return eventually to normal. The depressed brain gets stuck in stress mode. It fails to readjust. More specifically, danger or threat stimulates the hypothalamus to produce a signal to the pituitary to send a signal to the adrenal glands, just over the kidneys. The adrenal glands secrete cortisol to produce fight or flight reactions. Prolonged stress for vulnerable people results in a "stuck switch." The adrenal glands continue to pump out cortisol, damaging the brain, killing neurons, shrinking the hippocampus, stressing other organs, and producing the typical affects of depression. The neural circuits mediating positive emotionality—circuits engaging the nucleus accumbens and activating the dopamine reward system in the frontal cortex—shut down, producing anhedonia, and consequently a loss of motive and interest. The neural circuits transmitting negative emotionality—engaging the amygdala and the hypothalamic-pituitary-adrenal axis—lock into chronic activity, producing continuous anxiety and anguish.[27]

Shakespeare depicts in Hamlet a pathological condition—a mood disorder that in our current culture would be treated with antidepressant medication or electroconvulsive therapy. The intuitive psychological power that Shakespeare displays in depicting this condition is just one more piece of evidence supporting the legitimacy of the canonical status he holds. He holds that status not because he articulates patriarchal values or flatters British imperial pride, but for two chief reasons: his extraordinary verbal genius and his penetrating psychological insight.

But is *Hamlet* just a study in clinical depression? Bradley rightly raises this question, and rightly gives a negative answer to it. "It would be absurdly unjust to call *Hamlet* a study of melancholy," though "it contains such a study." What makes the difference between a study of melancholy and *Hamlet*? The other parts of Hamlet's mind and personality—his intellect and character. "A slower and more limited and positive mind might not have extended so widely through its world the disgust and disbelief that have entered into it. But Hamlet has the imagination which, for evil as well as good, feels and sees all things in one. Thought is the element of his life, and his thought is infected."[28] Again, Hamlet is not Everyman. He is one particular man, but a man with faculties that enable him to give his particular condition the broadest representative scope, to generalize from his condition to the human condition, and thus to give a habitation and a name to a major phase of human experience. Not all men and women have profoundly disturbed emotional relations with their mothers; not all men and women fall into severe clinical depression. But all men and women are vulnerable to those threats, and that vulnerability provides the basis of common understanding that makes it possible for most readers to feel with Hamlet, to empathize, to identify vicariously with his plight.

Analyzing Hamlet's Personality: The Five-Factor Model

Critics of *Hamlet* have given many good impressionistic accounts of Hamlet's personality. Using research developed over the past few decades, we can now systematize these common-language observations within an empirically established set of categories. These categories converge naturally with the common idiom, and indeed through one line of research they have derived directly from the common idiom. The "lexical" approach to personality begins with combing dictionaries for every term that has some reference to personality or temperament. The idea is that if a feature is sufficiently important to affect social interactions, it will become embedded in the common idiom. The categories that emerge from the lexical approach can be correlated with our understanding of human life history and can be causally linked with underlying neurobiological processes.[29] By locating Hamlet's personality within the current model of personality, we can get a better sense of the relations among the specific features of his personality, the circumstances of his life, and his emotional reactions. We should thus be able to give a more complete and adequate account of the thematic and tonal structure of the play to which Hamlet gives his name.

Personality researchers have reached consensus that within the Germanic language group (English, Dutch, the Scandinavian languages, etc.) and other language groups as well, there are five major factors of personality: Extraversion/Introversion, Agreeableness, Conscientiousness, Emotional Stability, and Openness to Experience. These five factors have emerged consistently from independent research teams using diverse methodologies. *Agreeableness* signals a pleasant, friendly disposition and a tendency to cooperate and compromise, versus a tendency to be self-centered and inconsiderate. *Extraversion* signals assertive, exuberant activity in the social world versus a tendency to be quiet, withdrawn, and disengaged. *Conscientiousness* refers to an inclination toward purposeful planning, organization, persistence, and reliability versus impulsivity, aimlessness, laziness, and undependability. *Emotional Stability* reflects a temperament that is calm and relatively free from negative feelings versus a temperament marked by extreme emotional reactivity and persistent anxiety, anger, or depression. *Openness to Experience* describes a dimension of personality that distinguishes *open* (imaginative, intellectual, creative, complex) people from *closed* (down-to-earth, conventional, simple) people.

When we speak of "human nature," we focus first of all on "human universals," on cognitive and behavioral features that everyone shares. We typically use personality, in contrast, to distinguish one person from another—for example, a friendly, careless extravert in contrast to a cold, conscientious introvert. In reality, personality factors are themselves human universals, integral parts of our common human nature. Each of the five factors has a common substratum. Individuals differ only in degree on each factor.[30] The underlying commonality in Extraversion/Introversion is the necessity to engage in some way with an external environment—the "approach" part of the basic "approach-avoidance" mechanism that links human reactive impulses with those of every species, even amoebas. Agreeableness is a measure of affiliative sociality, and since humans are social animals, most humans have some measure of affiliative sociality. Conscientiousness is a measure of any given person's disposition for organizing, planning, and carrying through on the tasks of life. Locating present action within a temporal continuum containing past and future is part of the specifically human cognitive apparatus.[31] Without some measure of conscientiousness, a person could not function at all. Emotional instability, sometimes labeled "neuroticism," is a measure of emotional reactivity in the range of negative affect (the "avoidance" half of the "approach-avoidance" mechanism). Emotional reactivity varies in intensity from individual to individual, but experiencing pain is normal and necessary. Without fear and sorrow, people would have no means of registering dangers or feeling the sense of loss. The ability to experience emotional pain, like the ability to experience physical pain, is an indispensable adaptive trait. Openness to Experience registers curiosity and a readiness to absorb experience of an imaginative, intellectual, and aesthetic character. Our readiness for culture—our disposition for producing emotionally charged symbolic forms—is the single most important feature of human nature that distinguishes us as a species from all other species.[32]

As Hazlitt, Bradley, and many others have recognized, Hamlet is both profoundly introverted and intellectual. He thus has a naturally meditative personality. He engages not directly with persons and situations but rather with his sense of them. He is conscientious and thus tormented by his own inability to function effectively. He is emotionally unstable, a trait that renders him particularly susceptible to depression—to being overwhelmed by stress, unable to cope. As a depressive, he is characteristically vacillating, indecisive, and ineffectual. In this respect, his emotional instability converges with his introversion. He is at one remove from direct action, and when it comes to action, indecisive. All of this is captured in the characterization of Hamlet offered by the protagonist of Goethe's *Wilhelm Meister's Apprenticeship*:

> "A lovely, pure, noble, and most moral nature, without the strength of nerve which forms a hero, sinks beneath a burden which it cannot bear and must not cast away. All duties are holy for him; the present is too hard. Impossibilities have been required of him; not in themselves impossibilities, but such for him. He winds, and turns, and torments himself; he advances and recoils; is ever put in mind, ever puts himself in mind; at last does all but lose his purpose from his thoughts; yet still without recovering his peace of mind."[33]

There remains the question of Agreeableness. Is Hamlet a nice, warm, friendly person? His admirers would like to think so. Hazlitt tries to palliate his behavior to Ophelia. I think Samuel Johnson is closer to the truth in speaking of Hamlet's "useless and wanton cruelty" to Ophelia.[34] And it isn't just Ophelia, embodiment of frail womanhood. More often than not, Hamlet is verbally caustic. He finds his vocation in witty put-downs. He delights in mocking Polonius, even after he has killed him. He sends Rosencrantz and Guildenstern to their deaths with no flicker of remorse or sadness. Quite the contrary, he exults in the success of his cunning stratagem. He tells Gertrude that he must be cruel only to be kind, but such rationalizations are common. Children readily detect the hypocrisy that so frequently lurks behind the phrase, "It's for your own good." Add all this up, and it seems unlikely that Hamlet would score even at the average on the factor "Agreeableness."

Protagonists tend to be agreeable, since readers do not readily cotton to disagreeable characters. But Hamlet never quite loses his audience, even when they flinch from his cruelty. There are at least five reasons for this. First, he is, after all, mightily put upon, struggling against crime and depravity that dwarf mere unpleasantness. Second, he preempts readers' resentment by being as brutally hateful to himself as he is to others. If in his accounting Ophelia is Representative Woman, fickle and false, Hamlet is himself Representative Man, "proud, revengeful, ambitious" (III. i. 122). Third, he is a satirist as well as a protagonist. He entices the audience to participate with him in exposing folly, wickedness, deceit, debauchery, treachery, venality, sycophancy, and foppishness. He is not merely depressed. He is angry, and because he is also driven to disguise, his anger finds vent in satirical wit. *Hamlet* is not a "tragicomedy," precisely, but it is a very funny tragedy. Ophelia fails to see the humor in her father's death, but most readers are irresistibly entertained by the patter of wicked puns that follow the good old man to his dinner, not where he eats but where he is eaten. Fourth, Hamlet never succumbs to mere egoism or cynicism. He is capable of filial affection, admiring friendship, and romantic love. And finally, perhaps most importantly, Hamlet's relations to other individuals are almost incidental to his central motive—to articulate his own imaginative sense of his situation. The high moments in *Hamlet*, the moments most remembered, are the soliloquies. Even in his tirade against Ophelia, she is scarcely more than a prop, an occasion for a monologue denouncing human nature. His one bosom friend, Horatio, is merely a sounding board for Hamlet's reflections. Hamlet speaks to himself, and we but overhear him.

Early evolutionary psychology deprecated the significance of individual differences and focused exclusively on human universals. This was a serious theoretical mistake.[35] Moreover, it lends support to the false charge that literary Darwinism cannot cope with individual texts because evolutionary psychology concerns itself only with human universals.[36] Individual variation is integral to the evolutionary process, and differences of personality allow individuals to occupy different niches within variable social ecologies.[37] *Hamlet* occupies a niche in the literary canon in good part because Hamlet's personality makes it possible for him to define a range of emotion—morbid, unhappy, bitter, angry, resentful, contemptuous, disgusted—that touches powerfully responsive chords in his audience. He articulates his condition as a general human condition, and while that representation is not the whole truth, it is enough of the truth to fix our attention and win our grave approval.

Just How Universal Is *Hamlet*?

Tooby and Cosmides are right, I think, in declaring that Hamlet's condition symbolizes an evolutionarily ancient adaptive problem: "the struggle for coherence and sanity amidst radical uncertainty." The way that problem manifests itself, though, depends very much on cultural, historical circumstance. Hamlet could not have existed either in Periclean Athens or in medieval Europe. His mind roams free over the whole scope of human experience, probing all questions, finding no clear answers, no firm structure of belief and value. Oedipus, in contrast, is always certain—first of his own rectitude, and then of his guilt. Socrates questions everyone else's beliefs and values, but Plato has the ideals of *The Republic* always comfortably in reserve for himself. Dante's inferno has its precise hierarchy of guilt and torment. *Hamlet* is different. Matthew Arnold registers this difference in describing Hamlet as a truly "modern" figure. In the 1853 preface to his *Poems*, Arnold explains why he has not included in the volume his one most ambitious poem, the closet drama "Empedocles on Etna." Though wearing ancient garb, Empedocles is a voice of Arnold's own time, expressing all the doubts and perplexities—religious, philosophical, moral, and social—that characterize the intellectual life of the Victorian period.[38]

> What those who are familiar only with the great monuments of early Greek genius suppose to be its exclusive characteristics have disappeared: the calm, the cheerfulness, the disinterested objectivity have disappeared; the dialogue of the mind with itself has commenced; modern problems have presented themselves; we hear already the doubts, we witness the discouragement, of Hamlet and of Faust.[39]

Doubt and discouragement do not first appear in human experience in the seventeenth century, much less the nineteenth, but there is no age before the Elizabethan in which doubt and discouragement achieve a supreme form of articulation, and no age before the Victorian in which they come to dominate the imaginative life of a whole culture. The three great philosophical poems of the Victorian period, Tennyson's *In Memoriam,* Arnold's "Empedocles on Etna," and Browning's "Bishop Blougram's Apology," are all meditations on religious and philosophical doubt, and to this canon one can add, as an appendix, Carlyle's *Sartor Resartus,* the collected poems of Arthur Hugh Clough, and Pater's *Marius the Epicurean.* In the postmodern period, we have stopped tormenting ourselves, for the most part, with religious doubt—not because we have solved the problems with which the Victorians struggled, but because we have given up on them and have resigned ourselves to the existential conditions they still hoped to avoid. The descendants of *Hamlet* in the modern period are works such as *The Waste Land, Long Day's Journey into Night, Waiting for Godot, La Nausée, The Seventh Seal,* and *Crow.*

One can hardly imagine what Sophocles or Dante would have made of *Hamlet,* or even what Chaucer would have made of it. We have made of it one of our very few most essential texts. We have taken it to heart and made it an anthem for our own imaginative lives. By assimilating the insights of the humanist tradition to an evolutionary understanding of human nature, we can now gain a better understanding of what that choice means.

Hamlet is a long, magnificently articulated cry of emotional pain and moral indignation. Mortally hurt in his inmost feelings, Hamlet clings to an imaginative ideal of courage, honor, dignity, and chivalrous love. That ideal is embodied in a ghost—"such a questionable shape" (I. iv. 44)—and that shape is almost all that stands between Hamlet and an actual world given over to bestial indulgence, false shows, treachery, and foolishness. He is slow to act, and when he does act, he brings cataclysmic ruin to himself and most of those who are closest to him. And yet, he is not a failure. He learns to look at death with clear and open eyes, accepting the frailty and transience of life. He is sensitive enough to register our worst fears in our most vulnerable moments and still in his own person give unmistakable proofs for the nobility of the human mind.

If this is not a tragedy for all times and seasons—not the kind of thing that would fulfill the deepest imaginative needs of Sophocles, Dante, or the Tiv—it nonetheless fulfills a tragic potential originating in the basic features of human nature. Perhaps at some point, possibly centuries from now, we shall no longer regard Hamlet's as one of the voices that speak most intimately to us, probing our fears, winning our fervent sympathy, voicing our outrage, making us laugh, and giving us an unsurpassed standard of meditative power. If that ever happens, we shall know that we have truly entered into yet another phase in the development of the human imagination.

PART THREE

Empirical Literary Study: An Experiment in Web-Based Research

CHAPTER 8

Agonistic Structure in Victorian Novels
Doing the Math

JOSEPH CARROLL, JONATHAN GOTTSCHALL,
JOHN JOHNSON, AND DANIEL KRUGER

Purposes

Three broad ambitions animate this study. Building on research in evolutionary social science, we aimed to (1) construct a model of human nature consisting of motives, emotions, features of personality, and preferences in marital partners; (2) use that model to analyze some specific body of literary texts and the responses of readers to those texts, and (3) produce data—information that could be quantified and could serve to test specific hypotheses about those texts.

Evolutionists are still in the process of constructing a full and adequate model of human nature. Evolutionary social scientists know much already about how human reproductive behavior and human sociality fit into the larger pattern of human evolution. They still have much to learn, though, about the ways literature and the other arts enter into human nature. Our model of human nature draws on our knowledge of imaginative culture, integrates that knowledge with evolutionary theories of culture, and produces data that enable us to draw conclusions on an issue of broad significance for both literary study and evolutionary social science: the adaptive function of literature and the other arts.

In order to make advances in knowledge, it is necessary to choose some particular subject. Genetics is a basic science that applies to all organisms, but geneticists first got an empirical fix on their subject by focusing minutely, with Mendel, on peas, and, with Morgan, on fruit flies. In place of peas and flies, we have taken as our subject British novels of the longer nineteenth century (Austen to Forster). As a literary topic, the subject is fairly broad, but our theoretical and methodological aims ultimately extend well beyond the specialist fields of British novels, the nineteenth century, British literature, narrative fiction, or even literary scholarship generally. This study is designed to engage the attention of literary scholars in all fields and also to engage the attention of social scientists. If it achieves its aims, this study would help persuade literary scholars that empirical methods offer rich opportunities for the advancement of knowledge about literature, and it would help persuade social scientists that the quantitative study of literature can shed important light on fundamental questions of human psychology and human social interaction. Our own research team combines these two prospective audiences. Two of us (Carroll and Gottschall) have been trained primarily as literary scholars, and two of us (Johnson and Kruger) primarily as social scientists.

The focal point for this study is "agonistic" structure: the organization of characters into protagonists, antagonists, and minor characters. The central question in the study is this: does agonistic structure reflect evolved dispositions for forming cooperative social groups? Suppressing or muting competition within a social group enhances group solidarity and organizes the group psychologically for cooperative endeavor. Our chief hypothesis was that protagonists and good minor characters would form communities of cooperative endeavor and that antagonists would exemplify dominance behavior. If this hypothesis proved correct, the ethos reflected in the agonistic structure of the novels would replicate the egalitarian ethos of hunter-gatherers, who stigmatize and suppress status-seeking in potentially dominant individuals. If suppressing dominance in hunter-gatherers fulfills an adaptive social function, and if agonistic structure in the novels engages the same social dispositions that animate hunter-gatherers, our study would lend support to the hypothesis that literature fulfills an adaptive social function.[1]

One of our chief working hypotheses is that when readers respond to characters in novels, they respond in much the same way, emotionally, as they respond to people in everyday life. They like or dislike them, admire them or despise them, fear them, feel sorry for them, or are amused by them. In writing fabricated accounts of human behavior, novelists select and organize their material for the purpose of generating such responses, and readers willingly cooperate with this purpose. They participate vicariously in the experiences depicted and form personal opinions about the qualities of the characters. Authors and readers thus collaborate in producing a simulated experience of emotionally responsive evaluative judgment. If agonistic structure is a main shaping feature in the organization of characters in novels, if agonistic structure engages evolved dispositions for forming cooperative social groups, and if novels provide a medium of shared imaginative experience on a large cultural scale, one could reasonably conclude that the novels provide a medium through which authors and readers affirm and reinforce cooperative dispositions on a large cultural scale.[2]

Agonistic structure clearly has a wide conceptual scope in its own right, but analyzing agonistic structure also serves a deeper purpose. By constructing a research design that correlates the features of characters with the responses of readers, we have sought to produce a first approximation to a universal set of categories for analyzing meaning structures in fictional narratives. In this context, "meaning," on one level, signifies the emotional and conceptual significance readers attribute to the organization of characters in the novels. On a second level, "meaning" consists in the psychological functions that organization fulfills. In order to identify those psychological functions, one can make inferences from actual effects. Consider an analogy with physiology. Saliva contains an enzyme that catalyzes the hydrolysis of starch into maltose and dextrin. On the basis of this effect, physiologists can reasonably infer that saliva functions to help digest food. So also, if one observes that agonistic structure has some definite psychological effect, one can formulate reasonable hypotheses about the function this effect fulfills. That function need not be consciously recognized by readers. In *Middlemarch*, speaking of Mrs. Bulstrode and her friend Mrs. Plymdale, George Eliot says that they were "well-meaning women both, knowing very little of their own motives." Like Eliot and like most other novelists, we are assuming that people can be moved powerfully by forces they do not always fully understand.[3]

Taking both attributed significance and psychological function as the referent of the term "meaning," our study presupposes that literary meaning can be reduced to constituent parts, measured, and located precisely within the causal network of nature. In this context, "nature" signifies both the physical world and the biocultural world that forms so large a portion of the environment humans create and inhabit. From the evolutionary perspective, culture does not stand apart from the biologically grounded dispositions of human nature. Culture is the medium through which humans organize those dispositions into systems that regulate public behavior and inform private thoughts.

When scholars and scientists suppose that literary meaning can be objectively, scientifically understood, they adopt a stance that contrasts sharply with the belief, common in the humanities, that literary meaning is illimitably complex and contains irreducible elements of the qualitatively unique. No one study could definitively confirm that all literary meaning can be objectively analyzed, but individual studies can provide strong evidence that major features of meaning can be effectively reduced to simple categories grounded in an evolutionary understanding of human nature. Quantifying literary meaning translates a naturalistic interpretive vision into empirical evidence that literary meaning is determinate, delimited in scope, and consilient with the knowledge of evolutionary biology.

The Game Plan

Collecting the Data and Sorting Characters into Sets. We created an online questionnaire, listed about 2,000 characters from 201 canonical British novels of the nineteenth and early twentieth centuries, and asked respondents to select individual characters and answer questions about each character selected. Potential research participants were identified by scanning lists of faculty in hundreds of English departments worldwide and selecting specialists in nineteenth-century British literature, especially scholars specializing in the novel. Invitations were also sent to multiple listservs dedicated to the discussion of Victorian literature or specific authors or groups of authors in our study. Approximately 519 respondents completed a total of 1,470 protocols on 435 characters. (A copy of the questionnaire used in the study can be accessed at: http://www-personal.umich.edu/~kruger/carroll-survey.html. The form is no longer active and will not be used to collect data.)

The questionnaire contains three sets of categories. One set consists in elements of personal identity: age, attractiveness, motives, the criteria of mate selection, and personality. (The sex of the characters was a given.) A second set of categories consists in readers' subjective responses to characters. Respondents rated characters on ten possible emotional responses and also signified whether they wished the character to succeed in achieving his or her goals. The third set consists in four possible "agonistic" role assignments: (1) protagonists, (2) friends and associates of protagonists, (3) antagonists, and (4) friends and associates of antagonists. Respondents were free to fill out questionnaires on any individual characters from the list. For each character selected, respondents assigned scores on each category of analysis and also assigned the character to one of the four possible agonistic roles.

Dividing the four agonistic character sets into male and female sets produces a total of eight character sets. The organization of characters into these eight sets forms an implicit empirical hypothesis—the hypothesis that agonistic structure, differentiated by sex, is a fundamental shaping feature in the organization of characters in the novels. We predicted (1) that each of the eight character sets would be sharply defined by a distinct and integrated array of features, that these features would correlate in sharply defined ways with the emotional responses of readers, and that both the features of characters and the emotional responses of readers would correlate, on the average, with character role assignments; (2) that characters identified as protagonists and their friends and associates would have attributed to them, on average, the features to which readers are most attracted and that they most admire; (3) that characters identified as antagonists and their friends and associates would have attributed to them, on average, the characteristics for which readers feel an aversion and of which they disapprove; (4) that protagonists would most completely realize the approbatory tendencies in reader response; and (5) that antagonists would most completely realize the aversive tendencies.

Averaging Scores. Comparing scores on characters who were coded by more than one respondent enables us to determine the levels of agreement among the respondents. The levels were quite high, well above the level that is considered acceptable in standard psychological research. When multiple readers did not agree on role assignments, characters were assigned to the role designated by the majority of the respondents. For characters who received multiple codings, scores of all the codings are averaged, so each character is counted only once in the total set of scores that produce averages for the whole data set. For instance, the most popular character, Elizabeth Bennet of Austen's *Pride and Prejudice*, received 81 codings, but those 81 sets of scores are averaged into one score, and that one averaged set of scores counts just the same, in the total data set, as the set of scores for John Dashwood, from Austen's *Sense and Sensibility*, who received only one coding. (Eighty of 81 respondents identified Elizabeth as a protagonist, and the measure of agreement on her scores among all her respondents was very high indeed. The one dissenting score on role assignment might have been a simple clicking error.)

Psychologists presuppose that when multiple respondents agree about features of people, those features actually exist. The subjects in this study are imagined people rather than actual people, but the principle is the same. Our design presupposes that the features identified by the respondents actually exist in the characters. Correlating emotional responses with attributed features enables us to assess the degree to which emotional responses are constrained by these attributed features. As it happens, there is a high degree of correlation between attributed features and the emotional responses of readers. Now, if the features readers identify in characters actually exist, those features are determined by authors. Authors stipulate a character's sex, age, personality, motives, and criteria for selecting mates. Readers largely agree in recognizing and identifying those features. If readers' emotional responses to characters show a high degree of correlation with attributed features (and they do), one can reasonably infer that authors have a high degree of control in determining readers' emotional responses to characters. Insofar as "meaning" consists in the two levels previously described—the significance readers attribute to the organization of characters, and the psychological functions fulfilled by this organization—one could reasonably conclude that authors have a high degree of control in determining meaning.

Condensing the Results. This chapter offers a condensed version of our findings—comparing only protagonists and antagonists (leaving out minor characters), and displaying the results only for motives, long-term mating, personality, and emotional responses. These results bring out the main tendencies in the data. Since our intended audience includes humanists not familiar with the technical idiom of statistics, the results are summarized in largely discursive form. Readers interested in obtaining more information on the statistical details can consult a more technical overview of the study.[4]

The next section explains the categories used to analyze character attributes and readers' responses. The explanations are followed by brief descriptions of the main results. In the final section, we draw conclusions from the results delineated in "Categories of Analysis."

Categories of Analysis

Motives. Motives are basic life goals. They are the chief organizing principle in human behavior. The categories for motives take account of the features of human life history that have been preserved from our mammalian and primate lineage; the specifically human reproductive characteristics that involve long-term pair-bonding, differing male-female mate-selection strategies, paternal investment, and the existence of extended kin networks; evolved human dispositions for forming coalitions, dominance hierarchies, and in-groups and out-groups; and the peculiarly human dispositions for acquiring and producing culture. Analyzing these topics generated a list of twelve basic motives: (1) Survival (fending off imminent physical danger or privation); (2) Finding a short-term romantic partner; (3) Finding or keeping a spouse; (4) Gaining or keeping wealth; (5) Gaining or keeping power; (6) Gaining or keeping prestige; (7) Obtaining education or culture; (8) Making friends and forming alliances; (9) Nurturing/fostering offspring or aiding other kin; (10) Aiding non-kin; (11) Building, creating, or discovering something; and (12) Performing routine tasks to gain a livelihood.[5]

We predicted (1) that protagonists would be generally affiliative in their motives—concerned with helping kin and making friends; (2) that antagonists would be chiefly concerned with acquiring wealth, power, and prestige; and (3) that protagonists would on average be much more concerned than antagonists or minor characters with acquiring education and cultural knowledge.

To bring the motives into a compact form, we conducted a statistical procedure known as "factor analysis." This procedure analyzes the elements that correlate with one another either negatively or positively. Any such cluster of correlated elements is called a "factor." For instance, in this data set, wealth, power, and prestige are very highly correlated with one another, and they are negatively correlated with helping non-kin. In other words, characters who scored high on seeking wealth also tended to score high on seeking power and prestige. Those same characters also tended to score low on helping non-kin. Seeking wealth thus has a positive correlation with seeking prestige and power and a negative correlation with helping non-kin. These clustered correlations form a factor that is here designated "Social Dominance." Social Dominance strongly distinguishes characters assigned to roles as antagonists.

In addition to Social Dominance, factor analysis produced four distinct motive factors: Constructive Effort, Romance, Nurture, and Subsistence. Constructive Effort most strongly characterizes protagonists, both male and female. It consists of two prosocial elements (helping non-kin and making friends) and two cultural elements (seeking education and building or creating something). The other three motive factors are distributed among male and female protagonists. Male protagonists are motivated by Subsistence, that is, by survival and by doing routine work to earn a living. Female protagonists are motivated by Romance (mating efforts) and by Nurture (caring for offspring and other kin). The distribution of these last three factors corresponds to the distribution of social roles in the period. Males were obligated to provide an income, and for most females "career" options were limited to marriage and family. Males were also more likely to encounter physical danger.

Figure 8.1 displays scores on motive factors for male and female protagonists and antagonists. The scores are displayed in standardized form, in units of standard deviation. The horizontal line at the zero point is the average score for all the characters on any given factor. Bars rising above the horizontal midline indicate scores above the average, and scores falling below the midline indicate scores below the average. A score of 1 would constitute a score one standard deviation higher than the average. A score of –1 would constitute a score one standard deviation lower than the average. For instance, female antagonists score .97 on Social Dominance (just .03 short of a single standard deviation above the average), and they score –.5 on Constructive Effort—exactly half a standard deviation below the average. (The same system is used in the other scores displayed further on.)

A standard deviation is the average distance from the mean. Units of standard deviation correlate with percentiles. For instance, on any given factor, for either a character set or an individual character, if a score is one standard deviation above the average, that score is higher than about 84% of all other scores on that factor. Half a standard deviation (.5) is higher than about 69% of all other scores. One and a half standard deviations (1.5) is higher than all but about 93% of all other scores. For negative scores (below average rather than above average), the percentiles are the same, with the direction reversed. For instance, a score of one standard deviation below average is lower than about 84% of all other scores.

To give just two quick examples of individual characters, Dorothea Brooke in George Eliot's *Middlemarch* is an exemplary female protagonist. She scores low on Dominance (–.9), very high on Constructive Effort (1.39), somewhat above average on Romance (.19), and fairly high on Nurture (.52). Mrs. Norris, in contrast, an antagonist from Austen's *Mansfield Park*, scores very high on Dominance (1.46) and low on Constructive Effort (–.76).

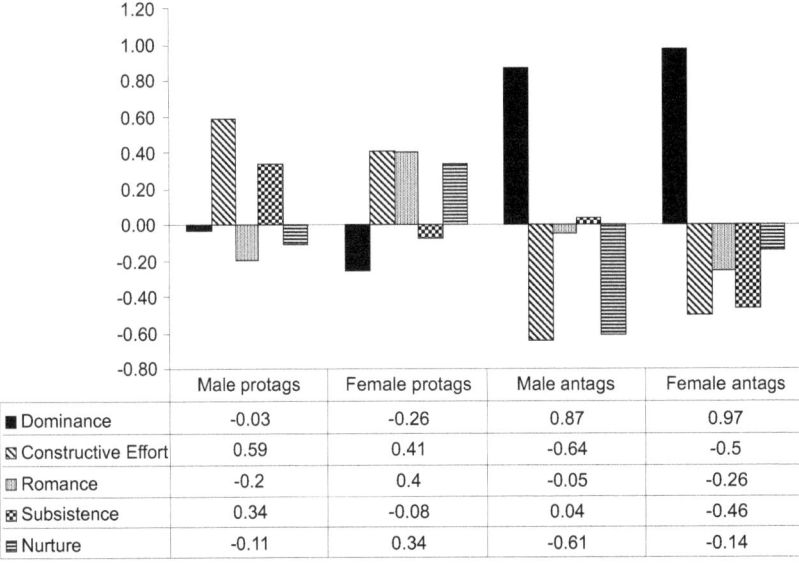

Figure 8.1 **Motive factors in protagonists and antagonists**

Criteria for Selecting Mates in the Long Term (Marital Partners). Studies in evolutionary psychology have identified general differences between the mating preferences of males and females. Males and females both are predicted to value intrinsic qualities such as kindness, intelligence, and reliability in mates, but males are predicted preferentially to value physical attractiveness in a mate, and females preferentially to value extrinsic attributes (wealth, prestige, and power) in a mate. These contrasting preferences are rooted in the logic of reproduction. Physical attractiveness in females serves as a proxy for youth and health in a woman—hence for reproductive potential—and extrinsic attributes enable a male to provide for a mate and her offspring. The seven terms just listed thus include criteria in which male and female preferences are expected to overlap and also criteria in which they are expected to differ.[6]

As in analyzing motives, statistical procedures compressed the criteria for selecting mates into clusters or "factors," that is categories that contain closely correlated elements. There are three clearly defined mate selection factors: a preference for (1) Intrinsic Qualities (reliability, kindness, intelligence); (2) Extrinsic Attributes (wealth, power, prestige); and (3) Physical Attractiveness (by itself).

Figure 8.2 displays the scores on mate selection. Male protagonists display a strongly marked preference for Physical Attractiveness in a mate. Female protagonists display a moderate preference for Extrinsic Attributes and a strong preference for Intrinsic Qualities. Female antagonists display a pronounced and exclusive preference for Extrinsic Attributes. Male antagonists score at or below average on all preferences. Both male and female antagonists score very far below average in preferences for Intrinsic Qualities.

Elizabeth Bennet from Austen's *Pride and Prejudice* offers an exemplary instance of criteria for selecting mates in female protagonists. She scores moderately high on seeking Extrinsic Attributes in a mate (.32), very high on seeking Intrinsic Qualities (1.15), and just about average on seeking Physical Attractiveness (-.03). In contrast to Elizabeth, Augusta Elton, an antagonist from Austen's *Emma*, scores very high on seeking Extrinsic Attributes (1.45) and very low on seeking Intrinsic Qualities (-1.15). Elizabeth's eventual marital choice, Fitzwilliam Darcy, deviates somewhat from the average male protagonist. He scores fairly high on seeking Physical Attractiveness

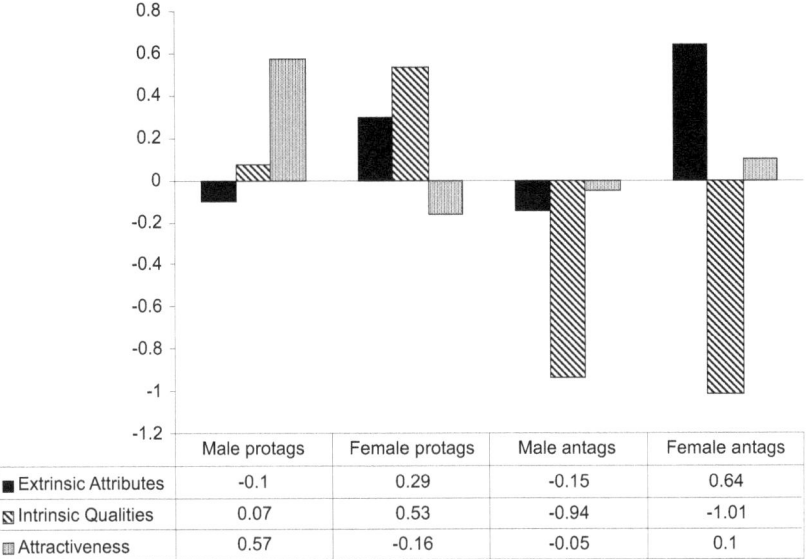

Figure 8.2. Criteria for selecting long-term mates in protagonists and antagonists

(.59) but also high on seeking Extrinsic Attributes (.60) and exceptionally high, for a male, on seeking Intrinsic Qualities (.81).

If a protagonist or antagonist deviates from the average, that deviation enters deeply into the imaginative qualities that distinguish one novel from another. If a pattern of such deviations emerges across an author's whole body of work (as is the case with Austen's protagonistic males), those deviations can help us to define the imaginative qualities that distinguish that author from other authors.

Personality Factors. The standard model for personality now is the five-factor or "big five" model. (See the essay on *Hamlet* in part 2 of this book.) *Extraversion* is a measure of engagement with the external world; *Agreeableness* a measure of prosocial warmth; *Conscientiousness* a measure of organization and reliability; *Emotional Stability* a measure of even temper; and *Openness to Experience* a measure of curiosity and aesthetic responsiveness.[7]

We predicted that (1) protagonists and their friends would on average score higher on the personality factor Agreeableness, a measure of warmth and affiliation; and (2) that protagonists would score higher than antagonists on Openness to Experience, a measure of intellectual vivacity.

Figure 8.3 displays the scores on personality. Male and female protagonists are both somewhat introverted, agreeable, conscientious, emotionally stable, and open to experience. Female protagonists score higher than any other set on Agreeableness, Conscientiousness, and Openness, and they score in the positive range on Stability. In personality, male protagonists look like slightly muted or moderated versions of female protagonists. Male and female antagonists are both relatively extraverted, highly disagreeable, and low in Stability and Openness. On each of the five factors, the protagonists and antagonists pair off and stand in contrast to one another.

Charlotte Brontë's Jane Eyre has a personality that is unequivocally protagonistic but that also has a distinctive cast common to Charlotte Brontë's protagonists and to those of her sister Anne: very low on Extraversion (–1.14), well above average on Agreeableness (.47) and Emotional Stability (.38), and high on Conscientiousness (.98) and Openness to Experience (.81). Bertha Rochester, in contrast, the madwoman in *Jane Eyre*, has a personality that is unequivocally antagonistic and that also reflects the character of her insanity: low on Agreeableness (–.80) and Openness to Experience

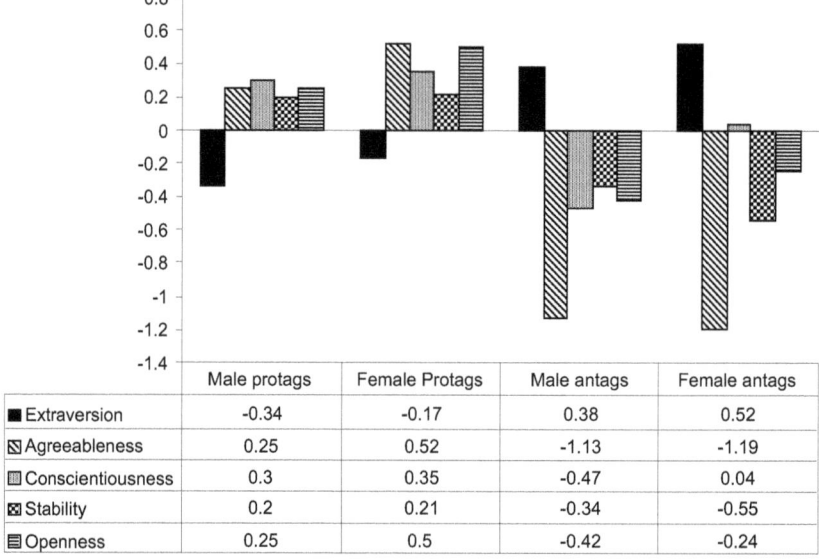

Figure 8.3 **Personality factors in protagonists and antagonists**

(−.46), and ultra-low on Conscientiousness (−1.46) and Emotional Stability (−1.61).

Emotional Responses. As noted earlier, scores on emotional responses do not signify the emotions *in* the characters but rather the emotional responses readers had *to* the characters. Our aim was to identify emotions that are universal and that are thus likely to be grounded in universal, evolved features of human psychology. The solution was to use Paul Ekman's influential set of seven basic or universal emotions: anger, fear, disgust, contempt, sadness, joy, and surprise. These terms were adapted for the purpose of registering graded responses specifically to persons or characters. Four of the seven terms were used unaltered: anger, disgust, contempt, and sadness. Fear was divided into two distinct items: fear *of* a character, and fear *for* a character. "Joy" (or "enjoyment") was adapted both to make it idiomatically appropriate as a response to a person and also to have it register some distinct qualitative differences. Two terms, "liking" and "admiration," served these purposes. "Surprise," like "joy," seems more appropriate as a descriptor for a response to a situation than as a descriptor for a response to a person or character. Consequently, in place of the word "surprise," we used the word "amusement," which combines the idea of surprise with an idea of positive emotion. One further term was included in the list of possible emotional responses: indifference. Indifference is the flip side of "interest," the otherwise undifferentiated sense that something matters, that it is important and worthy of attention.[8]

We predicted (1) that protagonists would receive high scores on the positive emotional responses "liking" and "admiration"; (2) that antagonists would receive high scores on the negative emotions "anger," "disgust," "contempt," and "fear-of" the character; (3) that protagonists would score higher on "sadness" and "fear-for" the character than antagonists; and (4) that major characters (protagonists and antagonists) would score lower on "indifference" than minor characters.

As in motives and mate selection, statistical procedures compressed the elements of emotional response into a smaller number of "factors." There were three clearly defined emotional response factors: (1) Dislike, which includes anger, disgust, contempt, and fear *of* the character, and which also includes negative correlations with admiration and liking; (2) Sorrow, which includes sadness and fear *for* the character and a negative correlation with amusement; and (3) Interest, which consists chiefly in a negative correlation with indifference.

Figure 8.4 displays the scores on emotional responses. The antagonists score very high on Dislike, low on Sorrow, and somewhat above average on Interest. Male and female protagonists both score low on Dislike and high on Sorrow. Female protagonists score high on Interest, but male protagonists, contrary to our expectations, score below average on Interest.

Count Dracula, from Stoker's *Dracula*, offers an unmistakably antagonistic profile: a very high score on Dislike (1.06), a respectable score on Interest (.33), and—despite having his head lopped off with a bowie knife—an only average score on Sorrow (-.06). In contrast, Anne Elliot, the protagonist of Austen's *Persuasion*, scores low on Dislike (-.76), high on Interest (.59). and moderately high on Sorrow (.40).

The Value of Systemic Analysis

The data from the questionnaire could have either confirmed or falsified the existence of agonistic structure. If the character sets had been indistinct, if they had displayed no particular patterns, if the

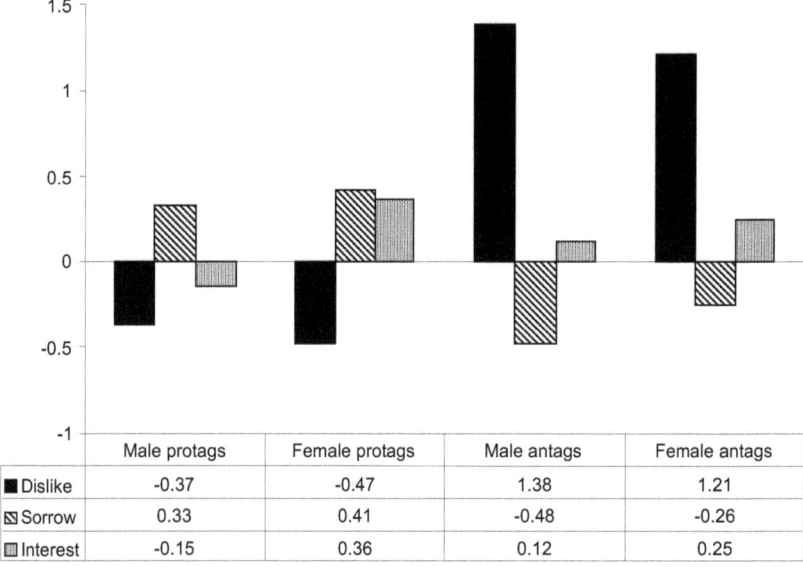

Figure 8.4 Emotional responses to protagonists and antagonists

content of character had not correlated with the emotional responses of readers, or if the responses of readers had not correlated with agonistic role assignments, the hypotheses built into the research design would have been falsified. As it turned out, the hypotheses were robustly confirmed. The character sets are sharply defined and contrasted through a correlated suite of characteristics: motives, mate selection, personality, age, and attractiveness. And that suite of characteristics correlates strongly with the emotional responses of readers. The use of an evolutionary model of human nature gives us the basis for a quantitative delineation of agonistic structure, and the clear delineation of agonistic structure in turn supports the analytic utility of an evolutionary model of human nature.

Agonistic structure in these novels displays a systematic contrast between desirable and undesirable traits in characters. Protagonists exemplify traits that evoke admiration and liking in readers, and antagonists exemplify traits that evoke anger, fear, contempt, and disgust. Antagonists virtually personify Social Dominance—the self-interested pursuit of wealth, prestige, and power. In these novels, those ambitions are sharply segregated from prosocial and culturally acquisitive dispositions. Antagonists are not only selfish and unfriendly but also undisciplined, emotionally unstable, and intellectually dull. Protagonists, in contrast, display motive dispositions and personality traits that exemplify strong personal development and healthy social adjustment. Protagonists are agreeable, conscientious, emotionally stable, and open to experience. Protagonists clearly represent the apex of the positive values implicit in agonistic structure. Both male and female protagonists score high on the motive factor Constructive Effort, a factor that combines prosocial and culturally acquisitive dispositions. Their introversion, in this context, seems part of their mildness. The extraversion of antagonists, in contrast, seen in the context of their scores on other personality factors and on motives, seems to indicate aggressive self-assertion.

There are of course exceptions to the large-scale patterns that prevail in the data—a small but distinct class of agonistically ambiguous characters such as Becky Sharp in Thackeray's *Vanity Fair*, Catherine and Heathcliff in Emily Brontë's *Wuthering Heights*, the Monster in Shelley's *Frankenstein*, Lucy Graham in Braddon's *Lady Audley's Secret*, and Dorian Gray in Wilde's *The Picture of Dorian Gray*. Such characters tend to score low on Agreeableness but also high on Openness to Experience, high on Dislike but also high on Sorrow. Such exceptions are extremely interesting but do not subvert the larger pattern. The larger pattern stands out clearly *despite* the blurring produced by the exceptions. An analogy might clarify this issue. When social scientists select a population of humans and score them on sexual orientation, a small percentage of their subjects have scores that are sexually ambiguous or that reverse heterosexual dispositions. The average scores for the total population nonetheless display clear patterns of heterosexual polarization—men preferring women, and women preferring men. Once one begins thinking statistically, one no longer gives undue prominence to special cases and exceptions. One thinks instead in terms of population averages. Within those population averages, one can make good analytic sense of the special cases and exceptions.

At the level of discrete observations from within the common language, some of our specific findings might seem fairly obvious. It might not, for instance, seem terribly surprising that readers dislike antagonists or that readers feel more sorrow, on average, for protagonists than for antagonists. Taken collectively, such findings nevertheless advance our knowledge in three distinct ways. Each finding serves as evidence for a large-scale hypothesis about the existence of agonistic structure; each forms part of a network of theoretically rationalized categories about human nature and literature; and each contributes to a total set of relations from which one can draw inferences not readily available to common observation.

Most people would acknowledge, in a casual way, that "readers dislike antagonists." But casual acknowledgments do not go very far toward providing empirical support for the proposition that novels are organized into systematic patterns of opposition between protagonists and antagonists. Topics of this sort are highly speculative; they admit of much ambiguity in definition; and appeals to specific cases, taken singly, could be manipulated in such a way as to support virtually any thesis on the subject. By producing data from many novels in which the word "dislike" correlates with specific attributes of characters, one limits the range of speculation and brings the subject within the scope of empirical knowledge.

In our view, observation at the level of discrete and fragmentary impression is less valuable, as knowledge, than observation lodged within theoretically rationalized categories. The word "dislike" is a common language term, but in our usage, it is also the product of a statistical analysis of ten emotional responses derived from the systematic empirical study of universal human emotions. Similar considerations apply to the other categories used to delineate character sets. The personality factor "Agreeableness" is a common language term, but it is also part of a model, derived from the statistical analysis of thousands of lexical items, that organizes personality into five superordinate factors. The motives and criteria of mate selection used in the questionnaire are couched in the common language, but they are also part of an integrated set of principles lodged within the explanatory context of evolutionary social science. Dislike correlates negatively with Agreeableness, positively with Social Dominance, and negatively with a preference for Intrinsic Qualities in a mate. Such correlations provide evidence for the existence of agonistic structure; the clear patterns of agonistic structure testify to the robust quality of the categories; and evolutionary social science provides a larger explanatory context both for the categories and for agonistic structure.

If one presupposes that agonistic structure exists, this or that finding in our study, taken singly, might not seem surprising, but for many readers in the humanities, the central premise of this study will probably be not only surprising but deeply disturbing. Our central premise is that both human nature and literary meaning can be circumscribed, reduced to finite elements, and quantified. We reduced human nature to a set of categories and used those categories to trace out quantitative relationships in responses to a large body of literary texts. This procedure tacitly negates the idea—nebulous and pervasive, Protean in its varieties—that literature and the experience of literature occupy a phenomenological realm that is separate and qualitatively distinct from the realm that can be understood by science.

Agonistic and Sexual Polarity

We made a number of detailed predictions about the relation between categories and character sets, and most of these predictions were confirmed, but some of our findings have been surprising to us, and all of our findings collectively have enabled us to draw inferences we could not have formulated before analyzing the data derived from the model of agonistic structure. Two findings seem to have an especially broad import for the organization of meaning and value in the novels: a bias for female-centered values, and the subordination of differences between males and females to differences between protagonists and antagonists.

In the world of these novels, males hold positions of political, institutional, and sometimes of economic power denied to females, but females hold a kind of psychological and moral power that is exemplified in their status as paradigmatic protagonists. The most important distinguishing features of antagonists, both male and female, are high scores on the motive factor Social Dominance (the desire for wealth, power, and prestige), low scores on the personality factor Agreeableness, and low scores on a preference for Intrinsic Qualities (intelligence, kindness, and reliability) in a mate. Female protagonists score lowest of any character set on Dominance and highest on Agreeableness and on preferring Intrinsic Qualities in mates. They also score highest in the typically protagonistic personality factors Conscientiousness, Emotional Stability, and Openness. In these important ways, female protagonists hold a central position within the normative value structure of the novels. The ethos of the novels is in this sense feminized or gynocentric.

Once one has isolated the components of agonistic structure and deployed a model of reading that includes basic emotions as a register of evaluatively polarized response, most of the scores on emotional response factors are predictable. There is, however, one surprising and seemingly anomalous finding that emerges from the scores on emotional responses—the relatively low score received by male protagonists on Interest. This finding ran contrary to our expectation that protagonists, both male and female, would score lower on indifference than any other character set. The apparent anomaly can be explained by the way agonistic polarization feeds into the psychology of cooperation. Male protagonists in our data set are relatively moderate, mild characters. They are introverted and agreeable, and they do not seek to dominate others socially. They are pleasant and conscientious, and they are also curious and alert. They are attractive characters, but they are not very assertive or aggressive characters. They excite very little Dislike at least in part because they do not excite much sense of competitive antagonism. They are not intent on acquiring wealth and power, and they are thoroughly domesticated within the forms of conventional propriety. They serve admirably to exemplify normative values of cooperative behavior, but in serving this function they seem to be diminished in some vital component of fascination, some element of charisma. They lack power, and in lacking power, they seem also to lack some quality that excites intensity of interest in emotional response.

We did not anticipate either that male protagonists would be so strongly preoccupied with Physical Attractiveness relative to other qualities or that male antagonists would be so relatively indifferent to Physical Attractiveness. The inference we draw from these findings is that the male desire for physical beauty in mates is part of the normative value structure of the novels. Male antagonists' relative indifference to Physical Attractiveness seems part of their general indifference to interpersonal relations.

If one were to look only at the motive factors, one might speculate that male antagonists correspond more closely to their gender norms than female antagonists do. Male antagonists could be conceived as personified reductions to male dominance striving. The relative indifference male antagonists feel toward any differentiating features in mates might, correspondingly, look like an exaggeration of the male tendency toward interpersonal insensitivity. Conceived in this way, male antagonists would appear to be ultra-male, and female antagonists, in contrast, would seem to cross a gender divide. Their reduction to dominance striving would be symptomatic of a certain masculinization of motive and temperament. They would be, in an important sense, de-sexed. Plausible as this line of interpretation might seem, it will not bear up under the weight of the evidence about male antagonists' relative indifference to Physical Attractiveness in a mate. Like female antagonistic dominance striving, that also is a form of de-sexing. Dominance striving devoid of all affiliative disposition constitutes a reduction to a core element of sex-neutral egoism. The essential character of male and female antagonists is thus not a sex or gender-specific tendency toward masculinization; it is a tendency toward sexual neutralization in the general isolation of an ego disconnected from all social bonds.

In the past thirty years or so, more criticism on the novel has been devoted to the issue of gender identity than to any other topic. The data in our study indicate that gender can be invested with a significance out of proportion to its true place in the structure of interpersonal relations in the novels and that it can be conceived in agonistically polarized ways out of keeping with the forms of social affiliation depicted in the novels. In this data set, differences between males and females are less prominent than differences between protagonists and antagonists. If polarized emotional responses were absent from the novels, or if those polarized responses covaried with differences between males and females, the differences between male and female characters might be conceived agonistically—as a conflict. The differences between male and female characters in motives and personality could be conceived as competing value structures. From a Marxist perspective, that competition would be interpreted as essentially political and economic in character, and from the deeper Darwinian perspective, it would ultimately be attributed to competing reproductive interests. The subordination of sex to agonistic role assignment, though, suggests that in these novels conflict between the sexes is subordinated to their shared and complementary interests. In the agonistic structure of plot and theme, male and female protagonists are allies. They cooperate in resisting the predatory threats of antagonists, and they join together to exemplify the values that elicit the readers' admiration and sympathy. Both male and female antagonists are massively preoccupied with material gain and social rank. That preoccupation stands in stark contrast to the more balanced and developed world of the protagonists—a world that includes sexual interest, romance, the care of family, friends, and the life of the mind. By isolating and stigmatizing dominance behavior, the novels affirm the shared values that bind its members into a community.

Comparing Male and Female Characters by Male and Female Authors

In the total set of 435 characters, characters by male authors outnumber characters by female authors by nearly two to one (281 vs. 154). Nonetheless, because a greater percentage of characters by female authors are major females (protagonists or antagonists), 47% of the major females in the whole data set are from novels by female authors (45% of female protagonists and 52% of female antagonists). Female authors contribute close to half of all major females (47%), of all good females (protagonists and their associates, 47%), and of all minor females (associates of both protagonists and antagonists, 45%).

In order to determine whether the sex of the author significantly influenced the depiction of sex-specific features in the characters, we compared the depiction of male and female characters by male and female authors. Male and female authors converge in most of the ways they describe similarities and differences between male and female characters. (We found one statistically significant difference in criteria for selecting a partner for short-term mating.) However, they also display biases. Though falling short of statistical significance, those biases tend in a clearly discernible direction. Both male and female authors tend to mute differences between their male and female characters. Male and female characters by male authors tend to resemble one another, and male and female characters by female authors tend to resemble one another. Male characters by female authors tend to look more like females. Female characters by male authors tend to look more like males. Male and female authors concur more closely in the depictions of motives in female characters than in the depictions of motives in male characters. That is, male characters by female authors look more like female characters than female characters by male authors look like male characters.

When we compare male and female characters by male and female authors in the major character sets (protagonists and antagonists), three features reach statistical significance. Male protagonists by female authors score significantly lower on valuing Extrinsic Attributes in a mate (wealth, power, and prestige). Male protagonists by female authors score significantly higher on Nurture. Female protagonists by female authors score significantly higher on Constructive Effort.

With respect to seeking Extrinsic Attributes in a long-term mate, male protagonists by female authors are less demanding than male protagonists by male authors. This result is a specific instance of a general tendency: compared to authors of the other sex, authors of each sex tend to depict characters of the sex different from the author as less demanding in selecting mates.

In novels by female novelists, male protagonists are more domestic (more nurturing), and female protagonists occupy a more prominent place in the public sphere to which Constructive Effort gives access. In novels by female authors, then, the difference in male and female sociosexual roles—relative to the difference in novels by male authors—is diminished from both directions: by differences in male motives and by differences in female motives.

The constituent elements of Constructive Effort are seeking education or culture; creating, building, or discovering something; making friends and forming alliances; and helping non-kin. Female protagonists by female authors score higher than female protagonists by male authors on making friends, seeking education, and helping non-kin. Out of all twelve motives that enter into the motive factors, the one motive with the largest difference for female protagonists by male and female authors is seeking "prestige." Across the whole set of 435 characters, prestige loads very strongly on the motive factor Social Dominance, where it clusters with seeking wealth and power. For female protagonists by female authors, in contrast, prestige separates out from the pursuit of wealth and power and clusters instead with the elements of Constructive Effort.

Characters with motivational profiles like those of female protagonists by female authors would scarcely be contented with purely domestic social roles. They want more education, a more active life in the public sphere, and greater public standing. In the advanced industrial nations, the social roles of women have of course changed dramatically in the past one hundred years. The depictions of female protagonists by female authors give evidence of the undercurrents that ultimately helped to produce these changes. Male authors also contribute to this movement—female protagonists by male authors score moderately high on Constructive Effort—but female authors clearly take the lead.

Despite differences in cross-sexed depictions—male characters by female authors and female characters by male authors—male and female authors fundamentally concur on the motivational tendencies that distinguish male and female characters. In novels by both male and female authors, male characters score higher than female characters on Dominance, Constructive Effort, and Subsistence, and they score lower than female characters on Romance and Nurture. In novels by both male and female authors, male characters choosing long-term mates score higher than female characters on preferring Physical Attractiveness, and they score lower than female characters on preferring Intrinsic Qualities and Extrinsic Attributes. In all these factors, it is only the magnitude of the differences that vary in male and female characters by male and female authors, and the magnitude of that difference reaches statistical significance in none of the factors.

Besides motives and mate selection, the one largest category of analysis for the content of character, in this study, is personality. With respect to personality factors, male and female characters by male and female authors score virtually the same. (The differences in their scores range between 0 and .2.)

The Adaptive Function of Agonistic Structure

Is it feasible to reason backward from our findings to formulate hypotheses about functions fictional narratives might have fulfilled in ancestral environments? By identifying one of the ways novels actually work for us now, can one produce evidence relevant to hypotheses about the evolutionary origin and adaptive function of the arts? Yes. Agonistic structure is a central principle in the organization of characters in the novels. Taking into account not just the representation of characters but the emotional responses of readers, one can identify agonistic structure as a simulated experience of emotionally responsive social interaction, and that experience has a clearly defined moral dimension. Agonistic structure precisely mirrors the kind of egalitarian social dynamic documented by Boehm in hunter-gatherers—our closest contemporary proxy to ancestral humans. As Boehm and others have argued, the dispositions that produce an egalitarian social dynamic are deeply embedded in the evolved and adapted character of human nature. An egalitarian social dynamic is the most important basic structural feature that distinguishes human social organization from the social organization of chimpanzees. In chimpanzee society, social organization is regulated exclusively by dominance. In human society, social organization is regulated by interactions between impulses of dominance and impulses for suppressing dominance. State societies with elaborate systems of hierarchy emerged only very recently in the evolutionary past, about ten thousand years ago, after the agricultural revolution made possible concentrations of resources and therefore power. Before the advent of despotism, the egalitarian disposition for suppressing dominance had, at a minimum, a hundred thousand years in which to become entrenched in human nature. In highly stratified societies, dominance assumes a new ascendancy, but no human society dispenses with the need for communitarian association. It seems likely, then, that agonistic structure in fictional narratives emerged in tandem with specifically human adaptations for cooperation and specifically human adaptations for creating imaginative constructs that embody the ethos of the tribe.[9]

Agonistic structure in these novels seems to serve as a medium for readers to participate vicariously in an egalitarian social ethos. If that is the case, the novels can be described as prosthetic extensions of social interactions that in nonliterate cultures require face-to-face interaction. If that face-to-face interaction fulfills an adaptive function, and if agonistic structure is a cultural technology that fulfills the same adaptive function, one could reasonably conclude that agonistic structure fulfills an adaptive function. We hope to see further empirical research that opens up new ways of probing this important issue.

We have suggested that the novels provide a medium of shared imaginative experience through which authors and readers affirm and reinforce egalitarian dispositions on a large cultural scale. At least one possible challenge to this hypothesis could readily be anticipated. Could it not plausibly be argued that the novels merely depict social dynamics as they actually occur in the real world? If that were the case, one would have no reason to suppose that the novels mediate psychological processes in the community of readers. The novels might merely serve readers' need to gain realistic information about the larger patterns of social life. To assess the cogency of this challenge, consider the large-scale patterns revealed in our data and ask whether those patterns plausibly reflect social reality:

The world is in reality divided into two main kinds of people. One kind is motivated exclusively by the desire for wealth, power, and prestige. These people have no affiliative dispositions whatsoever. Moreover, they are old, ugly, emotionally unstable, undisciplined, and narrow minded. The second kind of people, in contrast, have almost no desire for wealth, power, and prestige. They are animated by the purest and most self-forgetful dispositions for nurturing kin and helping non-kin. Moreover, they are young, attractive, emotionally stable, conscientious, and open-minded. Life consists in a series of clear-cut confrontations between these two kinds of people. Fortunately, the second set almost always wins, and lives happily ever after. This is reality, and novels do nothing except depict this reality in a true and faithful way.

In our view, this alternative hypothesis fails of conviction. The novels contain a vast fund of realistic social depiction and profound psychological analysis. In their larger imaginative structures, though, the novels evidently do not just represent human nature; they embody the impulses of human nature. Those impulses include a need to derogate dominance in others and to affirm one's identity as a member of a social group. Our evidence strongly suggests that those needs provide the emotional and imaginative force that shapes agonistic structure in the novel.

CHAPTER 9

Quantifying Agonistic Structure in *The Mayor of Casterbridge*

JOSEPH CARROLL, JOHN JOHNSON, JONATHAN GOTTSCHALL,
DANIEL KRUGER, AND STELIOS GEORGIADES

Models of Tragedy in the Interpretive History of *Mayor*

We set up a website for *The Mayor of Casterbridge* separate from the larger website listing about two thousand characters from about two hundred novels. To distinguish the two sites, we referred to the larger website as the "multi-novel website." Our aim in setting up an individual site for *Mayor* was to collect data on enough characters from a single novel to give a comprehensive analysis of the organization of characters and reader responses in that one novel. We chose *Mayor* in part because it is relatively compact, has only a few major characters, and has characters who are very distinctively marked in motives and personality. We listed six characters from *Mayor*: Henchard (the title character); his wife Susan; his stepdaughter Elizabeth-Jane; his rival Donald Farfrae; Lucetta, the woman for whose favors Henchard and Farfrae enter into competition; and Newson, the sailor who, at the beginning of the novel, buys Henchard's wife and daughter from him. Another reason for selecting *Mayor* as a case study is that it has an unusual agonistic and tonal structure. By using the average scores of the multi-novel website as a frame of reference, we hoped to tease out the structural peculiarities of *Mayor* and draw illuminating interpretive inferences from those peculiarities.

Interpretive commentary, and especially the interpretation of tone, is often regarded as a form of study too subjective and impressionistic ever to be brought within the range of quantification and empirical analysis. By giving a quantitative analysis of the tone in a single novel, we aimed to demonstrate that there need be no aspect of literary study inaccessible to empirical study, and further, that quantification could confirm, refine, correct, and develop the insights of traditional interpretive criticism.

We solicited participation in the *Mayor* study by directly contacting scholars who had published on Hardy and particularly on *Mayor* or on other Hardy novels. We also advertised the study on the listserv of the Thomas Hardy Association and listservs associated with the study of Victorian literature. All participation was anonymous, but we collected information about respondents' age, sex, level of education, when and why they read the novel, and whether they had published on *Mayor* or other works of Hardy. By analyzing this information, we determined that a total of eighty-five individual coders responded to the survey. Fifty-one were males, thirty-four females. The youngest respondent was twenty-three, and only eight respondents were under the age of thirty. All had college degrees. Nine had a bachelor's degree, twenty-one a master's, and fifty-five a doctorate. Twenty-five had published on *Mayor*; another twenty-three had published on some other novel by Hardy; and another ten had published on some other aspect of Hardy's work. Thus, a total of fifty-eight out of the eighty-five (68%) had published on some aspect of Hardy's work. Sixty-seven respondents reported having read the novel within the past five years, and thirty-one within the past year. Fifty-five read it either for teaching a class or for "professional purposes." In sum, almost all the respondents were very familiar with the novel. A number of respondents completed more than one protocol, and a total of 124 protocols were completed.

To assess the level at which respondents agreed in their assessments of the characters, we conducted "alpha reliability estimates." In most psychological research, alpha values around .70 are considered acceptable, and alphas in the .80 to .90 range are considered good. Values above .90 are normally achieved only by trained professionals. The average alphas across all categories for the *Mayor* respondents is .84. The lowest alpha values were for a minor character (Newson), who received only five codings. If we exclude Newson's alpha values, the average alpha values across all categories is .89. In other words, there was a high level of consensus among the respondents on all the substantive categories of analysis. Agonistic role assignments are a different matter. We discuss those at the beginning of the next section.

To orient readers who have not read *Mayor* or have not read it recently, we shall concisely summarize the plot. The actions and events in *Mayor* are like a roller coaster ride of wildly changing fortunes—especially the fortunes of Henchard, Susan, and Lucetta. In the opening chapter, Henchard is twenty-one years old. Embittered at being held back and burdened by family responsibilities, he gets drunk at a country fair and sells his wife and baby daughter. Within the next twenty years, he becomes a wealthy and respected corn merchant and is elected mayor of the market town Casterbridge. Meanwhile, his wife Susan has lived with Newson, the man who bought her. Her child from the marriage with Henchard has died, but she has had another child with Newson. Both children are named Elizabeth-Jane. Newson is lost at sea, and Susan returns to Henchard, deceiving him by telling him that Newson's child, now grown, is his child. He remarries her, but she dies soon after. Shortly after her death, Henchard tells Elizabeth-Jane that she is his daughter and asks her to take his name, but almost immediately after that he discovers that Elizabeth-Jane is not in fact his daughter. He does not tell her he had been deceived in believing himself her father, but he becomes cold and hostile toward her. Since her arrival in Casterbridge, Elizabeth-Jane has been romantically interested in Henchard's young protégé Farfrae, who had come to Casterbridge without place or prospect, but Farfrae loses interest in Elizabeth-Jane and takes up instead with Lucetta, who previously, unbeknownst to him, was Henchard's mistress. Henchard began his relationship with Farfrae by being overbearingly friendly, but he becomes jealous of Farfrae's popularity. Henchard becomes bitterly antagonistic to Farfrae, and they become competitors in business. After Susan's death, Henchard also becomes Farfrae's rival for Lucetta, and her preference for Farfrae embitters Henchard still further. Farfrae and Lucetta marry. In the period of just a few years after Susan's return, Henchard's fortunes have declined drastically, and Farfrae's fortunes have steadily risen. Henchard eventually loses both his wealth and his social position and is compelled to work as a lowly employee for Farfrae, who now dominates the corn trade and also becomes the new mayor of Casterbridge. Henchard attempts to kill Farfrae by throwing him out of a hay loft, but relents and breaks down in remorse. Lucetta has become pregnant with Farfrae's child, but her past with Henchard is made public. She becomes hysterical, has a seizure, and dies through complications with the pregnancy. Having lost his worldly position, Henchard seeks solace in establishing a bond with Elizabeth-Jane. They live together companionably for a while, but Elizabeth-Jane secretly renews her romantic relations with Farfrae, and then her biological father Newson reappears. Fearing to lose her, Henchard tells Newson that Elizabeth-Jane is dead. When his lie is about to be discovered, Henchard leaves Casterbridge to take up laboring work in a far district. He returns for Elizabeth-Jane's wedding, but she rejects him. He falls into despair, declines to eat, and dies.

As this summary should make clear, *The Mayor of Casterbridge* is in basic ways an unusual novel. Its protagonist, Michael Henchard, has personality traits and motivational dispositions that are more typical of antagonists than of protagonists. Moreover, Hardy's own perspective on the events seems remote and detached, thus discouraging the reader's emotional involvement in the story. Because of these peculiar features, *Mayor* constitutes an especially difficult challenge to interpretive criticism, and it is a challenge that previous criticism has been only partially successful in meeting.

Most commentators who seek to interpret the tonal and perspectival structure of *Mayor* use one of three distinct models of tragedy, or, with whatever cost to consistency, some combination of the three: (1) a model of retributive justice, (2) a model of Promethean Romantic heroism, or (3) a model of redemptive change. John Paterson offers a transcendental version of the model of retributive justice. In his view, tragedy depends on "moral and religious universals" and reaches resolution in vindicating "the existence of a moral order, an ethical substance, a standard of justice and rectitude, in terms of which man's experience can be rendered as the drama of his salvation as well as the drama of his damnation." The role of the tragic protagonist in this scheme is that of acknowledging this transcendent ethical order. Henchard offends against the cosmic order, which destroys him, but he also "stands for the grandeur of the human passions." He is thus the tragic agent of a "heroic imagination."[1] Like the model of retributive justice, the Promethean Romantic model focuses on the assertion of heroic though destructive grandeur. George Levine, for example, identifies "the romantic hero" as a figure of "large aspirations" and "uncontrollable energies that destroy with the force of an Alpine torrent." These heroic figures "desire beyond the limits of nature" and they thus exemplify qualities that are "quintessentially human." The tragic hero achieves "a new freedom of imagination" and represents "a new conception of human dignity."[2] In contrast both to the model of retributive justice and to the Promethean Romantic model, the model of redemptive change deprecates the idea of heroic passion and emphasizes instead the deplorable and contemptible aspects of the protagonist's career. Advocates of the redemptive model, like advocates of retributive justice, require that the protagonist feel contrition for his various misdeeds. As R. H. Hutton conceives it, Henchard's "tragic career of passionate sin, bitter penitence, and rude reparation" serves ultimately to bring him "to a better and humbler mind." In this model, the purpose of tragedy is to exemplify the way in which "circumstance" can serve "to chasten and purify character."[3] Elaine Showalter offers a modern feminist version of the redemptive model. In her reading, Henchard undergoes a transformation "from a romantic male individualism to a more complete humanity." By becoming less male, Henchard becomes more fully human, and he thus becomes "capable of tragic experience."[4]

These three models of tragedy have persisted over decades in which seemingly fundamental changes have taken place in the ideological and philosophical orientation of literary studies, and they have retained their basic structural character through numerous metamorphoses in theoretical concepts and vocabularies—old fashioned humanist, New Critical, archetypal, Marxist, Freudian, deconstructive, feminist, and the various hybrid blends of postmodernism. The persistence of these models suggests that in important ways the models function at imaginative levels deeper and more general than the various fashions through which they have retained their basic form. Each of these three models appeals to some historically conditioned articulation of a fundamental disposition in human nature. The model of retributive justice has an affinity with the ethos of the Old Testament, and its proponents are wont also to cite antecedents from Greek tragedy. The model of redemptive change, with its emphasis on salvation through moral transfiguration, has an obvious affinity with the Christian ethos. Like the model of retributive justice, the Promethean Romantic model operates in a cosmic sphere, but it repudiates the justice of the cosmic order and, like the redemptive model, locates its resolution within the affirmation of specifically human qualities. As its name suggests, the Romantic model is closely associated with the spiritual defiance of a certain phase of Romanticism, a phase identified more closely with Byron and Shelley than with Wordsworth or Keats. Each model appeals to a specific emotional range and finds its resolution in the gratification of some deep emotional need—the desire for justice, the claim for self-abnegating affiliation, or the assertion of individual power. The assertion of power and the claim for affiliation constitute the two basic forms of human social interaction. Justice mediates between these two forms.

Despite the archetypal scope of the three models of tragedy, none of the models is sufficiently deep and general to give a thoroughly cogent account of the tonal and perspectival structure of *Mayor*. The three models overlap in certain ways but conflict in other ways, and the inadequacies of each, as interpretive models, help to explain the persistence of its rivals. The model of retributive justice eliminates the element of chance in Hardy's vision of the world and adopts a stance of vindictive satisfaction incompatible with his tolerant humanity. The model of Promethean Romantic heroism glamorizes Henchard's character and strikes a note of vainglorious triumphalism incompatible with Hardy's shrewd irony. And the model of redemptive change blurs the essential continuity of Henchard's character and posits a sentimental resolution alien to Hardy's tragic austerity.

At about the time that he was writing *The Mayor of Casterbridge*, Hardy wrote a note in which he formulated a concept of tragedy that contains none of the distorting impedimenta of the three models that are typically invoked to account for the generic and tonal structure of the novel. "Tragedy. It may be put thus in brief: a tragedy exhibits a state of things in the life of an individual which unavoidably causes some natural aim or desire of his to end in a catastrophe when carried out."[5] This definition covers a broad spectrum of works typically regarded as tragic, and it is fully adequate to the situation of *The Mayor of Casterbridge*. It involves no commitment to a principle of poetic justice; it does not require us to derive affirmations of an essential human nobility from the struggles of the tragic protagonist; and it does not presuppose a morally uplifting transformation in the moral constitution of the protagonist.

We need not accept any of the main assumptions that have animated the standard tragic models used to interpret *Mayor*—that the novel must involve passional involvement with a heroic protagonist, that the protagonist must himself achieve an adequate interpretive perspective on his own experience, that the events of the story must affirm the existence of a morally meaningful order, that the story must culminate in the production of sublime affects, that it must exemplify moral improvement, or that it must provide some reassuring image of human goodness or nobility. If we reject these assumptions, we can avoid romanticizing or sentimentalizing the tragic protagonist. Henchard is a powerful, commanding personality, deeply flawed, often misguided, inadvertently self-destructive, and ultimately pathetic. Hardy does not himself feel that Henchard's career is a sublime or ennobling spectacle, and he does not invite the reader to feel that. The spectacle of "The Life and Death of the Mayor of Casterbridge"—the full main title of the novel—challenges Hardy to devise a perspective adequate to the contemplation of destructive passions and the mischances of life. Henchard himself can attain to no such perspective. He is not a reflective man, and to achieve a philosophic view of his own experience would require powers of detachment and of generalization that are alien to his nature.

One of Hardy's most perceptive critics, Lord David Cecil, observes that while Hardy had rejected Christian beliefs, his ethos remained deeply imbued with Christian values. "The Christian virtues—fidelity, compassion, humility—were the most beautiful to him."[6] In *Mayor*, those qualities are most fully exemplified by Elizabeth-Jane, but the qualities are not gender specific. In other Hardy novels, they are exemplified by both male and female characters—for instance, by Gabriel Oak in *Far from the Madding Crowd*, Diggory Venn in *The Return of the Native*, John Loveday in *The Trumpet Major*, Giles Winterborne in *The Woodlanders*, and Tess in *Tess of the d'Urbervilles*. Hardy himself regards all of these characters with affectionate respect, but in his more developed powers of reflective contemplation, he also stands apart from them, and above them. In the final chapter of *Mayor*, Hardy evokes Elizabeth-Jane's widest views in her mature life, and in that evocation, her perspective intermingles indistinguishably with Hardy's own:

> Her strong sense that neither she nor any human being deserved less than was given, did not blind her to the fact that there were

others receiving less who had deserved much more. And in being forced to class herself among the fortunate she did not cease to wonder at the persistence of the unforeseen, when the one to whom such unbroken tranquillity had been accorded in the adult stage was she whose youth had seemed to teach that happiness was but the occasional episode in a general drama of pain.[7]

Because she thus also stands apart and above, Elizabeth-Jane is not herself a passional protagonist. So far as the passional drama is concerned, she is only a good minor character. Within the perspectival drama—the struggle to attain an interpretive view adequate to the spectacle of Henchard's life—she is the central character. It is in her mind, and not in that of the protagonist, that Hardy locates his own sense of resolution.

"Interest" as a Key to the Tonal Structure of *Mayor*

Had we started with *Mayor*, and studied it alone, we could never have derived a clear idea of the standard agonistic structure of the novels of the period. The consensus level (including missing values) for assigning characters to roles in *Mayor* is low (69%, in contrast to 81% for all 206 multiply-coded characters in the multi-novel website), and the assignment of roles puts strong pressure on the standard agonistic logic articulated in the relations among personality, motives, mate-selection criteria, and emotional responses. The consensus rating on Henchard, the title character, is fairly high (88%). Fifty-six out of sixty-four respondents identify Henchard as the protagonist. But, compared to the profiles from the website for all the other novels in this study, Henchard's profile in motives and personality is more like that of an antagonist than that of a protagonist. His predominating motives are those of achieving wealth, power, and prestige; his scores on affiliative behavior and affiliative personality traits are low; he does not score high on cultural interests; he is highly unstable emotionally; and he receives high scores on the emotional response factor Dislike. Henchard comes into sharp conflict, in one way and another, with Farfrae and with Newson, and as a result, those two characters are identified as antagonists, but their scores on motive factors and personality factors are not like those of standard antagonists. Newson's profile is that of a good minor character. In motive factors, Farfrae's profile combines protagonistic and antagonistic features, but his personality profile is emphatically that of a protagonist.

In the multi-novel website, the one central motive factor that distinguishes protagonists, both male and female, from all other character sets, is a factor we label "Constructive Effort." It consists in two chief elements: affiliative and altruistic social behavior, and creative and culturally acquisitive intellectual interests. Henchard's stepdaughter Elizabeth-Jane displays a high level of Constructive Effort, and her personality also reflects features typically associated with protagonists. (See figures 9.1 and 9.2.)

She scores low in Extraversion and high in Agreeableness, Conscientiousness, and Emotional Stability. Despite her protagonistic features, our respondents affirm that the success or failure of her hopes and efforts is not a main feature in the outcome of the story, and they identify her as a good minor character, that is, as the friend or associate of a protagonist. That role assignment corresponds to the assessment of her role in most of the critical commentary on the novel. Elizabeth-Jane is clearly not a protagonist, but she nonetheless has a crucially important function in the story. She provides a point of view wider and wiser than that of any of the other characters. Her own success or failure is not central to the outcome of the story, but her perspective on the success or

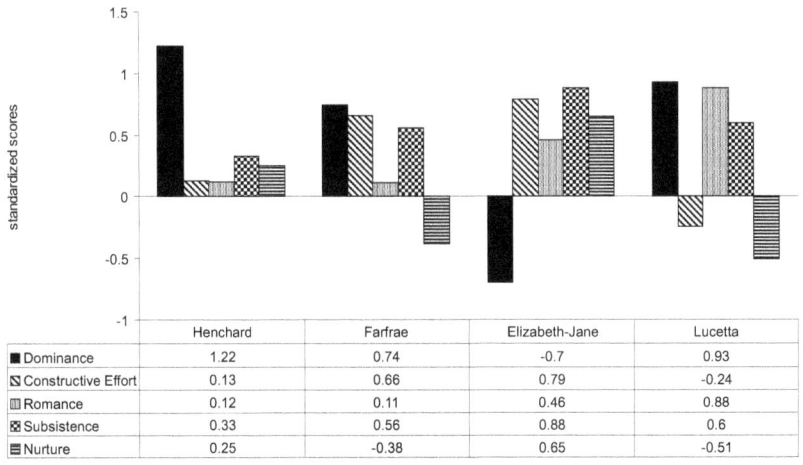

Figure 9.1 Motive factors in four main characters in *Mayor*

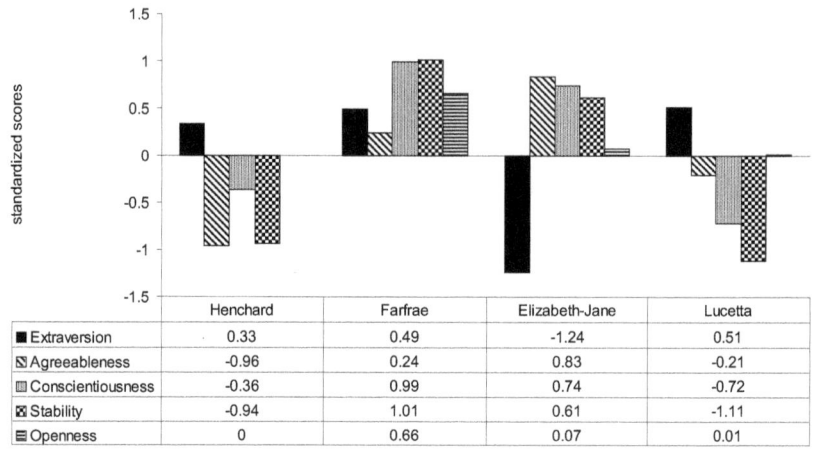

Figure 9.2 Personality factors in four main characters in *Mayor*

failure of other characters provides a standard of judgment that is close to Hardy's own. That standard modulates the emotional and tonal quality of the story and helps to guide the reader in gaining a perspective on the meaning of the story.[8]

With respect to personality, Farfrae is a very paragon of a male protagonist. He is right at average for male protagonists on Agreeableness, and far above average on Conscientiousness, Stability, and Openness. He is reliable in business, he is consistently cheerful and even-tempered, and he scores higher on Openness than any other character. He likes to read; he invents a process for restoring damaged wheat; and he introduces new agricultural technology into Casterbridge. And yet, nine of twenty-one respondents identified him as an antagonist; seven identified him as a good minor character; one as a bad minor character; and three said "other." Only one respondent identified him as a protagonist. Farfrae's motivational profile mixes protagonistic and antagonistic features, but his personality is overwhelmingly protagonistic. Despite his apparently appealing personality, readers are indifferent to him. He excites little Interest, and his scores on both Dislike (−.16) and Root For (−.14) are close to average.

Farfrae is a bright, cheerful, friendly, young man, ambitious and successful, but also constructive and open to new experiences.

He is a fortunate person, admirable, attractive, and successful, but within the emotional economy of this novel, that particular profile has no special claims on the interest or sympathy of the reader. The novel is designed around catastrophic losses and failures—those of Susan, first, and then of Lucetta, and ultimately of Henchard. Unlike a substantial portion of nineteenth-century novels, *Mayor* is not designed to align the reader's perspective with that of a Golden Youth, to engage the reader's sympathetic identification with that youth, to fulfill the reader's expectations concerning the hopes and fortunes of that youth, and to affirm the normative and central value of the personality and motives embodied in that youth. Within the perspectival and emotional economy of this novel, the concerns of a young man like Farfrae are relegated to marginal status, and the novel occupies itself instead in coping with forms of distress that remain outside the scope of Farfrae's empathic power.

The criteria that enter into mate selection typically differ among males and females in both good and bad character sets. Male protagonists tend to set a high value on Physical Attractiveness, some value on Intrinsic Qualities (intelligence, kindness, and reliability), and little value on Extrinsic Attributes (wealth, power, and status). Female protagonists tend to set the highest value on Intrinsic Attributes, a moderate value on Extrinsic Attributes, and little value on Physical Attractiveness. Male antagonists, curiously, score below average on all criteria for selecting mates. That is, they have no particular preferences. Female antagonists, in contrast, place the highest value on Extrinsic Attributes, slight value on Physical Attractiveness, and almost no value on Intrinsic Attributes. As the unusual structure of motives and personality in *Mayor* might lead us to anticipate, mate selection in *Mayor* disrupts these usual patterns. (See figure 9.3.)

Henchard and Farfrae vie for Lucetta, and in pursuing her they both mingle protagonistic and antagonistic features. They are strongly moved by her Physical Attractiveness but give no heed to her Intrinsic Qualities. They both display interest in her Extrinsic Attributes. In selecting Henchard, Lucetta displays the pattern of a typical female antagonist; she is interested only in his wealth, power, and prestige (External Attributes). In selecting Farfrae, in contrast, she is moved by all three criteria, though least by his Intrinsic Qualities. In a standard romantic comedy, the normative or model couple marries at the end of the story. Elizabeth-Jane

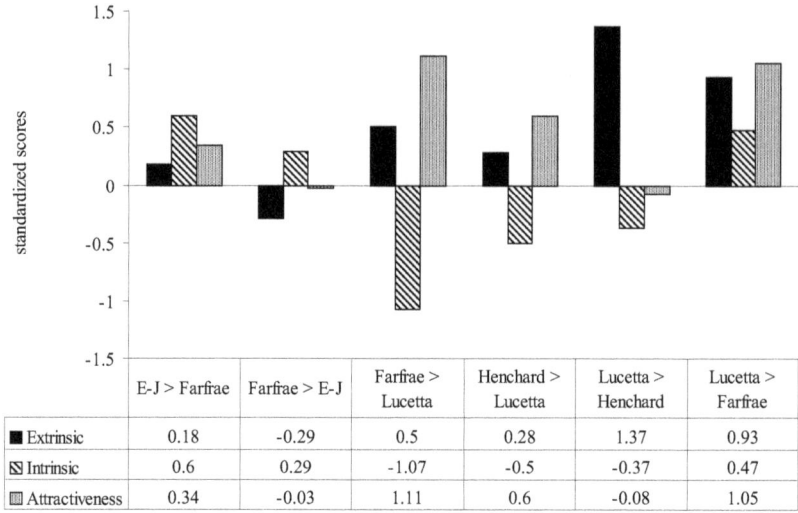

Figure 9.3 Criteria for selecting mates in *Mayor*

and Fafrae marry at the end of this story, but their mate selection pattern is unusual. In selecting Farfrae, Elizabeth-Jane gives highest priority to Intrinsic Qualities, but she also places a considerable emphasis on Physical Attractiveness. In selecting Elizabeth-Jane, Farfrae, in contrast, gives little heed to Physical Attractiveness. He gives some regard to Intrinsic Qualities, but contrasted with his interest in Lucetta, his interest in Elizabeth-Jane seems, in its romantic aspect, rather tepid.

Our research design does not aim directly at analyzing the complex interactions in point of view among the author, the characters, and the readers, but the elements of our design enable us to get at this perspectival dimension indirectly. The relations among the role assignments, the constitution of character, and the emotional responses of readers give the necessary clues to the peculiar perspectival and tonal structure of this novel, and by assessing that tonal structure we can make reasonable inferences about the specific kind of psychological work this particular novel is designed to accomplish both for the author and for the reader.

All novels perform some kind of psychological work. They activate the emotions and imaginative responses of readers and lead the readers through an integrated emotional process culminating in some kind of conclusion or point of rest ("resolution"). Most of the novels in our data set seek in a fairly simple and direct way to involve the reader in the story, to engage the reader's sympathetic identification with one or more main characters, or at least to activate the reader's sympathetic and appreciative responsiveness to the main characters. That sort of involvement is registered in part through one of the three emotional response factors that emerged from the factor analysis of emotional responses in the multi-novel website, the factor "Interest." The first emotional response factor is Dislike and is constituted by positive loadings on anger, disgust, contempt, and fear of the character, and by negative loadings on admiration and liking. (A factor "loading" indicates the weight given to each of the measurements used to define a factor.) The second emotional response factor is Sorrow and is constituted by positive loadings on sadness and fear for the character. The third emotional response factor is Interest. This factor has moderate positive loadings on admiration and liking, but the main element in Interest is a strong negative loading on indifference. Characters who score low on Interest have typically received very high scores on indifference. That is, our respondents have indicated that they are highly indifferent to the character. A high score on Interest suggests a strong degree of passional involvement with a character. Factor analysis, by design, identifies statistically independent themes. The factor analysis therefore reveals that the emotional response factor Interest is qualitatively distinct from the evaluatively charged response Dislike, which constitutes a measure of positive or negative emotional valence. Interest is also qualitatively distinct from Sorrow, which constitutes a measure of sympathy or compassion.

In one of the earliest responses to *Mayor*, an anonymous critic observed that the novel "does not contain a single character capable of arousing a passing interest in his or her welfare."[9] As the scores on Interest in our study indicate, this critic's observation of the fact is correct. (See figure 9.4.)

The inference the critic draws from that fact is, however, erroneous. The critic presupposes that some sort of passional involvement with characters is an indispensable requirement in all novels, so that the absence of interest is merely a defect, and a large one. Passional involvement is indeed a common way in which novels work, but it is not the only possible way, and it is not the way *The Mayor of Casterbridge* works. What Hardy is after

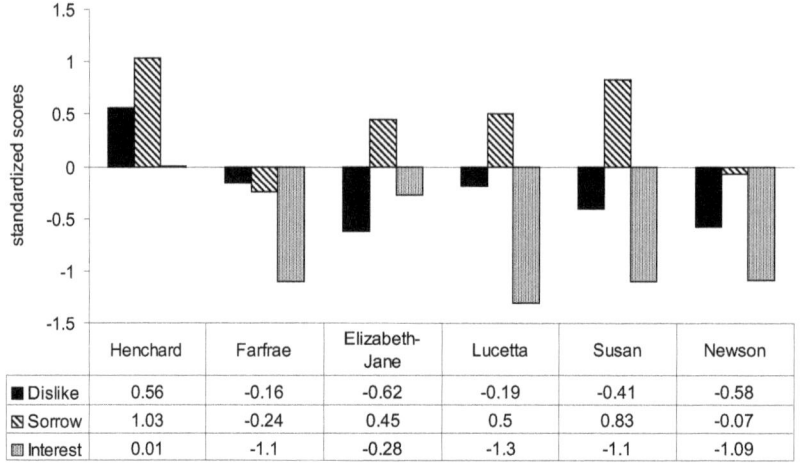

Figure 9.4 Emotional responses to characters in *Mayor*

in this novel is something rather different, and something fairly unusual, peculiar to Hardy, and perhaps more fully exemplified in this particular novel than in any other novel by Hardy. What Hardy is after is in fact something like the reverse of Interest. The kind of psychological work Hardy accomplishes in *Mayor* is that of gaining a reflective detachment from the story he depicts. He seeks himself to achieve a defensive, stoic stance against both passion and the vagaries of circumstance. Within the story itself, as a participant observer, Elizabeth-Jane embodies that stance.

Hardy worried about having cluttered the serial publication of the novel with sensational events, and he pruned and simplified the plot in the book version.[10] Even in its chastened form, the pace of the story is such that the rapidly shifting fortunes and love entanglements are like a spectacle seen through the wrong end of a telescope, a phantasmagoria of passion and folly, tinged with absurdity and futility. The most striking aspect of the emotional response to the novel as a whole is the extremely low level of Interest for the main characters. To get a comparative sense of the level of interest, we can line up the Interest scores for the forty-eight most frequently coded characters in the larger data set, add the six characters from *Mayor* to the list, and then sort the scores in descending order (high to low). Out of the fifty-four characters, the four lowest scores on

Interest are all from *Mayor* (Newson, Farfrae, Susan, and Lucetta). Henchard, though he excites strong emotional responses in Dislike and in Sorrow, nonetheless occupies the thirty-seventh position in the Interest scale, and Elizabeth-Jane, though she excites feelings of Admiration and Liking, occupies the forty-second position. Because the scores in the data set for the multi-novel website are standardized, the average score for all characters is zero. For the forty-eight most frequently coded characters, the average score on Interest is .17. For the six characters in *Mayor*, the average Interest score is –.81. Given the proportions of the normal curve, about 79% of all characters in the multi-novel website—major and minor together—have Interest scores higher than the average score for the six main characters in *Mayor*. If it is true that Hardy is seeking to damp down excitement, to discourage the emotional involvement of readers, he has evidently succeeded.

There is a good deal of Sorrow in the story—for Henchard, Elizabeth-Jane, Lucetta, and Susan—but the low scores on Interest suggest a low level of intensity in emotional response. Henchard is unequivocally the protagonist, but he scores high on Dislike. Farfrae and Newson come into conflict with Henchard or present obstacles to him, and they are thus assigned roles as antagonists, but neither scores in the antagonistic range on Dislike. In the scores for the larger data set, Root For—a measure of whether respondents want a character to succeed—correlates strongly and negatively with Dislike ($r = -.67$). For *Mayor*, the correlation is only –.03—essentially no correlation. The confounding of normal agonistic role assignments disrupts the usual relationship between liking or disliking characters and becoming emotionally invested in the outcome of the story. It seems likely that this disruption helps neutralize emotional responsiveness in readers and thus contributes to the low scores on Interest.

As narrator of *Mayor*, Hardy adopts a stance of reflective, stoic detachment. He seeks to gain a calm and distant perspective on the transient ambitions and passions of human life and the peripeties and contingencies of circumstance. Gaining detachment is not the most common kind of psychological work a novel accomplishes, but it is a common strategy for coping with life, and it is altogether consistent with Hardy's melancholy and philosophical temperament. Late in life, Hardy wrote a poem titled "For Life I Had Never Cared Greatly." This was not true, but it did reflect one of Hardy's persistent philosophical ambitions. He felt this ambition as an exceptionally keen need, because for life he had always cared very much, and he was thus vulnerable to all its travails.

Conclusions

Our data indicate that the agonistic structure of *Mayor* is very different from that of the average Victorian novel. It is not surprising, then, that *Mayor* has presented an especially difficult challenge to interpretive criticism. By quantifying the elements of tonal analysis, we can break up the prefabricated affective and conceptual structures that have shaped criticism on this particular novel. Reducing affective structures to their component parts can render interpretive analysis more flexible and more precise. Advances in flexibility and precision can refine common perceptions of exceptionally accessible authors such as Jane Austen, and they can also help to solve intractable interpretive problems in exceptionally difficult novels such as *Mayor*.

Adopting a quantitative approach need not render a critic less sensitive to nuances of character and tone. Quite the contrary. It can free us from distorting preconceptions, making it possible to see an old and familiar text with eyes newly opened. Many of the particular observations we make in this chapter converge closely with those of Hardy's other critics. It could hardly be otherwise. The questions in the questionnaire are couched in the common language and appeal to the common understanding. The data on which we base our conclusions have been contributed largely by professional scholars intimately familiar with Hardy's work. These scholars have not simply been blind to the attributes of the characters. They have only been unable to combine their particular observations and emotional responses into a coherent picture of the novel as a whole. Quantifying agonistic structure makes it possible to construct an interpretive model that corresponds more closely to the total structure of meaning in Hardy's work.

The organization of tonal and thematic elements in *Mayor* is unusual, but the elements themselves are common and familiar. They can be located on a continuum with the imaginative qualities in Hardy's other works. Those qualities include a sensually rich lyricism scarcely equaled in English outside the poetry of Shakespeare, Milton, and the Romantics. Hardy also has an extraordinarily high capacity for registering emotional pain. In his treatment of Tess in *Tess of the d'Urbervilles*, his proclivity for negative affect combines with tenderness and strength of mind. The result is a sublime elegy. In much of *Jude the Obscure*, Hardy's sensitivity to pain degenerates into neurasthenia, pure depressive affect of the sort monstrously personified in Jude's son "Father Time." *Jude* is written from a point of view morbidly fixated on the spectacle of sensitive human matter caught and mangled in destructive circumstance. Nonetheless, Jude's renunciation of life, at the end, has a ghastly magnificence that transcends self-pity. He passes beyond the reach of torment and achieves a final stage of utter indifference. In *Mayor*, Elizabeth-Jane succeeds in achieving a detached but compassionate perspective that does not involve losing all capacity for pleasure and all interest in life. In the final paragraphs of the novel, Hardy's perspective merges almost completely with that of Elizabeth-Jane. He commends her wisdom, and invites the reader to do the same.

Few critics have been able to elucidate the kind of psychological work Hardy accomplishes in *Mayor*, for himself or for them. But emotions can be powerfully active even when they are not fully understood or explained. Hardy's stance in *Mayor* has almost certainly exercised an emotional influence operating apart from critical efforts to explain it. Still, criticism is most satisfying when it both evokes and explains—evokes the feelings we have in reading a novel, and also stands apart from those feelings, analyzes them, and locates them within broader networks of explanation.

PART FOUR

Evolutionary Intellectual History

CHAPTER 10

The Power of Darwin's Vision

The Classic Status of *The Origin of Species*

The Origin of Species has special claims on our attention. It is one of the two or three most significant scientific works of all time—one of those works that fundamentally and permanently alter our vision of the world. At the same time, it is one of the few great scientific works that is also a great literary classic. It is written for the educated general reader and requires no specialized scientific training. It is argued with a singularly rigorous consistency, but it is also eloquent, imaginatively evocative, and rhetorically compelling. Although it is now nearly a century and a half old, it remains the single most comprehensive and commanding exposition of its subject, and its subject—the development of life throughout all of time—has a sublime scope and a unique significance.

Many fine scientists, scholars, and writers have now dedicated their lives to the subject of evolutionary biology, but none of this work has rendered Darwin's own contribution obsolete. Ernst Mayr, both a biologist and a historical scholar of the first rank, maintains that modern evolutionists differ from Darwin "almost entirely on matters of emphasis." Mayr himself is one of the main contributors to the "Modern Synthesis," that is, the integration of Darwin's theory of natural selection with Mendelian genetics. Despite the advances in modern technical understanding, he notes that "a modern evolutionist turns to Darwin's work again and again," and he observes, rightly, that "Darwin frequently understood things far more clearly than both his supporters and his opponents, including those of the present day." His summary judgment of Darwin's lasting historical significance is that "no one has influenced our modern worldview—both within and beyond science—to a greater extent than has this extraordinary Victorian."[1] In confirmation of these claims, we may look to a recent, comprehensive textbook of evolutionary biology, Mark Ridley's *Evolution*. Ridley informs his audience that "the classic case for evolution was made in Darwin's *On the Origin of Species*" and that Darwin's "general arguments still apply."[2] Further, on the specific and central topic of the evidence for evolution, Darwin gives "the classic account." Michael Ghiselin, another distinguished biological theorist and Darwin scholar, also affirms the enduring value of Darwin's commanding perspective. "To learn of the facts, one reads the latest journals. To understand biology, one reads Darwin."[3] Given such testimony as this, it would not be too much to say that if a student were to read only one book on evolution, the best book to read would still be *The Origin of Species*.

Plan of This Chapter

The extraordinary canonical position occupied by the *Origin* depends on three elements: the subject, the time, and the man. Darwin had a subject full of mystery and power, the one subject of the deepest possible significance for all living things; the time was right for the comprehension of that subject; and Darwin was the right man to achieve that comprehension. In this section of the chapter, I shall explain the sequence of topics for the chapter as a whole, and then in the next three sections I shall take up each of these three elements in turn. After commenting on Darwin's subject, the historical background to his work, and the character of mind that made it possible for him to discover and develop the theory of natural selection, I describe the one main evolutionary theory that stood as an alternative to Darwin's—the theory of Lamarck and Spencer. Turning then to the development of Darwin's own theory, I discuss the inception and gestation of the *Origin*. I also discuss Darwin's effort, in *The Descent of Man*, to incorporate human beings within the phylogenetic order—that is, within the classificatory system that derives from the common descent of all living things. ("Phylogeny" is the evolution of a genetically related group of organisms and is distinguished from "ontogeny," the development of an individual organism.) In locating Darwin in relation to both his sources and his successors, I use the idea of scientific revolutions as a leading theme. I compare Charles Lyell's revolution in geology with Darwin's revolution in evolutionary biology, and I examine the complex way in which Darwin assimilates his sources, incorporating some elements and using others as foils. Taking issue with Thomas Kuhn's notion of a simple paradigm shift or gestalt switch, I also examine the long delay before Darwin's own revolution was completed in the Modern Synthesis, a process that lasted from about 1920 through about 1950.

Darwin's Subject

Before commenting on the nature, sources, and development of Darwin's theory, it will be helpful to present a brief outline of the theory itself. Darwin's own summary of his theory, in the introduction to the *Origin*, is admirably succinct:

> As many more individuals of each species are born than can possibly survive; and as, consequently, there is a frequently recurring struggle for existence, it follows that any being, if it vary however slightly in any manner profitable to itself, under the complex and sometimes varying conditions of life, will have a better chance of surviving, and thus be *naturally selected*. From the strong principle of inheritance, any selected variety will tend to propagate its new and modified form.[4]

The logic of reasonable presuppositions and conditional inferences in this formulation is luminously clear and simple—so clear and simple that Darwin's young colleague, T. H. Huxley, first responded to the theory by exclaiming to himself, "'How extremely stupid not to have thought of that!'"[5] The apparent simplicity of the theory is deceptive and is in fact a measure of its extraordinary depth and power. Huxley was a man of exceptionally quick and sharp intellect, but he lacked Darwin's deep consistency and his power of formulating and sustaining a wholly original vision of the world. (When we speak of "genius," it is to such power that we refer.) Huxley saw instantly into the logic of the case, but he did not see instantly, and perhaps never fully and consistently grasped, the all-encompassing character of the theory, the way it implied and was implicated in every conceivable aspect of the structure and function of all living things—both of their internal organization and of their external relations to the physical world and to other living things. The significance of a scientific theory can be measured in good part by the ratio between simplicity of causal explanation on the one side and the extent of explanatory scope on the other. The most significant theories bring the largest range of phenomena within the smallest compass of causal explanation. Judged by this criterion of significance, the theory that Darwin squeezes succinctly but adequately into the few lines just quoted ranks with the theories of Copernicus, Newton, Einstein, and Crick and Watson. That is, it is one of the few most successful efforts at scientific explanation in the history of science.

This succinct summary has a dry, logical, almost arithmetical character. Such is the nature of simplicity in causal explanation. But the full scope of the world of phenomena encompassed by the theory has a magnitude that staggers the imagination, and this magnitude has a specific aesthetic character. Since the time of Longinus, the definition of the sublime has been that of a grandeur that expands the imagination to its limits and then escapes those limits. Darwin himself understood this effect and had the imaginative capacity simultaneously to explain and to wonder. In his autobiography, he complains that in his later life he had lost all capacity for aesthetic pleasure, for poetry and art, though he still much enjoyed novels. For this apparent atrophy of the aesthetic faculties, he blames his exclusive concentration on scientific work.[6] A perhaps more just apprehension of the case is to say that Darwin's aesthetic and imaginative energy had gradually become wholly absorbed into the creative vision that became his life's work. In the final sentence of the *Origin*, after once again summarizing succinctly the presuppositions and logical linkages of his theory, he gives full rhetorical and emotional expression to his imaginative apprehension of his subject:

> There is grandeur in this view of life, with its several powers, having been originally breathed into a few forms or into one; and that, whilst this planet has gone cycling on according to the fixed law of gravity, from so simple a beginning endless forms most beautiful and most wonderful have been, and are being, evolved.

Darwin's imagination poises over two kinds of tension. One is the tension between the simplicity of life's origin and the multifarious complexity into which natural forms have evolved and continue to evolve. The other kind of tension is that between the invariable, unitary law of gravity, producing only an endless, cyclical repetition of planetary motion, on the one side, and on the other the perpetual change of living forms through time. Darwin envisions the whole progression of life on earth within the cosmic scope of Newton's celestial mechanics, and in his imaginative response to this progression he integrates the austere intellectual satisfaction of causal explanation with a luxuriant delight in the complexity of life.

In his chapter on geological succession, Darwin observes that through his theory "we can understand how it is that all the forms of life, ancient and recent, make together one grand system; for all are connected by generation" (*OS*, 309). The *Origin* is full of impressive rhetorical passages, but its larger imaginative effect derives from the quietly meditative and methodical exposition of this whole "grand system" in its diverse aspects. The first and most apparent aspect is that of systematics, the classification of all living things within a hierarchical order. In the century before Darwin, the great Swedish systematist Linnaeus (Carl von Linné, 1708–1778) had created a workable system of classification. He had thus provided an indispensable platform for Darwin's own work, but Linnaeus' system had no causal mechanism and no temporal dimension. As Ghiselin observes, in Darwin's theory, "classification ceased to be merely descriptive and became explanatory."[7] The hierarchical order of systematic classification exists *because* all living beings are connected in a phylogenetic line. They are connected, as Darwin says, by "generation."

Darwin's insight into the phylogenetic basis of systematic order is a true insight—an insight that peers into the reality of things. Consequently, once he has in possession this one central clue to the organization of life, he can use it as a guide to every other aspect of natural history. It becomes his golden thread through the mazes of anatomy, development (ontogeny), reproductive interactions, ecology, paleontology, and the geographical distribution of living things. Darwin gives extended expositions to his findings in all these fields of inquiry. These expositions serve as evidence for his core theory, and his core theory serves as an explanatory hypothesis for the organization of the evidence.

In the generation before Darwin, in rough parallel to Linnaeus, the distinguished French anatomist and paleontologist Georges Cuvier had established two grand principles of anatomy: (1) that all animals can be arranged into a few basic body plans (in Cuvier's system, vertebrates, mollusks, *articulata* [e.g., insects], and *radiata* [e.g., jellyfish); and (2) that the internal organization of all animals displays an integrated functional order; one part or organ requires and implies another part of a particular kind. For instance, an animal with the teeth and claws of a predator will also predictably have a digestive tract designed for the digestion of meat. An animal with wings designed for flight will also have a heart that beats fast enough and a skeletal structure light enough for sustained flight. The same logic of phylogenetic connection that explains the classificatory order described by Linnaeus explains also the organization of life under a few distinct body plans. And the idea of adaptation by means of natural selection explains the integrated functional order in the internal organization of organisms. Animals make sense as integrated functional wholes not because they have been created in that way, once and for all, but because they have evolved in adaptive relation to the conditions of their existence.

The theory of special creation and the theory of natural selection are both compatible with the integrated functional organization of animals—and indeed integrated functional organization is the primary evidence put forth in the argument of natural theology: the argument that "design" implies a designer. But the theory of special creation, in contrast to the theory of natural selection, is not a causal explanation so much as a simple appeal to divine intervention—the *deus ex machina* of biology. More importantly, from the standpoint of causal logic, while special creation can account for integrated functional organization simply by claiming that such organization displays the wisdom and beneficence of the Creator, it cannot account for *imperfections* in functional organization. If God created animals to be perfectly adapted to their environments, why did he provide them with rudimentary organs such as the human appendix? Why did he provide upland geese with webbed feet that they never use for swimming? Moreover, why did he manage things in such a way that the same sequence of bones appears in the forelimbs both of reptiles and of mammals, and in the wings both of birds and of bats? Was this sequence of bones optimally efficient for the diverse activities of all these animals? To questions such as these, special creation can provide no answer. In contrast, Darwin's theory of descent with modification—the theory that all organisms have descended from previous organisms, and that in the course of descent the form of organisms has gradually become modified through a process of adaptation by means of natural selection—provides an answer. By invoking this theory, we can understand that all adaptive structures derive from previous structures; adaptation never begins from nothing, and inherited structure places necessary constraints on all functional organization.

By explaining both the internal organization of organisms and their classificatory order, Darwin enabled himself to give an intelligible account of life in three of its main dynamic aspects: (1) the internal development of individual organisms (ontogeny); (2) the distribution of species over space in time; and (3) the interactions of organisms within ecological systems.

By positing the selection of adaptive characteristics at differing points in the life history of an organism, Darwin was able to explain, correctly, the partial parallelism between embryonic development (ontogeny) and the place a species occupies within the generational sequence (phylogeny). The human embryo, for instance, at one stage contains gills, and at another a rudimentary tail. (Darwin comments on this topic in the *Descent of Man*.)[8] Darwin's disciple Ernst Haeckel later exaggerated this insight into embryonic development into the misleadingly overgeneralized claim that *ontogeny recapitulates phylogeny*, but Darwin's own observations (*OS*, 373–74) display the judicious precision and circumspection that usually characterize his work, and these observations form an integral part of the network of logic and evidence in his exposition. In his history of embryology, John Moore makes a point about Darwin's historical significance very similar to the point that Ghiselin makes about Darwin's place in the history of systematics. "The Darwinian paradigm shift of 1859 changed not only what biologists did but also provided an explanation for what they observed. The new paradigm was able to offer a satisfying explanation for much that had already been learned. In fact, the data themselves seemed to be awaiting some organizing theory, and Darwin's basic idea provided it." The theory of natural selection "did far more than make some otherwise confusing embryological phenomena understandable. It accounted for the grand phenomenon of organisms belonging to sets or taxonomic groups," and as a result it "gave embryologists a mission of first-rate theoretical importance—the search for lineages in the minutiae of development."⁹

By identifying phylogenetic organization, Darwin was able to situate plants and animals in distinct lineages and to observe the way these family groups have distributed themselves geographically through migration and the dispersal of seeds. In the later chapters of the *Origin*, Darwin reconstructs the changes over geological time in the ecosystems that have occupied specific portions of the earth. It is worth emphasizing that the "one long argument" of the *Origin* is not just the exposition of a *theory*. It is also a geographical and ecological *history*. The theory provides the crucial clues for the history, and the history in turn supports and illustrates the theory. The historical portions of the *Origin* constitute a dramatic narrative that is immense in its scope and detail. In the first of his two chapters on "Geographical Distribution," for instance, Darwin gives a masterful and compelling account of the flow of life forms over vast continental land masses, driven by the advance and retreat of ice sheets and the oscillations of the earth's crust, over millions of years.

Darwin's understanding of adaptive form links the internal organization of animals with their environmental conditions—conditions consisting both of the physical environment and also (a point to which Darwin gives special emphasis) of the other organisms with which they interact. This interdependency of organisms is the subject of ecology, and the *Origin* offers both a classic exposition of the principles at work in ecological analysis and also a series of narrative and rhetorical evocations that are among the most striking and memorable passages in the book. In these passages, logical argument interlinks symbiotically with naturally poetic imagery, each expanding and supporting the other. Here is an instance:

> A corollary of the highest importance may be deduced from the foregoing remarks, namely, that the structure of every organic being is related, in the most essential yet often hidden manner, to that of all other organic beings, with which it comes into competition for food or residence, or from which it has to escape, or on which it preys. This is obvious in the structure of the teeth and talons of the tiger; and in that of the legs and claws of the parasite which clings to the hair on the tiger's body. But in the beautifully plumed seed of the dandelion, and in the flattened

and fringed legs of the water-beetle, the relation seems at first confined to the elements of air and water. Yet the advantage of plumed seeds no doubt stands in the closest relation to the land being already thickly clothed by other plants; so that the seeds may be widely distributed and fall on unoccupied ground. In the water-beetle, the structure of its legs, so well adapted for diving, allows it to compete with other aquatic insects, to hunt for its own prey, and to escape serving as prey to other animals. (*OS*, 142)

Darwin envisions the world as a ceaseless process of biotic interactions leading to transformations of organic form. The driving force behind these interactions is the drive toward survival and reproduction.

Natural theologians such as William Paley had examined ecological interactions from within the constraining need to affirm a beneficent order aimed at the production of the highest happiness for the highest number. Darwin's ecological vision, in contrast, takes a radically naturalistic turn. In one celebrated passage, contemplating "the plants and bushes enclosing an entangled bank," Darwin insists that the appearance of randomness or "chance" in their distribution is delusory and that in reality all happens according to "definite laws." These laws are those of a struggle and conflict that in terms of human moral order are mere anarchy and chaos:

> What war between insect and insect—between insects, snails, and other animals with birds and beasts of prey—all striving to increase, and all feeding on each other or on the trees or their seeds and seedlings, or on the other plants which first clothed the ground and thus checked the growth of the trees! (*OS*, 141)

Darwin takes full account of symbiotic relationships, and also of cooperation among social animals such as bees, ants, and wolves (and, in *The Descent of Man*, human beings), but his vision is fundamentally one of competitive struggle. He repeatedly uses phrases such as "the great battle of life," and the "war of nature." As one of the several possible empirical findings that could falsify his theory, the most striking and decisive that Darwin cites is his contention that if even one instance could be found of a species having developed an adaptation solely for the benefit of some other species, "it would annihilate my theory"(*OS*, 220). The vision of nature Darwin offers is not that of some broad, abstract, intellective pattern, but that of living impulse, eager, frantic, animating every single organism, vast and minute, in inconceivable numbers, everywhere on earth, persisting throughout all the time of organic life. In his own generation, this vision was startling in its novelty and strangeness, but it was also massively convincing. It was as if the fog had been dispersed, and for the first time people saw the living world as it really is and said, "Of course, yes, that's it."

Darwin's vision of nature can be disturbing in its recognition of ferocity and ruthlessness, but it can also be ennobling in its response to beauty and power. Darwin's own response to his subject has a quality of adult realism that, in retrospect, makes the fantasies of anthropomorphic providentialism seem puerile and sentimental. In the concluding sentence to the central chapter "Natural Selection," Darwin finds a poetic image that recapitulates his phylogenetic diagram of branching lineages and that also captures the tonal extremes in the subjective qualities of his vision:

> As buds give rise by growth to fresh buds, and these, if vigorous, branch out and overtop on all sides many a feebler branch, so by generation I believe it has been with the great Tree of Life, which fills with its dead and broken branches the crust of the earth, and covers the surface with its ever branching and beautiful ramifications. (*OS*, 177)

Death and destruction are inseparable parts of the organic process. No small part of Darwin's achievement is to have had the strength of mind necessary to rise above our partial human identifications and to stand, clear of eye and unabashed, before the total order of nature.

The Historical Moment of the *Origin*

In the latter half of the eighteenth and the first half of the nineteenth century, natural history had made huge strides in a wide array of specialized disciplines. The tradition of "natural theology," or the study of adaptive structure and ecological relations—interpreted as indications of providential order or "design"—had already extended from the time of John Ray (1627–1705) to that of William Paley (1743–1805). In the course of the eighteenth century, Linnaeus had for the first time set taxonomy—the classification of living things in coherently related groups—on a sound footing. The Comte de Buffon (1707–1788), the greatest of the naturalists among the French encyclopedists, had pioneered the study of geographical distribution—the designation of distinct groups of plants and animals in different parts of the world. Explorer naturalists, typified for Darwin by Alexander von Humboldt (1769–1859), had transformed the genre of travel writing into a medium of research into biogeography and ecology. (Darwin's own research as a naturalist during the voyage of the *Beagle* made important contributions to this tradition.) The deliberate breeding of animals and plants had of course gone on since before the beginning of recorded history, but the methodical study of breeding—of hybridism and of variation and the inheritance of variations—had emerged only within the previous century. The development of modern scientific embryology can be dated from Karl Ernst von Baer's discovery of the mammalian egg in 1827. The embryological researches of Darwin and his contemporaries provided important new insights into the structure and development of organisms and opened the way for Darwin's speculation into the relations between the individual development of a single organism (ontogeny) and the development over generations of species and higher *taxa* (phylogeny). (*Taxa* is the plural of *taxon*. *Taxa* are distinct groups of organisms of any rank. The taxonomic system locates all organisms within a classificatory hierarchy, thus: kingdom, phylum, class, order, family, genus, species, individual organism.) In the decades immediately preceding Darwin's maturity, Cuvier had established a strong scientific foundation both for comparative anatomy and for paleontology and had thus provided an indispensable basis for using anatomical structure to analyze the phylogenetic relations among organisms. In the work of Thomas Malthus (1766–1834), economics had turned its attention to the elemental biological interaction between population pressure and the supply of food. Most importantly of all, in a period from about 1750 to 1830, geology had emerged from the realm of fantastic speculation, established itself as a progressive empirical science, extended the scale of geological time from thousands of years to thousands of millions of years, and provided a model for the idea of massive alterations in structure resulting from the accumulation of changes so minute as almost to escape notice within the scale of a human lifetime.

Darwin's academic career at Cambridge was undistinguished, but while in university he had pursued natural history as a hobby and had made scientific friends. One of these friends, the botanist John Henslow, recommended Darwin for a post as unofficial naturalist on board *H.M.S. Beagle*, assigned to take soundings along the coast of South America and circumnavigate the globe for the purpose of making chronometric calculations. The *Beagle* set sail late in 1831 and returned to England nearly five years later. Much of that time, while his shipmates went about their different duties, Darwin was ashore, exploring, geologizing, and collecting specimens in natural history. Just before Darwin set out on the voyage of the *Beagle*, Robert Fitzroy, the captain of the ship, gave him what was probably the single most important intellectual gift of his life, the newly published first volume of Charles Lyell's *Principles of Geology*. (Volumes 2 and 3 reached Darwin in the course of the voyage.) Darwin assimilated Lyell's geological vision and used it as the main guide to his own geological observations on the voyage, and indeed for the rest of his life.

From Lyell Darwin inherited both substantial intellectual property and a burden of intellectual debts—unsolved scientific problems—that helped give direction to his own work. Lyell was both a geologist and a species theorist. In the former field, he achieved lasting distinction and provided an indispensable basis for the development of Darwin's own theories. As a species theorist, Lyell's creative work merely registered the stresses and perplexities of the species problem as it had developed up to his own time.

As a geologist, Lyell adopted the uniformitarian views of James Hutton. In *Theory of the Earth* (1795), Hutton had described the earth as a homeostatic system in which the slow and perpetual building up and wearing away of land over billions of years maintains a large-scale equilibrium. In the four decades between Hutton's work and Lyell's publication of the *Principles of Geology*, empirical geologists had made major advances in reliable knowledge about volcanic activity, erosion, stratigraphy, fossils, and other aspects of practical geology. Lyell assimilated this information, supplemented it with his own original and highly perceptive observations about crustal movements, and organized it within the basic framework of Hutton's theory of a homeostatic equilibrium between erosion and the formation of new land masses.

The principles of geology propounded by Lyell were fundamentally sound and relatively comprehensive. Since the time of his initial synthesis, there have been only two really fundamental additions to geological theory. The first was the theory of glaciers that was developed by Louis Agassiz during Lyell's own lifetime and that Lyell assimilated into later editions of the *Principles of Geology*. The second was the theory of continental drift, eventually expanding into the theory of plate tectonics, that was first sketched out by Alfred Wegener in the first and second decades of the twentieth century. The theory of plate tectonics has solved many interesting puzzles in geology, paleontology, and the geographical distribution of animals and plants, but this theory has only extended and expanded Lyell's synthesis, not replaced it. Darwin's theory of evolution by means of natural selection, in contrast, decisively demonstrated that Lyell's hypotheses as a species theorist were fundamentally wrong. Darwin could incorporate certain aspects of Lyell's thinking about species, and especially the idea that species become extinct through failure to adapt to environmental change—but the larger structure of Lyell's theory was erroneous, and what Darwin did was not to assimilate it but rather to reject it and to replace it with an alternate, better theory.

At the time that Lyell took up the species problem, the main alternatives for explaining the distribution of species over time were those of Lamarck and Cuvier. In his *Philosophie Zoologique* and other works, Lamarck had formulated a radical alternative to the idea that species had been created in their fixed and final form. He had proposed that species evolve over time, driven forward by some mysterious internal impetus toward ever-increasing complexity of structure and directed ultimately toward transformation into the supposedly highest of all anatomical forms—the human. Along the way, Lamarck speculated, the pure impulse of complexification leading to anthropomorphic perfection is deflected and distorted by the need of organisms to adapt to the various stresses of local environmental conditions. Cuvier had flatly rejected Lamarck's evolutionary theses and had affirmed that all species are specially created in a single, primary act of creation, that of Genesis. Since he also recognized the reality of extinction, he necessarily supposed that the actual number of species is declining over time.

Cuvier did important work in assessing fossils and strata in the Paris basin, but he drew erroneous conclusions from the discontinuities in both the sedimentary sequences and the fossil record. He did not recognize the basic stratigraphic principle with which William Smith (1769–1839) can be credited. (Stratigraphy is the branch of geology that concerns itself with analyzing the sequence of sedimentary strata.) Smith was the founder of English stratigraphy, and through his stratigraphical map of England, he established the principle that strata are complete in no one area; sediments have been both laid down and eroded over widely dispersed areas, with the result that a complete stratigraphical column—a complete sequence of all actual strata in their chronological order—can be obtained only by collating strata from different regions. This act of comparison and collation was the main business of practical geology from the time of Hutton and of the stratigraphical pioneer Abraham Werner (1750–1815) through the time of Adam Sedgwick (1785–1873), the Cambridge stratigrapher who gave Darwin his first lessons in practical geology (and who later, incidentally, violently opposed his theory of evolution on religious grounds). Since Cuvier's formative period came a little too early for him to assimilate Smith's principle, he was both a geological and a biological catastrophist. He accounted for discontinuities in the fossil record by supposing that a series of catastrophic floods or sudden ice invasions had eliminated the biota in any given area multiple times. He accounted for the different fossils in more recent strata by supposing that after each catastrophe somewhat different (but already existing) plants and animals had migrated in from neighboring regions. (In the most extravagant of all theories of special creation, Cuvier's catastrophist disciple Louis Agassiz, among others, supposed that whole new biotas—new assemblies of plants and animals—had been specially created after each successive catastrophe.)

Both Lamarck and Cuvier perceived certain aspects of the species problem correctly, but neither formulated a plausible total theory. Cuvier believed that some species represented in fossils had become extinct, and Lamarck believed they had evolved into currently living forms. Thus far each was at least partly correct (not all extinct species have evolved into living species), and their views were complementary, but other aspects of their views prevented the complementarity from being perceived as part of the whole, larger picture. Lamarck denied the possibility of extinction (except in rare cases of human depredation), did not adequately recognize basic differences of body plan among distinct groups of animals, and provided no plausible mechanism for evolutionary development. Through his paleontological researches into the extinct fauna of the Paris basin, and especially the large extinct mammals such as mammoths, Cuvier decisively established the reality of extinction, but he rejected Lamarck's correct contention that species could change over time.

Lyell rejected Cuvier's view of geological history as a series of catastrophic floods and sudden massive invasions of ice, but he also reacted with alarm to Lamarck's evolutionary speculations. These speculations seemed to conflict with his uniformitarian convictions that the current state of the earth has been the result of past actions very similar to the actions we see occurring around us at the present time. An even more important objection was that the speculations implied continuity between human beings and the primates (specifically, "orang-outangs"). In reaction to Lamarck, Lyell sought to modify the idea of special creation in such a way that he could acknowledge the reality of extinction and still integrate special creation with his own uniformitarian geology. Lyell hypothesized that only a small proportion of the biota existing at any given geological period becomes extinct and that the extinct species are replaced with species that have been newly created. Replacements of the biota would thus be slow, gradual, and continuous. Lyell suggested that species went extinct because of a failure to adapt to changing environmental conditions, but he offered no causal mechanism for the introduction of new species. Like Cuvier, and in stark conflict with the paleontological evidence, he denied any "progressive" character—any increase in morphological complexity—in the stratigraphic column. He suggested instead that newly introduced species always replaced, with some slight differences, other species within the same general class. Certain species of birds or fish, say, would become extinct, and new but not too dissimilar species of birds and fish would be created to replace them. As a species theorist, then, what Lyell mainly offered Darwin, apart from his exposition and critique of Lamarck, was a set of puzzles and perplexities that it became the chief occupation of Darwin's life to solve.

Darwin himself was always generous in his appreciation of Lyell's achievements and open in his avowals of how much his own work had benefited from Lyell's influence. In the *Origin*, Darwin speaks of the *Principles of Geology* as a work that "the future historian will recognise as having produced a revolution in natural science" (*OS*, 271). For the purpose of understanding the nature of the Darwinian revolution, and of scientific revolutions generally, it is instructive to compare the relation of two great revolutionaries to their chief predecessors: Lyell's relation to Hutton, and Darwin's relation to Lyell. At the time Lyell produced the first edition of *Principles of Geology*, Hutton was largely in eclipse. His views on the importance of volcanic activity in creating land masses had received less credence than the Neptunist theory of Abraham Werner, who had hypothesized that all the sediments had been precipitated from a universal ocean. At the time of Lyell's work, the dominant, received view in geology was the somewhat different version of catastrophism—the theory of successive, relatively local floods and ice invasions—propounded by Cuvier. In this climate, given Cuvier's daunting prestige, to advocate a Huttonian view—the idea of continuous, relatively slight changes in the earth's surface over incalculable immensities of geological time—required considerable boldness and originality. Nonetheless, within just a few years, Lyell had established his new, Huttonian synthesis as the dominant, mainstream view. Toulmin and Goodfield lucidly characterize Lyell's historical position in relation to his predecessor.

> Lyell's position differed from Hutton's in only two serious respects. Firstly, he sets less store on the providential character of geological change.... Secondly, where Hutton's account of geological development had inevitably been only schematic, his own could be elaborate and detailed. The intervening forty years had left their mark.... [Lyell] had at his disposal a much larger and more varied range of examples, and the range of mechanisms he could illustrate and establish was correspondingly larger and more varied. Instead of the earlier crude opposition between fire and water, he could demonstrate the geological effects of a dozen different agencies, acting either in combination or against one another; and it was the marginal balance between all these agencies, at any one place and time, which determined whether the Earth's crust was being built up or worn down at that point.[10]

Lyell established uniformitarianism as a historical phenomenon and as a methodology. The method is that of reasoning from present causes to past events, and the factual presupposition that justifies the method is the idea that geological change results from small natural changes working over vast periods of time. (Historians of biology designate the factual presupposition as "actualism.") Both the fact and the method presupposed a time scale greater by orders of magnitude than that of any previous human imagining. Hutton must be accorded credit for the original conception of deep geological time, but Lyell gave it definitive confirmation and made it fully available to the empirical imagination.

By integrating empirical information within a Huttonian theoretical framework, Lyell established geology firmly as a science. He thus brought to a close the long phase of fanciful speculation in geology, a phase that included all ancient myths—preliterate, classical, and biblical—and also the quasi-mythic cosmogonic speculations of theorists such as Burnet, Buffon, and Werner. (Thomas Burnet [1635–1750] was a clergyman and author of *Sacred Theory of the Earth*, an extravagant exercise in geological fantasy.) Cuvier had made major advances in paleontology, and Lyell assimilated those to his system, thus correcting Hutton's failure to take adequate account of the fossil record, but Cuvier's catastrophism also constituted the last major flutter of the old speculative fancy in the construction of stories about the earth. Lyell himself acutely diagnoses this fancy as an inevitable imaginative consequence of a radically foreshortened time scale in which to compress the titanic transformations in the earth's crust.

Lyell was overwhelmingly the single most important influence on Darwin's work. Through his uniformitarianism—his vision of change as the consequence of small natural changes working over vast periods of time—Lyell provided the basis for Darwin's formulation of a scientifically correct theory of the development of life on earth. Darwin began his career on the *Beagle* voyage at least as much a geologist as a biologist. His first mature work of scientific discovery—what Ghiselin rightly calls his first great synthesis—was his theory of coral reefs. This theory deploys a chief principle of Lyell's geology—the perpetual rising and falling, the slow undulation, of the earth's crust. On the basis of this theory, Darwin corrected Lyell's own erroneous theory about the formation of coral reefs. (Lyell hypothesized that they grew upward from stable undersea mountains; Darwin correctly surmised that most of them grew at the edge of mountains that were gradually sinking beneath the surface of the ocean, and that others grew at the fringes of land masses undergoing elevation.) By revising Lyell's specific theory about coral reefs, Darwin solved a geological puzzle, and he thus also brilliantly confirmed Lyell's larger principle.

Lyell and Darwin may be envisioned as a triumphal scientific succession—one great monarch succeeding another. Lyell definitively shifted geology from the realm of fanciful speculation to that of science, and Darwin did the same for biology. Among the fanciful speculations that Darwin replaced, we may count Lyell's own theory about the extinction and succession of species. In one respect, Darwin's achievement looms larger than that of Lyell. Though a geologist of genius, Lyell was not a theoretical discoverer of the first magnitude. His work was that of synthesis and integration. He adopted Hutton's basic scheme and used it to assimilate the more recent work in vulcanism, paleontology, and stratigraphy. Darwin both discovered the basic theory of descent with modification by means of natural selection and also produced the synthesis of empirical disciplines that confirmed it. If we compare *The Origin of Species* with the *Principles of Geology*, we can still greatly admire Lyell's achievement, and we can perceive its vital importance to Darwin, but in comparison we can also appreciate all the more fully the singular, world-historical character of Darwin's book—the magnitude of its scope and the depth of its significance, its originality, the grandeur of its design, the intricate unity of its argument, and the sustained, symphonic power of its exposition.

Darwin's Intellectual Character

The middle of the nineteenth century was the right time for the formulation of the theory of natural selection because this whole network of naturalistic research had finally produced all the elements that were necessary to it. Darwin was the right man to undertake the formulation for several reasons. He had the rare capacity for original, creative thinking about elemental realities. He had both the training and the depth of mind that were necessary to recognize the significance of his subject, and he had the ambition that drove him to seize the unique opportunity history had given him. He had the social and material conditions that made it possible for him to dedicate himself to his project. He had an extraordinary capacity for sustained, detailed, multifarious inquiry oriented to one large, synthetic aim. He had an exceptional gift for insight into the mechanisms of living things, and an equally exceptional gift for integrating all his observations and inquiries into a unified theoretical vision. As Matthew Arnold said of the Greek dramatist Sophocles, Darwin saw life steadily, and he saw it whole.

In one important respect, Darwin's virtue was negative: he was not able *not* to conceive his subject in a profoundly coherent way. He had no capacity for evasion or equivocation. His mind did not admit of getting lost in details or of becoming stymied in inconsequential implications. He did not respond to the allurements of specious inferences. Partisan bias and special pleading were wholly alien to him. He weighed counterevidence or arguments that told against his views not simply as a matter of obligation, grudgingly, or as a strategy of argument. He cared for the full weight of an argument. As Henry James might put it, he wanted its full value, and he understood instinctively, as a part of his intellectual personality, that the weight of an argument consists of the conclusions that emerge from the combined force of all the evidence and all the reasons that can be brought to bear on a subject.

Darwin's integrity enters crucially into the standing of the *Origin* as both a scientific and a literary classic. Michael Ghiselin has registered the importance of intellectual integrity as a source of strength in the construction of Darwin's theories. "Darwin's success may readily be explained by a very simple hypothesis which seems not to have occurred to his critics: he thought. He reasoned systematically, imaginatively, and rigorously, and he criticized his own ideas."[11] Such qualities were apparent also to the most astute among Darwin's contemporaries. Speaking on the occasion of Darwin's death, T. H. Huxley, one of the most effective public proponents of Darwin's views, described "a certain intense and almost passionate honesty by which all his thoughts and actions were irradiated, as by a central fire."[12] Huxley explains how this "rarest and greatest of endowments" worked in Darwin both as a productive force and as a disciplinary constraint. As a productive force, it led him to undertake "prodigious labours of original investigation and of reading," and it drove him "to obtain clear and distinct ideas upon every topic with which he occupied himself." As a disciplinary constraint, it "kept his vivid imagination and great speculative powers within due bounds" and "made him accept criticisms and suggestions from anybody and everybody." One of the people from whom Darwin most eagerly sought criticism was his closest personal friend, the eminent botanist Joseph Hooker. In his review of the *Origin*, Hooker describes Darwin's integrity not merely as a feature of intellect but also as a social quality. It is an index of civilization that reveals itself in the tone and manner of Darwin's writing. "Whatever may be thought of Mr. Darwin's ultimate conclusions, it cannot be denied that it would be difficult in the whole range of the literature of science to find a book so exclusively devoted to the development of theoretical inquiries, which at the same time is throughout so full of conscientious care, so fair in argument, and so considerate in tone."[13]

Much has been made of Darwin's supposed dullness, exemplified, it is thought, by his respectable but undistinguished performance as a student. In his *Autobiography*, Darwin himself contributed to this tradition. He assessed his own abilities with unfeigned modesty and with the dispassionate weighing of pros and cons that had long since become the governing habit of his mind. He concluded, "With such moderate abilities as I possess, it is truly surprising that thus I should have influenced to a considerable extent the beliefs of scientific men on some important points." The surprise is perhaps justifiable on the grounds, as he himself explained, that he had not the "quickness of apprehension or wit" of a sort that distinguished Huxley. He acknowledged further that his memory was "extensive, yet hazy," but since his scholarly methods were highly organized, memory was not a significant handicap. When he needed information, he could "generally recollect where to search." In counterweight to his deficiencies or limitations, Darwin credited himself with methodical habits of observation and with an instinct of reasoning. "From my early youth I have had the strongest desire to understand or explain what I observed,—that is, to group all facts under some general laws." Darwin's mind readily generated hypotheses—this is what we mean by creativity in science—but Darwin made a conscientious and effective effort to submit all hypotheses to dispassionate scrutiny. "I have steadily endeavoured to keep my mind free, so as to give up any hypothesis, however much beloved (and I cannot resist forming one on every subject), as soon as facts are shown to be opposed to it."[14]

Darwin's combination of inventive fertility and self-critical rigor is singular, and singularly efficacious, particularly when it is combined with the power of sustained inquiry—"the patience to reflect or ponder for any number of years over any unexplained problem."[15] Although he puts the case modestly and fairly, such qualities as Darwin ascribes to himself actually contain the whole organon of the scientific ethos. Darwin could succeed as an independent and original thinker of the first magnitude because he encompassed within his own method and character all the necessary phases or aspects of generating and testing hypotheses that are normally distributed throughout the scientific community and that constitute a long-range institutional process.

Biographers not infrequently speak of Darwin's easy circumstances with a measure of resentment or disdain, as if he was somehow cheating or taking undue advantage of an unearned and illegitimate privilege deriving from an unjust distribution of wealth. In *The Descent of Man,* Darwin himself observes that the existence of a leisure class is an absolute prerequisite to the achievements of high civilization. "The presence of a body of well-instructed men, who have not to labour for their daily bread, is important to a degree that cannot be over-estimated; as all high intellectual work is carried on by them, and on such work material progress of all kinds mainly depends."[16] In our own time, such work is done by a professional class trained and commissioned for it. Darwin himself notes that in his own class many people no doubt made no very good use of their privileged circumstances. But resentment is an appropriate response to privilege only if the opportunities are wasted. Darwin did not waste them.

Darwin's supposed dullness as a student needs to be assessed with some care. His dispositions were not toward classical scholarship and language study but toward natural science. He was thus never a prize student, but whole generations of prize students can now be recollected only by digging deep into the decaying documents that register forgotten names. A few prize students have presumably been genuinely animated by the conventional curriculum of the Greek and Roman classics; they were fortunate in that this field just happened to answer to their real talents and interests. Probably most prize students, though, in achieving prizes, have given evidence less of inherent interest in philology than of social ambition and a willingness to accept conventional guidelines of activity. As a boy, Darwin pursued natural science purely as a hobby, and because it was a hobby he and others could not help but regard it, with some mild disapproval, as a form of dissipation, as a leisure pursuit to be indulged a little guiltily as a distraction from the serious work of deciphering the same standard texts in classical languages that many generations of students had already deciphered. In some biographical accounts, too little is made of the fact that to pursue natural science at all, with delighted if guilty eagerness, gave important evidence of spontaneous curiosity and intellectual animation.

As a boy and a young man, Darwin read widely in the course of liberal studies that was common for educated gentlemen of his age. And again, too little is often made of this phase of Darwin's education. Independently and for purely personal pleasure, he read the major English poets. Even when just a boy, he read Shakespeare with rapt and absorbed attention. On the *Beagle*, when he went ashore, the one book he took for pleasure, when he could take only one book, was *Paradise Lost*.[17] Though he was not keen on the minutiae of language study, and though his own genius was oriented mainly to science and not to literature, Darwin had a large general intelligence that responded with spontaneous delight to the finest artistic language the English literary tradition could supply. Such responsive aptitudes are not so common as we might suppose, particularly if our sense of the norm for literary intelligence is derived almost exclusively from reading biographies of novelists and poets.

In assessing Darwin's academic career, we would do well to recall other men who were regarded, in their student days, as rather dull dogs, but who consoled themselves for their mediocrity in conventional academic performance by reading widely and with absorbed delight in the liberal arts and by pursuing artistic or intellectual activities of their own devising. By answering to this description, Darwin joins a company that includes, among others, the Duke of Wellington and Winston Churchill. It would perhaps be too much to say that success in conventional academic pursuits in one's youth is a certain sign of ultimate mediocrity, but such success is by no means incompatible with mediocrity, and the reverse proposition—that mediocrity in school work is an unpassable barrier to originality and to greatness of achievement—is quite certainly false.

The one scholar who has most adequately grasped the nature of Darwin's intellectual character is Michael Ghiselin. Carefully weighing the evidence of Darwin's aptitudes in various areas, he concludes, "By the conventional indices, his intelligence quotient would probably indicate intellectual superiority, but not genius. Yet such standards can have little meaning in judging a unique individual with such unusual talents. Whatever may have been Darwin's intellectual resources, he used them with almost superhuman effectiveness." Balancing off raw IQ against intellectual character, Ghiselin suggests that "perhaps we should attribute his accomplishment less to intelligence than to wisdom." By the word "wisdom," Ghiselin means something more than and rather different from moral and emotional judiciousness, though Darwin also possessed these qualities in abundance. He means the whole array of characteristics that enable a scientist to get at the truth of his subject. Darwin's curiosity and inventiveness, his caution and circumspection, his ambition, objectivity, and patient determination all play a part in his success, and as Ghiselin rightly observes, "in the final analysis, the real criterion of greatness in such matters is success."[18] Had he not possessed a mind of a truly extraordinary quality, he could not have succeeded as he did.

The nature of Darwin's success has puzzled some scholars in part because each of his mental characteristics, though admirable, is not in itself unique or even extraordinary. Insofar as Darwin's achievement depends on the quality of his own mind, what accounts for his greatness is the way all of his mental characteristics enter into combination. The combination of Darwin's characteristics was truly exceptional, and the evidence for the exceptional nature of this combination is that it enabled him to grapple effectively with the problem with which history presented him. Ghiselin concisely summarizes Darwin's career as a theoretical biologist. "On seeing that there was evidence for evolution, in spite of what others had concluded, he had the courage and ability to seek out and to discover its mechanism. Grasping the potentialities of his discovery, he had the audacity to develop a comprehensive system of biological ideas on a scale which has scarcely been appreciated." Courage, audacity, and sustained constructive energy are all virtues on a heroic scale, but in Darwin's case they were made effective by being placed under the command of a characteristic that seems rather quiet, mild, and modest. Of all the characteristics that contribute to Darwin's achievement, Ghiselin and Darwin concur in believing that the one most important characteristic was a simple matter of disposition or preference—the disposition "to prefer having an opinion which is true."[19] As simple and even modest as such a disposition might seem, Huxley is right in designating it as the "rarest and greatest of endowments."

The Lamarckian and Spencerian Alternative to Darwinism

The Chevalier de Lamarck (1744–1829) has received more respectful attention from his modern commentators than he received from most of his contemporaries. Mayr notes that Lamarck was the first scientist to propose a consistent theory of gradual evolution, and Simpson argues that Lamarck is historically important because he explicitly described evolution as "a general fact embracing every form of life in a single historical process."[20] In his own day, as an evolutionary theorist, Lamarck had little standing among reputable biologists. He was overshadowed and overborne by the great Cuvier, who proved, contrary to Lamarck's own views, that extinction was a reality of the paleontological record. Lamarck envisioned a progressive transformation of species that began with the spontaneous generation of simple microorganisms. Over evolutionary time, driven by an internal need for a complexification of structure leading ultimately to the perfected human form, these microorganisms gradually moved up the scale of nature. Along the way in this progression, species were deflected and a little distorted by being compelled to adapt to specific environmental conditions, but their finely graded variations of structure nonetheless ultimately constituted the unbroken links in a temporalized Great Chain of Being.

The quasi-Lamarckian intimations of evolutionary development in the work of Darwin's grandfather, Erasmus Darwin (particularly in *Zoonomia*) were only slight and undeveloped poetic fancies and had little or no impact on subsequent biological theory. In contrast, Darwin's near contemporary Robert Chambers made a great popular sensation with the publication of his quasi-Lamarckian theory of evolution in *Vestiges of Creation* (1844), but Chambers was a journalist, not a serious scientist, and his fanciful speculations drew little sympathetic scientific attention. Darwin and other commentators have sometimes suggested that Chambers prepared the public mind to be more favorably receptive to Darwin's evolutionary theory, but as Huxley's violently hostile reaction to Chambers suggests,

Chambers might have done the cause of evolutionary theory more harm than good by casting it into the range of fantastic pseudoscience. Lamarck had been little known in England until Lyell gave an exposition of his views in the second volume of *Principles of Geology* (1832), and Lyell had expounded these views only for the purpose of repudiating them. When Darwin himself took up the cause of evolution, he was consistently eager to distance himself from Lamarck and emphatic in his expressions of disdain.[21] In the *Origin*, Darwin seldom cites Lamarck as a source of valid observation, and he does not even take him as a primary foil or alternative. The main polemical foil against which Darwin constructs his own positive arguments is that of special creation.

Lamarck's theory of evolution is progressive and teleological. That is, like almost all theorists of historical development in the nineteenth century, Lamarck believed that historical change was a form of improvement and that these improvements were directed toward some ultimate goal. In Lamarck's theory, as in most such theories, the movement is animated by some internal dynamic, but this internal dynamic is simply the mechanism for the realization of an essentially providential design instituted by a beneficent deity. Familiar versions of teleological progressivism include widely divergent ideological constructions: Hegel's absolute idealism—the idea of a World Spirit manifesting itself in the dialectical progression of culture and particularly well disposed to the Prussian State; Marx's dialectical materialism, a theory of class-based social interactions leading inevitably to the egalitarian utopia of communism; the utilitarian utopianism of Comte and St. Simon, in which all culture progresses reliably through the stages of supernatural and metaphysical development finally to come to poise in the "scientific" humanitarianism of positivism; and the schemes of various British Victorian constructors of cultural theories, including Carlyle, Mill, and Arnold.[22]

Darwin's theory of natural selection has so successfully eliminated teleology from the pool of common metaphysical ideas that many casual modern commentators forget that the most fundamental element of Lamarck's evolutionism was orthogenic progressivism—that is, an innate tendency to development along some "straight line" directed toward a determinate end. The inheritance of acquired characteristics was for Lamarck a secondary or subsidiary mechanism. Darwin himself conceded some limited scope to this latter principle, and progressively more in later editions of the *Origin* as he sought to hedge his bets against criticisms based on problems in the theory of inheritance and the extent of geological time. (See the section titled "The Nature of the Darwinian Revolution" later in this chapter for comments on Jenkin's critique of blending inheritance and on Kelvin's arguments about the age of the earth.) The idea of the inheritance of acquired characteristics was largely disconfirmed by the protogeneticist August Weismann in the late nineteenth century and was dealt decisive blows by the consolidation of the Modern Synthesis and by the discovery of DNA, but it appears as at least a minor issue as late as 1982.[23]

When Darwin scoffs contemptuously at Lamarck, he is not attacking the one idea with which Lamarck is now most familiarly associated—the inheritance of acquired characteristics. He is attacking two things. One is a subsidiary mechanism for the inheritance of acquired characteristics, the mechanism of "willing" (Darwin notes this could hardly apply to plants); he replaces this with simple "use and disuse," as in the loss of sight by moles or cave-dwelling animals. The other and more important object of Darwin's scorn is the idea of an inherent tendency to progress. Lamarck's theory is heavily inflected by the spirit of theodicy—that is, the effort to explain away or rationalize the existence of evil and thus to reconcile it with the existence of an omnipotent and benign deity. It is a biologized version of the Leibnitzian idea that our world is the best of all possible worlds. (Gottfried Wilhelm Leibnitz [1646–1716] was a German polymath and optimist philosopher.) As in the case of Paley's natural theology, the need to affirm the benevolence of Providence is a central underlying motive and a primary regulative principle for the formulation of the theory. In that respect, it is not a scientific theory but a theological theory. In fundamental and decisive contrast, Darwin's theory is scientific both in motive and in character. Its motive is to provide a causal explanation that makes the best sense of the total body of available evidence, and its character is mechanistic. Darwin recognizes various metaphysical implications of his theory, with their attendant emotional sensations, but in seeking confirmation of the theory, he appeals not to the consonance of the theory with a preconceived metaphysical order but rather to the explanatory adequacy of the causal mechanism he has identified. These differences of orientation have a correlative in method and manner. Darwin was given over heart and soul to the scientific method, and he responded with almost instinctive disgust to the general slackness of argument in Lamarck and to the license he gave to his unconstrained speculative fancy.

Lyell gives a vivid and compelling summary of Lamarck's thesis, and on the descriptive level this thesis has a lot to say for itself—it takes in the progressive development of species and the lability of the species form. In contrast, Lyell's own views on this issue are relatively weak; they look like desperate countermeasures, and they were never widely accepted. In expounding Lamarck's theory, rather than approaching the problem, as Lamarck does, as a set of theological propositions, Lyell concentrates on the biological problem of the instability of species. As a result, Lyell makes Lamarck's theory seem more attractive and plausible than Lamarck himself makes it seem. He does not make it plausible to Darwin, but he does provide a point of entry into Lamarck's general vision for the second most prominent English evolutionist of the nineteenth century—Herbert Spencer (1820-1903). As Spencer notes in his *Autobiography*, he first became acquainted with Lamarck, when he was only about twenty years old, through Lyell's exposition.[24] Despite Lyell's rejection of Lamarck, Spencer himself found the ideas immediately attractive. He absorbed them into the innermost fiber of his intellectual life, and from that time forward he never deviated in his devotion to them.

Spencer was primarily a social philosopher, and his first book, the *Social Statics* (1851), is an exercise in integrating Lamarckian teleological progressivism, utilitarian ethical theory, and an extreme form of libertarian individualism associated with laissez-faire economics. Excerpts from this work were included in my contextualized edition of Darwin's *Origin* because they provide particularly vivid and virulent instances of the social ideology that is commonly mislabeled "social Darwinism." As the date of publication for *Social Statics* makes clear, this ideology was formulated before the publication of *The Origin of Species* in 1859; the ideology was in no way influenced by Darwin's own work. Spencer envisions an ultimate utopian social order that will be achieved by gradually eliminating social undesirables and perfectly synchronizing the symbiotic interactions in a population of maximally efficient egoists. In rough parallel to the idea of natural selection, he envisions the gradual perfecting of the human type through the elimination of relatively unsuccessful human organisms and the transmission of acquired improvements, by Lamarckian inheritance, in the offspring of the strong and successful members of the population. That is, in contrast to Darwin's view, Spencer's view is that the members of a species adjust to their circumstances with varying degrees of success. These adjustments bring about structural modifications in their constitution, and the more successfully "adapted" pass on their improved constitutions to their offspring. Darwin allows for Lamarckian adaptation only as a minor, subsidiary process. For him, adaptation occurs not through behavioral changes within a single life cycle but rather through random variation and the differential survival of offspring over many generations.

The evolutionary theory propounded in *Social Statics* does not present the evolutionary process as a mechanism for the transmutation of species but rather as a means for the perfecting of the latent ideal form within a species, particularly the human species. In this important respect, Spencer is still presupposing the Aristotelian concept of species as an "archetype," that is, an ideal form with an unchanging essence. Spencer has used Lamarck as a means of modifying this Aristotelian concept, but only in such a way as to extend the concept over evolutionary time. That is, the ideal form of the species gradually fulfills or realizes itself not in a single generation but in the course of many generations.

The moral and social theories propounded in Darwin's *Descent of Man*—his one substantial essay in the field of evolutionary psychology and social ideology—are very different from the "social Spencerism" of *Social Statics*. Darwin's moral psychology is founded on the principle not of egoistic competition among isolable units in a social group but on the principle of evolved social sympathy. Spencer has almost unconsciously incorporated into his social psychology an idea derived from utilitarian economics and utilitarian ethics: that humans are in origin and essence nonsocial, that they are self-contained units designed to maximize individual pleasure, and that they incorporate into social groups only as a matter of convenience or necessity. Darwin, in contrast, with his far greater intuitive penetration into human nature, perceives that human beings are social animals and that their whole motivational and emotional organization is geared toward interdependent interaction with other humans. The evolved basis for that interdependence is social sympathy. Accordingly, Darwin's own vision of a utopian perfecting of the human social order consists not in maximizing the egoistic efficiency of individuals but in a gradual expansion of social sympathy so that it includes first all other human beings, of all nations and races, and then finally all living things.[25] It is still a utopia—Darwin was to that extent bound within the ideological constraints of his age—but it is a utopia of enlightened humanitarian ecologists, not of finely honed utilitarian egoists.

Darwin and Spencer run parallel courses, with neither having any substantial influence on the other. Spencer formulated his main ideas and wrote some of his foundational works while Darwin was still a relatively obscure naturalist working in specialized areas such as the geology of coral reefs and the classification of barnacles. Darwin formulated his own core theory in 1838, when Spencer was only eighteen and Darwin had never heard of him. After his first foray into social ideology, Spencer developed a much larger, more grandiose theory of evolution on a cosmic scale. This general theory of cosmic evolution first appeared, in nucleus, in an essay of 1852 entitled "The Development Hypothesis" and was then given an elaborate, full-dress formulation in his definitive philosophical work, the *First Principles* (1862). The cosmic theory depends on intrinsic formal processes abstracted from any specific field of action; it is basically a theory that uses abstract terms to describe a process of increasing organizational complexity—the simultaneous proliferation of smaller units of organization and their incorporation into ever larger systemic units. This is a descriptive pattern that Spencer mistakenly regarded as a causal mechanism and hence as a form of explanation. He could make this mistake because he presupposed an intrinsic principle of progress as a first principle and needed only to deduce, as he believed, the logical order through which that principle would necessarily articulate itself. It is to this order of problem that Darwin refers when he complains that what Spencer trades in are not explanations but only "definitions."[26]

After formulating his grand scheme of cosmic evolution, Spencer dedicated the rest of his life to using it as a pattern within which to organize every field of knowledge. He wrote books giving what he and many of his admirers took to be definitive formulations of all the knowledge that could possibly be contained within the fields of astronomy, geology, biology, sociology, psychology, and ethics. Each field was passed through the abstract formula of complexification—of "an advance from a diffused, indeterminate, and uniform distribution of Matter, to a concentrated, determinate, and multiform distribution of it," that is, "from a confused simplicity to an orderly complexity."[27] In passing through biology, or passing biology through this filter of preconceived ideas, Spencer pauses long enough to incorporate, as he believes, Darwin's theory of natural selection, but he never so much as glimpses the way in which Darwin's theory actually supplants and cancels his own merely formal exposition.[28]

In his *Autobiography*, Darwin declares, rightly, that while he was sometimes impressed with Spencer's apparent brilliance, he never derived much of value from him in the way of scientific propositions.[29] It nonetheless remains the case that Spencer is responsible for coining the one phrase, "the survival of the fittest," which is most often used as a kind of shorthand code phrase for the theory of natural selection. This phrase was first used by Spencer in *The Principles of Biology* in 1864. On Wallace's advice, Darwin adopted it in the fifth edition of the *Origin* in 1869. The advice was ill-considered, both in the giving and the receiving of it. From that one phrase has emerged a persistent pseudo-issue in the philosophical critique of Darwinism. It has been a source of unnecessary, purely semantic confusion. The argument runs thus: if fitness is defined by survival, "the survival of the fittest" means only that survivors survive. The phrase offers a good instance of the way in which the "definitions" that were Spencer's stock-in-trade incline toward "tautologies." But the putative problem is not in the concept the phrase is meant to encapsulate but in the phrase itself. Darwin's own formulations of the idea of natural selection have nothing tautological about them. Organisms vary in the characteristics that enable them to survive and reproduce; such variations are heritable; and the differential transmission of heritable variations leads over many generations to fundamental changes in adaptive structures, and hence, eventually, to speciation. Darwin adopted Spencer's phrase only on the tacit understanding that it would serve as a shorthand term implying all the content in his own concept of natural selection, but if one takes Spencer's phrase at face value, it strips out the elements of heritable variations and differential reproductive success. In order to avoid giving occasion for confusion, it is probably a good idea simply to avoid using the phrase.

Spencer was the most promising of Lamarck's offspring, but however splendidly he flourished in his own generation, Spencer's lineage long since faded into obscurity and has now sunk into extinction. In this respect, Spencer and Lamarck are to modern Darwinism what Neanderthals were to Cro-Magnons, not ancestors in a direct line of descent, but separate species running parallel to one another and (in all likelihood) interbreeding little or not at all. The Neanderthals survived for hundreds of thousands of years in Ice Age Europe and the Levant, but they coexisted with Cro-Magnons for only a few thousand years. Between forty and twenty-seven thousand years ago, as the ice retreated and the Cro-Magnons migrated in from the South, the Neanderthals disappeared from the earth, either directly exterminated or simply pushed out of viable habitats by the better equipped and more highly organized invaders who replaced them. The many volumes of Spencer's encyclopedia of universal knowledge are like the skeletal remnants of an extinct people, kept in cabinets as objects of antiquarian curiosity, a little dusty and strange, icons of an evolutionary dead end, and thus melancholy mementos of an ultimate failure and futility.

The Inception and Gestation of Darwin's Theory

The Place of Origin of Species *in Darwin's Career*

Before looking more closely into the development of Darwin's theory, I shall sketch out the familiar story of Darwin's career. Darwin was the son of a wealthy country doctor and the grandson of an Enlightenment scientist and poet—Erasmus Darwin. After an abortive effort at attending medical school in Edinburgh, he studied at Cambridge with the intention of taking orders and entering the Church. His appointment as unofficial naturalist aboard the *Beagle* rescued him from his clerical destiny and enabled him to find his vocation as a serious student of geology and natural history. He sent home large and valuable collections of flora, fauna, and fossils, along with letters containing scientific observations, and when he returned to England, he was met with scientific acclaim and welcomed warmly into the community of practicing geologists and naturalists.

Darwin did not discover natural selection or evolution while on his voyage, but as he himself observes in the first paragraph of the *Origin*, the biogeographical and paleontological observations he made on the voyage were the primary stimulus for the development of his theory. Shortly after returning to England, he began reading and meditating on the species question and jotting down his reflections in a series of notebooks. As he explains in his autobiography, the catalytic event in the formulation of his theory was his reading of Thomas Malthus' *Essay on Population*.[30] Malthus' mathematical conception of the way birth rates inevitably exceed the food supply—what we might call the carrying capacity of the environment—formed the final, essential link in the chain of reasoning that constituted Darwin's theory. Darwin "discovered" or formulated the theory of descent with modification by means of natural selection in 1838. In 1842, he wrote a sketch of the theory at about the length of a standard scholarly article. The sequence of topics in this sketch was essentially the same as that which he used for the sequence of chapters in the *Origin*. In 1844, he expanded this sketch into a book-length manuscript, which he set aside with instructions that in case of his death his wife should find someone to edit and publish it.

From 1844 until 1858, Darwin worked on a variety of projects. For the first few years after his return, he was occupied mainly with the materials from the voyage. In addition to the travel narrative itself, he published commentaries on the collections from his voyage and important works of geological inquiry that had resulted from his observations on the voyage. From 1846 to 1854, Darwin devoted himself to mastering the classification and anatomical structure of both the living and extinct species of a single class of animals, that of the cirripedes or barnacles. The eight years Darwin spent on the study of barnacles made important contributions to several developing fields of inquiry—to systematics, paleontology, embryology, and comparative anatomy—and it gave Darwin himself a firm professional grounding in all these areas. Moreover, by exploring the intricate variations on hermaphroditism and sexual polarity among related species of barnacles, Darwin opened an entirely new field of inquiry into the evolution of sex. During all this time, he never ceased collecting information, conducting experiments, and reflecting on the origin of species. At various points along the way, he confided his ideas to a few close associates—to his friend and botanical colleague Hooker, to his geological mentor Lyell, to his young admirer and anatomical colleague Huxley, and to his main American correspondent, the botanist Asa Gray.

Having finally completed his exhaustive study of cirripedes, in 1856, at Lyell's urging Darwin finally began writing his big book on species. He planned the work on such a massive scale, so extravagant in its detail and so circumspect in its consideration of sources and facts that it seems to have been intended to forestall and overwhelm all conceivable objection. In the *Origin* itself, Darwin often speaks with regret of having to pass over the extensive catalogues of facts that he promises to make available at some future time. However dense and concise the *Origin* itself might be, it is still quite a long and hefty book, and I think it safe to assume that most readers do not share Darwin's regret at not being able to linger over the massive documentation he has had to pass by.

Darwin had completed several hundred pages of his "big species book" when, in June of 1858, he was suddenly given a rude shock. Alfred Russel Wallace (1823–1913), fourteen years Darwin's junior, was one of Darwin's many scientific correspondents. Wallace was in the Malay Archipelago studying natural history much as Darwin had done in South America. While recovering from an attack of malaria, he recalled his reading of Malthus from years before, and this recollection precipitated in his mind the theory of descent with modification by means of natural selection—just as reading Malthus had precipitated the theory in Darwin's mind. (Wallace was not one of the people to whom Darwin confided his own ideas on evolution.) Wallace sent a short paper to Darwin sketching out his ideas on the subject. In the accompanying letter, with some diffidence, he asked Darwin to assess the paper, and if he saw any merit in it, to publish it.

Darwin's consternation was extreme. He had been working on the theory of natural selection for nigh on twenty years. He had already elaborated his theory at book length, had amassed huge quantities of evidence, and was in the process of producing a tome that was to have been simultaneously original and definitive. And now he was being scooped by a young colleague who had, during a fit of malaria, had a sudden insight into the same logic that had animated all of Darwin's efforts. Darwin was determined to do nothing mean or dishonorable, but he was understandably anxious not altogether to lose the credit for priority in the discovery of his theory. He turned for advice to the two men he trusted most, Lyell and Hooker. They proposed simultaneous publication and suggested that Wallace's paper be presented, side by side with a paper by Darwin, at a meeting of the Linnean Society. This solution was acceptable to everyone concerned. Darwin's paper consisted of two separate pieces stitched together for the occasion: one a chapter on natural selection from the manuscript of 1844, and the other an excerpt from a letter of 1857 to Asa Gray in which Darwin had given a complete outline of his theory.

After the shock of receiving Wallace's paper, Darwin decided to postpone completion of his "big species book" and instead to produce a shorter, denser work, devoid of footnotes, an "abstract" of the larger project. (Darwin initially proposed to the publisher that the book be entitled "*An Abstract of an Essay on the Origin of Species and Varieties through Natural Selection*," but the editor sagely dissuaded him from so tentative and cumbersome a title.) The big species book was never taken up again, but the "abstract," ultimately titled *On the Origin of Species by Means of Natural Selection, or the Preservation of Favoured Races in the Struggle for Life*, became the definitive and original work Darwin had wished to produce. As he himself says in his *Autobiography*, "It is no doubt the chief work of my life."[31] It was published on November 24, 1859. It was an immediate success, rapidly sold out, and a second edition, with a few revisions mainly of a copyediting character, was published six weeks later, on January 7, 1860. Four other editions followed, the last in 1872, and the book gradually expanded in size, as Darwin incorporated new information and included responses to some of the criticism that had been published. The sixth edition is nearly a third again as long as the first edition.

In the remaining two decades of his life, in addition to these revised editions of the *Origin*, Darwin published a long sequence of monographs and papers on specialized topics in the field of evolutionary biology that he had himself invented—notably on variation and inheritance, botanical adaptations of many kinds, sexual dimorphism and sexual selection, human moral psychology, the anatomy of emotional expression, and the ecology of earthworms. These more particular studies partly fulfilled the promises, made repeatedly in the *Origin*, that Darwin would provide more supporting evidence on particular points in some later work. Darwin continued doing original research until the end of his life, and his later works incorporated the results both from his studies and from his own experiments in botany and ecology.

The notebooks reveal that Darwin had gained the essential insights of the *Origin* two decades before it was published, and the essays of 1842 and 1844 demonstrate that he was already at that time able to give a coherent exposition of the basic theory of descent with variation by means of natural selection. What then, if anything, did Darwin gain through waiting for fourteen years before writing the final version of his work? There were three main forms of gain: (1) vastly more detail both in apt illustration and in considered inference, (2) an extended compositional process that resulted in an extraordinary density, coherence, and clarity in the exposition; and (3) one new idea, or at least a latent idea rendered explicit and available for development. The process of composition consisted of alternating phases of expansion and condensation, of filling in details and then of abstracting and summarizing. The one new idea is described in Darwin's *Autobiography*. He explains that there was one basic problem he had not adequately formulated in 1844—the problem of "divergence" or branching speciation, as opposed to linear descent.[32]

Darwin's Discovery of Divergence

There is some uncertainty about what Darwin's discovery of divergence means and what it amounts to. As Darwin's son and editor Francis Darwin observes in an introduction and a note for the 1844 manuscript, the idea of divergence—the gradual diversification of species from a parent stock—is strongly implied in the 1844 manuscript,[33] and indeed, Darwin's notebooks contain diagrammatic sketches of branching evolution similar to that which he presents formally in the *Origin* (*OS*, 168–69). What Darwin seems to mean by his discovery of divergence is the idea of ecological niches as a source of diversification. This meaning is obliquely apparent in the passage from the *Autobiography* in which Darwin describes his moment of insight about divergence, and it is much more clearly apparent in the chapter on divergence (chapter 7) in the unfinished big species book and in the letter of 1857 to Asa Gray (part 5) that became part of Darwin's Linnean Society paper of 1858.[34]

In Darwin's own account of the development of his theory, the idea of ecological niches hit him with the force of sudden revelation, as if it were the last of that whole series of brilliant flashes of insight that fill the notebooks. The problem with Darwin's account is that the idea of ecological niches as a means of speciation is already clearly present both in the 1844 manuscript and in the second, revised edition of the *Voyage of the Beagle*, published in 1845. In the *Beagle* passage, commenting on the variation of finches in the Galapagos, Darwin makes a statement that, in retrospect, seems to contain in nucleus the whole of the theory of the *Origin*. "Seeing this gradation and diversity of structure in one small, intimately-related group of birds, one might really fancy that from an original paucity of birds in this archipelago, one species had been taken and modified for different ends."[35] In the 1844 manuscript, in a passage on the "Variation of Organic Beings in a Wild State"—immediately preceding the passage that was included in the Linnean Society paper of 1858—Darwin asks the reader to envision natural selection as a sort of magnified human breeder seeking to form a new species on a volcanic island. The ecological conditions of the island would be somewhat different from those in the original home of the species, and the species would thus need to be "adapted to new ends." Starting from this example, obviously inspired by his own Galapagos findings, Darwin proposes that we take an expansive view of the principle involved. "With time enough, such a Being [that is, the Magnified Breeder] might rationally . . . aim at almost any result." Darwin takes mistletoe as an example. "Let this imaginary Being wish, from seeing a plant growing on the decaying matter in a forest and choked by other plants, to give it power of growing on the rotten stems of trees," and from there the plant may be supposed to develop the capacity for growing on "sound wood."[36] With naturalistic verve, Darwin describes the way in which, throughout its adaptive transformations, mistletoe would have evolved in coadaptive relation with the birds and insects that help it to propagate. It was a felicitous example, and in the *Origin* Darwin continued to use mistletoe to illustrate the coadaptation of species in an ecosystem (*OS*, 96). (Glick and Kohn suggest that it was only after 1844 that Darwin came to believe that "new species could be formed without geographic isolation," but as the mistletoe example indicates, Darwin had already recognized ecological speciation in 1844, and indeed the same example appears in the sketch of 1842.)[37]

Darwin himself clearly believed that at some point after 1844 he had suddenly received a new and important inspiration about ecological diversification. In order to reconcile this belief in the novelty of his insight with the evidence of what he had already written in 1844, we shall perhaps be compelled to make a distinction between having an idea available in latent form, on the one side, and making it present as an active and conscious source for further reflection, on the other. Making it explicit gave Darwin the impetus for some actual observations and experiments. For instance, it presented to him the idea that a single bit of ground could provide a large arena for adaptation and selection (*OS*, 164–65). More importantly, Darwin seems to have had a sudden, far-reaching insight into the scope and significance of ecological diversification as a central organizing principle within the whole economy of nature.

In 1844, Darwin had been thinking of ecological diversification mainly as a means of speciation, that is, as a causal mechanism that explains the morphological transformations and diversification of a species over time. His new inspiration might well have been the realization that this temporal dimension does not exhaust the explanatory significance of the mechanism. The idea of ecological diversification associates itself in Darwin's mind with the coadaptation of species, and it thus provides a point of entry into the intricate and dynamic interactions that take place *at any given moment in time* within a total network of biotic interdependencies. These interdependencies are in turn an integral part of the total environmental situation that conditions and constrains evolutionary change. Hence Darwin's strong and repeated emphasis, in the *Origin* itself, that among the factors regulating adaptive change interaction with other species is even more important than the physical environment. By articulating the idea of ecological diversification, Darwin extended the explanatory reach and the range of supporting evidence for his theory—and not just for the theory of descent with modification but specifically for the theory of natural selection. This idea was also the source and subject for some of the greatest rhetorical moments in the *Origin*. In passages like that of the "entangled bank," the rhetoric rises as a stylistic register for the scope and depth of Darwin's imaginative vision.

A Tale of Two Manuscripts: 1844 and 1859

Apart from the one problematic conceptual development involving divergence, the chief difference produced by the long gestation of Darwin's work lies in the composition. How much of a difference is that? If Darwin had died and his widow had fulfilled his wishes, getting Hooker, say, to edit and publish the 1844 manuscript, what effect would this book have had, compared to that of the *Origin*? The basic components of the core argument are all there, as indeed they are in Wallace's short and sketchy paper, but the *Origin* works out that argument in rigorous detail, with an accumulated wealth of information that places it at the apex of biological knowledge available in Darwin's own time. The *Origin* makes vividly clear the rigorous coherence of the argument as it is carried out through the essayistic or discursive equivalent of experimentation. Darwin's implied position with respect to the reader is something like this: "Here is my idea; let us see how it stands up against all the information now available in all the relevant fields of inquiry—classification, paleontology, geographical distribution, geology, comparative anatomy, embryology, the breeding experiments of hybridizers and domestic breeders, and the study of instinct in social insects. Let us pose all the most difficult questions we can, as if we were the devil's advocate—our own most serious and probing critics—and impartially weigh the evidence. Let us see whether the theory I have advanced can provide a coherent and reasonable explanation for all this information. And particularly, let us perpetually test my hypotheses by posing them as alternatives to the theory of special creation." The result of this experiment is of course that special creation breaks down repeatedly. It is unable to account for the information Darwin presents. His own theory not only accounts for that information piecemeal, in individual cases and in each field; it demonstrates that the information in each field forms a seamless web of interlocking explanation for the information in all the other fields.

In 1838 and 1839, Darwin had sudden brilliant bursts of insight into the essential mechanism and into its "metaphysical" or psychological implications. In 1842 and 1844, he formulated the theory in a discursively coherent form. But it was not until 1857 through 1859 that he was able to marshal his theories and his facts into a vision both massive and minute, saturated at every point with concrete evidence and matured reflective analysis. Probably, if only the 1844 manuscript had been published, other scientists would have taken up the project and in fragmentary, collective efforts have pieced together the whole puzzle. The logic and reality of the case would have led inexorably, though perhaps slowly, to that result. The final outcome for science might possibly have been little different from what it has been. The outcome for literature would have been utterly different, because the world would have been deprived of one of its great masterpieces—arguably the only work of this scientific magnitude that is also fully accessible as a work of literary imagination.

Both the 1844 manuscript and the *Origin* begin with a chapter on variation under domestication. This is an analogical argument based on homely and familiar associations, especially on the popular hobby of pigeon breeding, and it is designed not as a proof but rather as an illustrative parallel for Darwin's theory of natural selection. It is designed to open the reader's imagination to the general process of variation, selection, and divergence in organic form. The core theoretical argument of the *Origin*, the argument about actual causal processes in nature, occupies three central chapters (chapters 2, 3, and 4): "Variation under Nature," "Struggle for Existence," and "Natural Selection." In the 1844 manuscript, this core argument is outlined in just a few pages in part of chapter 2, and the other topics in that chapter are not very coherently related either to the core argument or to one another. More than half of chapter 2 is devoted to the topic of hybridism, to which Darwin devotes a whole chapter (chapter 8) in the *Origin*. In the 1844 manuscript, the discussion of variation in nature forms a small part of chapter 2. In the *Origin*, this topic occupies all of the second chapter ("Variation under Nature") and all of chapter 5, "Laws of Variation." The other chapter topics in the 1844 manuscript have counterparts in the *Origin* ("Instinct," "Geographical Distribution," and the rest), but the earlier versions are far less detailed and lack the tight consecutiveness of argument—the dense interweaving of observation and inferential reasoning—that distinguish the concordant discussions in the *Origin*. The topic of "Difficulties on Theory," a whole chapter in the *Origin*, appears in the 1844 manuscript as a patchy set of comments parceled out into sections of chapters 3 and 6.

The Style of Argument in Origin of Species: *An Instance*

It might be well to compare two passages in order to provide at least one detailed example of the kind of difference that waiting for fifteen years made in the work we read today. In chapter 2 of the 1844 manuscript, Darwin discusses the distinction between varieties and species. At the end of the chapter, under the heading "Limits of Variation," he makes a speculative inference on the stabilization of species in nature:

> I repeat that we know nothing of any limit to the possible amount of variation, and therefore to the number and differences of the races, which might be produced by the natural means of

selection, so infinitely more efficient than the agency of man. Races thus produced would probably be very "true"; and if from having been adapted to different conditions of existence, they possessed different constitutions, if suddenly removed to some new station, they would perhaps be sterile and their offspring would perhaps be infertile. Such races would be indistinguishable from species. But is there any evidence that the species, which surround us on all sides, have been thus produced? This is a question which an examination of the economy of nature we might expect would answer either in the affirmative or negative.[38]

This is the very end of the central theoretical chapter, but the conclusions are conjectural and tentative. The style is abrupt and choppy. The appeal to "the economy of nature" is vague and inconclusive. The succeeding chapter, "On the Variation of Instincts and Other Mental Attributes under Domestication and in a State of Nature, [etc.]" does not take up the question left hanging at the end of chapter 2.

One could contrast the fade-out at the end of this chapter to the rhetorical climax of the magnificent "Tree of Life" image with which Darwin concludes the chapter "Natural Selection" in the *Origin*. Here I wish to make a somewhat different comparison, a comparison not only of rhetorical effect but of argumentative style. I shall quote one long passage from chapter 5 of the *Origin*, "Laws of Variation." This passage is the conclusion not to a core theoretical sequence nor even to a whole chapter. It is the conclusion to a subsection of a labeled section of a chapter. The section is labeled by being introduced with an italicized proposition that itself constitutes a positive, unqualified affirmation. "*A part developed in any species in an extraordinary degree or manner, in comparison with the same part in allied species, tends to be highly variable*" (*OS*, 189). The subsection of this labeled section is introduced by a firm development of this proposition, and the development is supported by a confident appeal to a bit of established common knowledge in natural history. "The principle included in these remarks may be extended. It is notorious that specific characters are more variable than generic" (that is, that characteristics distinguishing species vary more than characteristics distinguishing whole genera) (*OS*, 192). This particular observation does not appear in the 1844 manuscript. In the *Origin*, it helps confirm and clarify the larger, sustained vision of morphological features emerging out of the flux of minute individual differences and assuming, over almost unimaginable expanses of geological time, ever higher rank in the classificatory hierarchy. Such features first stabilize as characteristics of species and then as characteristics of genera, and on up the hierarchy through families, orders, classes, phyla, and kingdoms. As each feature assumes higher classificatory rank, it becomes the shared property of a branching, diversified array of lower-ranked forms. In the passage I am about to quote, the particular point Darwin makes concerns only the relative variability of species and genera, but the tight logic of evidence and reasoning through which he makes this point invokes the core logic of natural selection, and this one point thus becomes yet another microcosmic confirmation of the total theory:

> Finally, then, I conclude that the greater variability of specific characters, or those which distinguish species from species, than of generic characters, or those which the species possess in common;—that the frequent extreme variability of any part which is developed in a species in an extraordinary manner in comparison with the same part in its congeners; and the slight degree of variability in a part, however extraordinarily it may be developed, if it be common to a whole group of species;—that the great variability of secondary sexual characters, and the great amount of difference in these same characters between closely allied species;—that secondary sexual and ordinary specific differences are generally displayed in the same parts of the organisation,—are all principles closely connected together. All being mainly due to the species of the same group having descended from a common progenitor, from whom they have inherited much in common,—to parts which have recently and largely varied being more likely still to go on varying than parts which have long been inherited and have not varied,—to natural selection having more or less completely, according to the lapse of time, overmastered the tendency to reversion and to further variability,—to sexual selection being less rigid than ordinary selection,—and to variations in the same parts having been accumulated by natural and sexual selection, and having been thus adapted for secondary sexual, and for ordinary specific purposes. (*OS*, 194–95)

This whole long passage consists of only two sentences, but Darwin has learned all the most important lessons about syntactic parallelism and subordination that more than a century of classic English prose had sought to teach. The first half of the passage consists of a sequence of precise factual observations tied together in parallel substantive phrases by the word "that." The pivot or hinge of the passage is the main clause of the sentence: "are all principles closely connected together." The main clause of the second sentence follows immediately and answers with almost colloquial ease to the main clause of the first sentence: "All being mainly due to. . . . " And then follows a second sequence of parallel subordinate phrases, introduced by the word "to" ("due to"), providing a causal explanation for the facts adduced in the first sentence. The causal explanation is presented not as a random list of causes but rather as a tightly linked causal sequence that outlines the actual historical sequence and that also incorporates, as fine and relevant embellishment, a reflection on the subordinate and ancillary character of sexual selection—that is, the selection of characteristics adapted for the purpose not of survival but of advantage in propagation.

The sustained symphonic power of Darwin's composition in the *Origin* depends on the mastery of many such local units of argument and exposition. As it happens, few historians of science are also trained analysts of rhetoric and composition, and few literary scholars have been sufficiently receptive to Darwin's subject matter to give adequate attention to the rhetorical and literary characteristics of the *Origin*. One consequence of this gap between what C. P. Snow called "the two cultures" is that the splendid literary quality of the *Origin* has never received its due meed of praise. A more serious consequence is that many otherwise competent readers have failed to grasp the sheer density and coherence of logical argument that is the foundation for that literary quality.

"Why the Delay?"

This one example could be replicated by many others, and it should make clear that the *Origin* represents an immense advance over the 1844 manuscript in the quality of composition, and it should be clear further that the word "composition" involves more than cosmetic or aesthetic qualities. It involves articulated argument interwoven with matured observation. Virtually every commentator on Darwin's career broaches the question, "Why the delay?" If Darwin had a book-length manuscript prepared in 1844, why did he wait another fifteen years before publishing his book? One common answer to that question is that he delayed because he was *afraid* to publish—afraid to offend the public, afraid to endanger his social and professional position, afraid even to upset his wife. In its most extreme form—as it appears for instance in the biography of Darwin by Desmond and Moore—proponents of this view attribute Darwin's severe, chronic gastrointestinal disorder to hysterical anxiety about the potential public reception of his work. This view of the case has a certain *National Enquirer* flavor of lurid headline sensationalism. *Radical Scientist Crippled by Terror of His Own Theory!* But even in its milder forms—as it appears for instance in Ruse's *The Darwinian Revolution*—this line of interpretation evinces a certain cynicism and betrays a basic deficiency in interpretive judgment. It fails to register the difference in the quality of argument between a lightly and not very coherently sketched outline, on the one side, and a dense, comprehensive, tightly woven fabric of argument on the other.[39]

The *Origin* manifests on virtually every page the results of intense study and absorbed reflection sustained over a period of nearly two decades by a scientific genius in the most robust phase of his development. Darwin could no doubt have spent two or three years polishing and refining the composition of the 1844 manuscript, and the published result would presumably have been a respectable contribution to scientific speculation. But no amount of attention devoted merely to polishing the 1844 manuscript could have produced a work even remotely so dense and thorough as the *Origin*. Darwin did not yet know enough, and had not thought enough, to produce the definitive work his theory had the potential to produce. From 1844 to 1859, the efforts that went into Darwin's studies in geology and natural history, and particularly his work on barnacles, enabled him to master entire fields of information in respect to which, in 1844, he was but a novice. In addition to his published work, over those years he collected an immense quantity of information—of facts accompanied by analytic reflection—that were slated for publication in the big species book. Though not directly cited, this information was an active force behind the momentum of argument that goes into passages like that which I quoted earlier. Some people write many books very quickly, but the speed and frequency usually result in thinness and repetition. Books like the *Origin* take time—just exactly as much time as Darwin did in fact take. There was no "delay," only a protracted preparation. Did the result justify all the time and effort that went into it? Most emphatically, it did.

Impact and Aftermath

In order to assess the value of Darwin's two decades of preparation for the public impact of his work, we can compare it with the impact made by the Linnean Society papers and the *Origin*. The papers had almost no impact; they went virtually unnoticed. In his summary of the society's activities for that year, the society's president, Thomas Bell, expressed regret that the year had not been "'marked by any of those striking discoveries which at once revolutionise, so to speak, the department of science on which they bear.'"[40] In a letter to Darwin in 1864, Wallace himself draws the appropriate inference. "As to the theory of Natural Selection itself, I shall always maintain it to be actually yours and yours only. You had worked it out in details I had never thought of, years before I had a ray of light on the subject, and my paper would never have convinced anybody or been noticed as more than an ingenious speculation, whereas your book has revolutionised the study of Natural History, and carried away captive the best men of the present age. All the merit I claim is the having been the means of inducing you to write and publish at once." The modern historian of biology David Hull seconds Wallace's opinion. "If all Darwin and Wallace had done was to publish their Linnean papers, it is very unlikely that biology would have been revolutionized. These papers were mere sketches." In the *Origin* itself, in contrast, Darwin "scanned the wide range of phenomena that his theory had to explain and showed which cases it could handle without any difficulty, which were doubtful cases, and which anomalies." The result was that he "converted a promising sketch into a scientific theory."[41]

The rhythm of composition for the *Origin* peaked in the first edition. After his two decades of preparation, there was one last phase of intensive editing. Darwin heavily rewrote the whole manuscript once it was in proofs. (Lyell's wife had read the penultimate version and complained of obscurities of expression.) The second edition, which appeared about six weeks after the first, offered only some minor editorial polishing and can reasonably be considered part of the compositional apex. After the second edition, the apex was clearly past, and in subsequent editions Darwin's further work on his manuscript became counterproductive. He became entangled with contemporary criticisms, and he undertook retrenchments and elaborated qualifications in response to scientific criticism based on inferences—about the mechanism of inheritance and geological time—now known to be erroneous. The result is a diffuse expansiveness and a slight blurring of the clear outlines of the argument. Accordingly, for my own edition of *The Origin*, I chose to use the first edition as a primary text, correcting it against only minor copyediting revisions in the second edition.

Darwin's Use of Malthus

Since Darwin's discovery predated that of Wallace by nearly two decades, one cannot precisely characterize the case as one of "simultaneous" discovery, but the timing is close enough, on a historical scale, to support the contention that the time was ripe for the discovery, that all the essential elements were in place. One of those elements, the firsthand experience of a practicing naturalist in the wild, would have been available to relatively few people at the time. The other elements were all publicly accessible and in wide possession among an educated lay public for whom natural history was a much more common and absorbing preoccupation than it is at the present time. (Many a country parson, like Mr. Farebrother in George Eliot's *Middlemarch*, had his private cabinet of prize specimens.) And yet, only Darwin and Wallace came across the idea, and they had an almost identical experience of revelation. In both cases, the crucial, crystallizing experience was that of reading Malthus' *Essay on Population*.

Why was Malthus so important? In a word: food. In our own day, in the affluent West, our main problem concerning food is that we have too much of it and thus have to make strenuous efforts of restraint and disciplined physical activity to avoid obesity. If we examine some of the contextual material assembled in the Broadview edition of the *Origin*, we shall realize how anomalous this present situation is. In passages from the Bible, Paley, Lamarck, Spencer, Malthus, Lyell, and Wallace, one is forcibly struck by the preoccupation with hunger and death converging on the question of population. Even God, at the end of the first chapter of Genesis, is concerned about food, and everyone else is preoccupied with the question of burgeoning and unfeedable masses of reproducing organisms. This preoccupation will be the more intelligible if we consider the conditions of life at the time. Famine was a regular feature of life in England itself up until the beginning of the nineteenth century. Famines recurred often on the continent into the 1830s, and in Ireland through the middle of the century. The 1840s in England are commonly referred to as "the hungry forties," and in Ireland, in the potato famines of the late 1840s, hundreds of thousands of people starved to death. In the early-middle decades of the nineteenth century, Dickens' novels register the prevalence of chronic hunger and malnutrition as a pervasive feature of life among large masses of the common people. Even well-fed people like Darwin and Wallace could not help but perceive the pressure of hunger in the population as a whole. Nonetheless, by presenting this phenomenon as an arithmetical calculation, Malthus made it vividly and dramatically apparent to them in a way simple observation had failed to do.

Darwin and Wallace both were fine and experienced observers of nature, but neither of them had actually, directly, observed the mass death that accompanies each generation. For each of them, this was an inference derived from the Malthusian calculation and transferred readily to animal populations. Darwin and Wallace both instantaneously saw how it applied across the whole animal kingdom. They simply had not registered it before: if animal populations remain stable, and the numbers born to sets of parents exceeds two, the excess must be presumed to have died. Since the numbers born do in fact regularly exceed the number of parents, the annual cycle necessarily involves a holocaust, a vast dying, as regular as the clockwork that serves Paley as a metaphor of benevolent providential design. If one factors in heritable variation and takes account of the way these variations affect survival and reproduction, the conclusion has a stunning simplicity: natural selection.

Darwin's use of Malthus presents us with an exemplary instance in the history of scientific discovery. Darwin adopted Malthus' specific arithmetical insight and incorporated it as a component of a much more complex theory. This is a classic instance in the growth of scientific ideas, and it provides us with an occasion to compare two fundamentally different views of science: (1) the social constructionist or Marxist view that any given scientific theory merely reflects the larger set of social, economic, and ideological forces at work in the scientists' world, and (2) the realist and objectivist view that science constitutes a developing knowledge of the actual world.

Malthus saw human population as a homeostatic system, that is, a system that sustained an equilibrium in numbers through an internal, self-regulating mechanism. If people become better off, they produce more offspring. If they produce more offspring, they starve, and the population remains stable. Darwin did not see evolution itself as a homeostatic system, but he affirmed Malthus' observation about the relative stability of population sizes, and he extended that observation from humans to all populations. The reproductive behavior of individual organisms has a predictable systemic effect on the population. If, as Malthus maintains, the population of a given species remains relatively stable in numbers over time, and if individual members vary in their ability to survive and propagate, and if those variations are heritable, then over time the population as a whole will change in adaptive structure. *Stability in the numbers of a population* thus becomes an integral logical component of an explanation for *change in adaptive structure in a population*. Both the stability of population numbers and the idea of a population as a self-regulating system are essential components of the whole argument. As Ghiselin observes, "Seeing in Malthus how the interaction of individuals in the same species may be affected by the intrinsic properties of each organism, and how there could be cumulative effects, Darwin and Wallace were able to conjoin all the disparate elements into a unitary system which constituted the theory of natural selection."[42] In Darwin's much larger formulation, that system includes both the element of stability in numbers, taken directly from Malthus, and also the element of adaptive change that results from the interaction of variation and differential reproductive success.

Context of Discovery and Context of Verification

In terms of the philosophy of science, this account of the relation between Darwin and Malthus is realist and objectivist in orientation. (In the modern philosophy of science, the most prominent proponent of this orientation has been Karl Popper.) In contrast, the social constructionist or Marxist conception of Darwin's theory presents that theory as an analogue to economic competition and takes Malthus as a primary inspiration for the formulation. Social constructionists treat Malthus as a source for an analogy rather than a component of a logical and empirical structure, and they typically do not directly assess the empirical validity of Darwin's theory. Instead, they cast doubt on Darwin's theory through a form of argument that we can describe as guilt by association. Social constructionists reject Malthus and capitalism on both moral and economic grounds; and they present Darwin's theory as itself a mere reflex of capitalist ideology. In this way, Darwin's theory can be presented as both scientifically arbitrary and morally retrograde. (For a prominent example of this approach, again see the biography of Darwin by Desmond and Moore.)

John Maynard Smith is both a confirmed Marxist in his political orientation and also a convinced Darwinian. Indeed, he is one of the two or three most creative and influential evolutionary theorists in the latter half of the twentieth century. His account of Darwin's relation to Malthus can serve to illustrate how a commentator can have an intelligent respect for the objective validity of science but still also reasonably and plausibly assess the way social context enters into the formation of scientific theories. He comments on the "subtle processes whereby ideas derived from a study of social relationships influence the theories of natural scientists," and he notes that "Darwin was consciously influenced by the ideas expressed by Malthus in his *Essay on Population*." Maynard Smith is by no means sympathetic to what he takes to be Malthus' own motive in writing his book—"to justify the existence of poverty among a considerable section of the population"—but he nonetheless acknowledges that Malthus' main thesis is correct: "that animal and plant species, including the human species, are capable of indefinite increase in numbers in optimal conditions."[43] Maynard Smith has thus tacitly invoked the distinction between "context of discovery" and "context of confirmation." Whatever Malthus' motives might have been, the only real question, from a scientific point of view, is whether his observation is sound.

The distinction between context of discovery and of verification shapes Maynard Smith's assessment of Darwin's relation to the larger social and economic context in which he worked. Maynard Smith observes that "Darwin must also have been influenced by the fact that he lived in the era of competitive capitalism, when some firms were improving their techniques, and increasing in size and affluence, while others were going bankrupt, and old crafts were dying out. It is unlikely that the concepts of competition and the struggle for existence in nature would have occurred to him so readily had he lived in a more static feudal society." The condition of society makes Darwin more receptive to certain observations than he otherwise might have been, but to say this much is not to say either that social conditions wholly cause or control the formulation of Darwin's theories, nor that those theories are incorrect. Maynard Smith invokes certain social conditions to explain, in part, how Darwin's imagination might have been primed or made ready for the observation of certain facts and the formulation of certain ideas. To determine whether those observations and ideas are in fact true, one must invoke specifically scientific criteria of judgment.

A serious effort to bring criticism to bear on the scientific validity of Darwin's theory must look either to the factual basis of his propositions, their logical connection, or their implications for a variety of empirical areas: genetics, biogeography, paleontology, comparative anatomy, and embryology. If it turned out, for instance, that Malthus, Darwin, and Wallace were wrong in assuming that population numbers tend to remain relatively stable over time, and if they were wrong in the inference that many more members of a given population are born than survive to reproduce—if a population could expand indefinitely, without limitation through the availability of resources—then one main element of Darwin's theory would collapse, and the whole complex of ideas would be invalid. Or, if modern genetics had proven that heritable variations could not be sustained beyond a few generations, again, the system would collapse. This was an implication of the faulty inheritance theory of Darwin's own time—the idea of "blending inheritance"—and it presented one of the most serious challenges to his system. And again, if Lord Kelvin had been right, and the earth were only somewhere between twenty million and forty million years old, the scale of geological time required by Darwin's system would have failed. If the catastrophist view of the fossil record had been vindicated because paleontologists had turned up cases in which one whole animal group was succeeded by a different group, suddenly and contiguously, without any intervening forms, and with no possibility of migration, the idea of "special creation" would have received strong vindication, and Darwin's system would have been seriously challenged. One can imagine, for instance, an Australian discovery in which marsupials were suddenly supplanted, in the fossil record, for a period of a few thousand or million years, then just as suddenly replaced by mammals of the more modern type, only to be succeeded once again by marsupials. This is not empirically impossible; it is consistent with the hypothesis of special creation; but it is altogether inconsistent with the theory of descent with modification by means of natural selection. And of course, it never happened.

Darwin's Evolutionary Psychology

The Descent of Man was published in 1871, twelve years after the publication of the *Origin*, but the implications for man are a distinct and powerful part of the initial inspirations that Darwin scribbled into his notebooks in the late 1830s. In the introduction to the *Descent*, Darwin explains that for diplomatic reasons he chose to avoid any extensive consideration of human beings in the *Origin*, but that for the sake of his integrity he had felt obliged to observe, toward the very end of the book, that his theory had human implications. "In the distant future, I see open fields for far more important researches. Psychology will be based on a new foundation, that of the necessary acquirement of each mental power and capacity by gradation. Light will be thrown on the origin of man and his history" (*OS*, 397). Following Darwin's lead, most commentators cite this one passage as the only reference to man in the *Origin*, but they thus overlook, as did Darwin himself, two references that are, in their own quiet way, even more effective. In the chapter "Struggle for Existence," Darwin cites "slow-breeding man" as an instance of potentially geometric reproductive rates, and in "Difficulties on Theory," Darwin casually takes human racial differences as yet another example of our ignorance concerning "slight and unimportant variations" such as skin color or hair (*OS*, 134, 218–19). Human beings are included in these passages along with examples drawn from wild species and from domesticated animals. The implication is that human beings are simply one more animal species and that their characteristics have the same causes and provide evidence for the same principles that enter into the discussion of plants, elephants, turkeys, vultures, and cattle (the other examples that are used in the passages).

In the *Descent*, Darwin carries the implications of these passages to their logical conclusion. He locates human beings in their phylogenetic heritage, as primates, and in support of his phylogenetic analysis, he brings forward arresting evidence from comparative anatomy and embryology. Regarding humans as social animals, he examines their forms of behavior and social organization as natural manifestations of their elementary biological dispositions for survival and reproduction, and he locates them within an ecological context that restricts all conditioning influences to those that also affect other animals.

The *Origin* succeeded in effecting a sudden and massive transformation in the educated public's view of descent with modification. Emboldened by this success, Darwin set out in the *Descent* to complete his survey of "the higher animals" (*OS*, 398). In some respects, this earliest of all essays in evolutionary psychology is still one of the best and most profound. It points the way toward a mature social science methodology that incorporates information from studies in anatomy, embryology, psychology, and anthropology. In company with his excellent methodology, Darwin brought to his subject a fine moral consciousness that helped direct his insights into the evolved social psychology of human beings. He gave a classic analysis of the two basic components of human moral psychology: (1) an evolved social sympathy, and (2) a capacity for reflective judgment that situates all present action in relation to longer temporal sequences and thus makes "conscience" possible. The incisiveness of this analysis has yet to be surpassed in even the most recent and sophisticated works of evolutionary psychology and evolutionary ethics. It nonetheless remains the case that it is the *Origin*, and not the *Descent*, that is "the chief work" of Darwin's life. Darwin brought to the study of man the same naturalistic intuition that he brought to the question of species, but the time was less ripe, and by the time he wrote his book, he was himself past the peak of his productive power.

Lyell effected a revolution in geology, and Darwin in biology. A similar revolution is now taking place in psychology and the other social sciences. After the first decade of the twentieth century, the leading figures in the social sciences instituted a long phase of ideological suppression in the service of an ideology of cultural autonomy, and we are only just now, in the past few decades, finally taking up again the naturalistic methodology that Darwin pioneered. Prediction involving timing in such a matter is of course risky, but I shall take the risk and affirm that we are just now on the verge of completing the Darwinian revolution in the social sciences. In order to understand what that revolution would entail, we can compare the *Descent* with the *Origin*. The *Descent* is full of fascinating observations and penetrating insights, but it lacks the deep systematic order that distinguishes the *Origin*. In the *Origin*, every fact and observation has a clear place within the tight logical structure outlined in the introduction and in the last paragraph of the book. The *Descent* is more casually organized, looser, more impressionistic. As is commonly observed, it is actually two separate books, awkwardly joined. One book is an anthropological essay on human nature, with some specific reference to evolved sex differences. The other book is a lengthy and highly detailed technical treatise on sexual dimorphism in the animal kingdom. Even apart from the question of awkwardly combining two distinct books, the anthropological essay in the *Descent* lacks the extraordinary logical rigor that distinguishes the *Origin*. The array of motives, emotions, and cognitive dispositions analyzed in the book have no tight, necessary relation to one another within a total system of motivational structures that are rooted in the elementary principles of natural selection. The level of conceptual organization in the *Descent* is less like that of the *Origin* than, say, that of John Locke's seminal but big and baggy *Essay Concerning Human Understanding*. It is a classic work, but not the kind of authoritative and definitive work Darwin achieved in biology proper.

In order to complete the Darwinian revolution in the social sciences, we shall have to integrate at least two schools of modern Darwinian psychology. One school is that of "sociobiology," which concentrates on the ultimate regulative principles of inclusive fitness and that can be criticized for too crudely or simply reducing human motives to the drive toward reproductive success. The second school is that of "evolutionary psychology," which aims at identifying a disparate array of genetically derived and physiologically based behavioral mechanisms targeted to the solution of specific adaptive problems. In order to extend this synthesis from the social sciences to the humanities, we shall also have to be able to take account of the adaptive functions of the arts and to understand the formal organization of the arts as prosthetic extensions of evolved cognitive aptitudes. What has been missing, until recently, is a complete causal integration of elementary biological principles with complex psychological structures, complex forms of social organization, and complex forms of cognitive activity. Human life history theory provides the concepts for that complete causal integration. (See the section titled "A Model of Human Nature" in the first chapter in this book.)

The Nature of the Darwinian Revolution

The history of evolutionary theory after 1859 can be divided into a few distinct phases. In the period from 1859 to that of Darwin's death in 1882, Darwin radically transformed the received view of evolution. Within just a few years, most reputable scientists came to accept that evolution, the transformation of species over time, had in fact occurred. But most scientists did not confidently accept natural selection as the primary mechanism through which those transformations took place. There was a long interregnum, lasting from about 1859 to about 1920, in which uncertainty over the mechanism of heredity and the extent of geological time placed the theory of natural selection in doubt. Fleeming Jenkin (1833–1885) pointed out that if inherited variations were "blended" in each successive generation, any variation would inevitably be swamped by the common characteristics of the species, and the physicist William Thomson (later Lord Kelvin, 1824–1907), on the basis of ingenious calculations about the dissipation of heat from the earth, argued that the earth was much younger than Lyell and Darwin had supposed, so that far too little time had passed for evolutionary change on Darwin's model. Around the turn of the century, Kelvin's theory of heat loss was corrected by the discovery of continuous, heat-producing radioactive decay from within the earth. Gregor Mendel (1822–1884) had discovered particulate inheritance in 1856—providing the solution for the problem of blending inheritance—but his theories were not recognized and assimilated until the turn of the century, and even then geneticists mistakenly believed that evolutionary change would require macromutational leaps, not the gradual accumulation of adaptive changes required by the theory of natural selection. Around 1920, three distinguished geneticists, Ronald Fisher (1890–1962), John Haldane (1892–1964), and Sewall Wright (1889–1988), began publishing the papers that reconciled Mendelian genetics with natural selection. In the period from about 1920 to about 1950, biological theorists from a wide array of specialized disciplines—natural history, systematics, paleontology, ecology, and other areas—integrated their work with that of the geneticists and thus produced the Modern Synthesis. The Modern Synthesis is the culmination of the Darwinian revolution and forms the basis for the authoritative current framework of scientific evolutionary theory. The discovery of the structure of DNA in 1953 has only confirmed and strengthened the basic theoretical structure of the Modern Synthesis. By identifying the specific molecular mechanisms that regulate variation, sexual recombination, mutation, and inheritance, the discovery of DNA has empirically validated key components of Darwin's theory, and has given decisive proof for his hypothesis that all of life on earth, through all its multifarious transformations in structure, forms a single, unbroken chain of hereditary transmission.

Mayr sets the Darwinian revolution in opposition to two distinct models. One is the idea that advance in science is "steady and regular." He attributes this view to no specific authority, and it is not clear that any serious theorist actually holds by it, though it might roughly describe some level of vague popular belief. The other model is that of Thomas Kuhn, which Mayr characterizes, fairly enough, as "a series of revolutions separated by long periods of steadily progressing normal science." Mayr describes a range of possible developments considerably wider than that envisioned within Kuhn's model:

> When we study particular scientific disciplines we observe great irregularities: theories become fashionable, others fall into eclipse; some fields enjoy considerable consensus among their active workers, other fields are split into several camps of specialists furiously feuding with one another. This latter description applies well to evolutionary biology between 1859 and about 1940.[44]

As Huxley explains, within ten years of publishing the *Origin*, Darwin had effected an almost complete transformation in the received view about one chief component of his theory—the contention that species had not been separately created and are not fixed and stable; that all species derive from descent with modification.[45] Huxley suggests also something of the uncertainty that still hovered over the other main component of Darwin's theory: natural selection as the central mechanism of change. Mayr is more emphatic than Huxley about the opposition to natural selection, and in this respect he reflects the now authoritative consensus. He maintains that "the opposition to natural selection continued unabated for some eighty years after the publication of the *Origin*. Except for a few naturalists, there was hardly a single biologist, and certainly not a single experimental biologist, who adopted natural selection as the exclusive cause of adaptation."[46]

In the concluding paragraph to the introduction of the *Origin*, Darwin himself tacitly anticipated a development not unlike that which Kuhn describes for scientific revolutions. Assuming provisionally that his main arguments would prove persuasive, he cautioned, "No one ought to feel surprise at much remaining as yet unexplained in regard to the origin of species and varieties, if he makes due allowance for our profound ignorance in regard to the mutual relations of all the beings which live around us" (*OS*, 98). The ignorance Darwin has in mind evidently concerns the detailed ecological knowledge about the relations of species to their habitats and the coevolutionary, interdependent adaptations of species connected to one another within an ecological web. Such problems, however important, would constitute, on the level of theory, details. That is, they would correspond to what Kuhn identifies as "puzzles" or matters of detailed inquiry wholly within the framework of an established theory. And puzzles of that sort there certainly have been. Darwinism has constituted an immense research program for naturalists—pointing the way toward their detailed inquiries into adaptive structure, embryology, geographical distribution, systematics, and ecological organization. But Darwin's own hopes for the completeness of his theoretical revolution were, as it turned out, too sanguine. One of the problems of detail within his theory, the nature of inheritance, proved so large a puzzle, with so many false leads and incomplete solutions, that one major element of his total theory—natural selection—remained in doubt for at least sixty years, until Fisher, Haldane, and Wright began to publish the papers on Mendelian genetics and natural selection that laid the foundations for the Modern Synthesis.

Mayr is correct in affirming that the Darwinian "paradigm shift" took several decades to complete. For the theory of natural selection, there was never any sudden "gestalt" switch. As Mayr describes it, giving the inside view of a major contributor to the process, there was instead a gradually accumulating body of theoretical genetic work that slowly converged with the work of "naturalists" (ecologists and systematists) and paleontologists. The combined weight of these different fields eventually convinced the majority of scientists qualified to judge in the case. Ridley confirms Mayr's account, and he concurs with Mayr in locating the consolidation of the Modern Synthesis in the late forties. He observes that by the mid-forties "the modern synthesis had penetrated all areas of biology. The 30 members of a 'committee on common problems of genetics, systematics, and paleontology' who met (with some other experts) at Princeton in 1947 represented all areas of biology. But they shared a common viewpoint, the viewpoint of Mendelism and neo-Darwinism. A similar unanimity of 30 leading figures in genetics, morphology, systematics, and paleontology would have been difficult to achieve before that date."[47]

The history of Darwinism offers an opportunity for assessing the most important epistemological issue raised by Kuhn's model: the question as to whether scientists are capable of reflecting critically on their own ideas and, on the basis of these critical reflections, modifying their views. Kuhn describes a paradigm as a total structure of ideas that regulates what scientists can think and even what they can actually see. In Kuhn's presentation, if certain phenomena do not fit within a paradigm, scientists are unable to perceive those phenomena. Mayr's description of the slow and messy progress of the Darwinian revolution subverts Kuhn's notion of a simple, total framework, a "paradigm" that scientists either accept, blindly, or reject and replace with another paradigm that they then also accept in an equally uncritical fashion. On this issue, Kuhn's most effective theoretical opponent has been Karl Popper, who identifies Kuhn's "Myth of the Framework" as "in our time, the central bulwark of irrationalism."[48]

Kuhn's model has an at best imperfect fit to the process of scientific theory formation in the period after Darwin presented his theory. How does it fit in the other direction? That is, how well does Darwin's own transformation accord with the notion of a sudden and radical gestalt switch? Darwin radically altered the prevailing view about the origin of species, and he proposed a mechanism that had never been considered as the central mechanism that regulated all of phylogenetic history. But the "switch" that occurred in his own thinking was no complete and total replacement of all previous ideas and information about species. As we have seen in considering the background to Darwin's work, he absorbed information and ideas from a very wide range of sources. His relations to both Malthus and Lyell are particularly instructive in this respect. Darwin assimilated Malthus' insight into population pressure and food supply as one central component of his own theory, but he also incorporated it into a much larger theory that involved adaptive structural changes of which Malthus had no inkling. Darwin assimilated major elements of Lyell's geology, developed them further (as in the theory of coral reefs), and used them to help explain essential points about geographical distribution and the fossil record. He rejected Lyell's general theory, if an idea so tentative and sketchy can be called a theory, about the origin of species, but he also incorporated Lyell's argument that a chief engine of extinction is the failure of a species to adapt to change of climate. In none of this do we see anything remotely like the sudden and total replacement of one structure of ideas by another. What we see instead is a steadily accumulating body of ideas and information that many individual scientists piece together into local groups—as if they were working out segments of a picture puzzle in which the segments remained, for the time being, disconnected from one another—until one scientist (or in this case two, if one also counts Wallace), sees the way in which all the partial segments fit into one total larger pattern.

To conceive of Darwin's achievement in this way need not diminish our sense of the importance of the creative, innovative power of the individual scientist. Ghiselin cites Kuhn's idea that "scientific revolution results from the failures and contradictions of the prevailing system," and while he grants that "such may well be the case for conventional scientists," he insists that "Darwin was an exception. He restructured traditional fields and erected new paradigms when the positive development of his ideas suggested something new."[49] Darwin's case thus confirms, he argues, that "the success of at least some revolutionary thinkers" may be attributed "not to sociological forces, but to an innovative mentality." This is a false antithesis. One can see the way it presents itself in Ghiselin's thinking, and indeed the Kuhnian model has often been taken, by its proponents, in the light in which Ghiselin sees it. The antithesis depends on an overly simple opposition between two kinds of productive force. In this overly simple formulation, either the whole social context produces a theory, or the individual genius of the scientist produces the theory. In reality, both elements are necessary parts of scientific discovery. Science is a collective, social enterprise. Darwin depended on the findings of an extended network of researchers—his scientific correspondents number in the hundreds—and a long tradition of geological and biological investigation. He did indeed observe the failures and contradictions of the prevailing system, and through the positive development of his ideas he provided solutions for them. If there had been no unsolved problems within the prevailing system, there would have been no need to formulate new explanations. To acknowledge that the existence of problems is a prerequisite for the formulation of solutions need not derogate from the "innovative mentality" of creative scientific genius. Maynard Smith gets this issue into proper focus. After listing the various kinds of information and inspiration that fed into the formulation of Darwin's theory, he observes that all these elements "provided Darwin with the necessary methods of attack and materials for study; it required his individual genius to weld them into a comprehensive theory of organic evolution."[50]

The fortunes of Darwin's theory in the period that lay between the publication of the *Origin* and the consolidation of the Modern Synthesis offer us one signal measure of the quality and magnitude of Darwin's genius. In his *Autobiography*, Darwin notes that "some of my critics have said, 'Oh, he is a good observer, but has no power of reasoning.'" Mildly but astutely, Darwin comments, "I do not think that this can be true, for the *Origin of Species* is one long argument from the beginning to the end."[51] What Darwin saw, and what almost everyone else (including Huxley) failed to see, for nearly a century, was that his theory consisted in interconnected and interdependent bodies of evidence and reasoning. The theory has a total logical structure, and that structure has a kind of tough validity that should have rendered it presumptively correct from the beginning.

Even Mayr, distinguished biologist and historian though he undoubtedly is, wrongly believes that the different elements of Darwin's theory are isolable. The evidence he brings forward to confirm that view is that in the minds of most of Darwin's contemporaries the idea of descent with modification was in fact isolated from the idea of natural selection. Mayr acknowledges that Darwin himself regarded the elements of his theory as "a unity," so that in disputing this claim he is placing the combined weight of Darwin's contemporaries and successors against the weight of Darwin's own judgment. In order to support the judgment thus rendered, Mayr asserts that "natural selection is dealt with in the first four chapters" but that "in the remaining ten chapters natural selection is not featured."[52] These latter chapters, Mayr maintains, deal only with descent, not with the mechanism of natural selection. In both this specific affirmation and the larger claim it is intended to support, Mayr is demonstrably mistaken.

Darwin saw clearly the logical necessity of all the parts of the theory fitting together. He understood that natural selection "almost inevitably induces extinction and divergence of character in the many descendants from one dominant parent-species" (*OS*, 363). Divergence itself "explains that great and universal feature in the affinities of all organic beings, namely, their subordination in group under group," and in this strict logical sense natural selection is an integral causal component of systematics or classification. The systematic classification of all living things involves both linkages and gaps; the gaps reflect extinction, and extinction is an effect of selection. As with classification, so also with geographical distribution. The radiation of dominant groups of flora and fauna within distinct geographical regions is a result not merely of descent but of the dominance of certain groups over other groups, and this dominance is the result of selection (*OS*, 131, 323–24, 330, 346). Even more apparently, ecology is a matter not of descent over time but of current interactions regulated by the adaptation of organisms to their environments—environments in which other organisms are at least as important as the physical features of the land, water, or air. Ecosystems are "systems" precisely because they involve elaborate interactions among organisms that have coevolved in adaptive relation to one another, as predator, prey, parasite, and symbiont.

If we shift our focus from large-scale populational interactions to the structure and development of individual organisms, selection remains central to Darwin's argument. Once we have set aside the idea of "design" as the result of divine fiat, adaptation can be explained only through natural selection, and imperfections in adaptive design are also a result of selection. One of the most important principles of morphology is the linkage of species through "homologous" structures. "What can be more curious than that the hand of a man, formed for grasping, that of a mole for digging, the leg of the horse, the paddle of the porpoise, and the wing of the bat, should all be constructed on the same pattern, and should include the same bones, in the same relative positions?" (*OS*, 364). This singularity or curiosity of the natural world cannot be explained through design or even merely through a form of descent like that envisioned by Lamarck. But it can be explained "on the theory of the natural selection of successive slight modifications" (*OS*, 364–65), and such modifications can be organized in adaptively functional ways only by natural selection. So also with embryology. Embryos share an initial phylogenetic commonality, and only in the course of ontogenetic development do they progressively differentiate into the characteristics of some one distinct species. At certain points in the phylogenetic history of all organisms, selection has activated morphological change for adaptive purposes, and those changes appear at specific points in the ontogenetic sequence of each organism (*OS*, 367–74).

In all of these major fields of evidence, then—in systematics, geographical distribution, ecology, morphology, and embryology—selection is central to Darwin's explanation of descent with modification. The field of variation and heredity was of course not yet established as "genetics," and it was the most mysterious and obscure part of Darwin's subject, but even in this field the logic of natural selection shed light for Darwin. It explained, for instance, why it is that "*a part developed in any species in an extraordinary degree or manner . . . tends to be highly variable*" (*OS*, 189). Darwin's solution for this conundrum is that "an extraordinary amount of modification implies an unusually large and long-continued amount of variability, which has continually been accumulated by natural selection for the benefit of the species" (*OS*, 191), and that this recent variability has remained active.

Darwin understood the integrity of his own argument, and he understood further, as most of his successors did not, the constraining force of that argument. The total logic of the argument pointed decisively to the necessity or inevitability of the existence of mechanisms of inheritance of a sort that were in fact eventually recognized in the synthesis of Mendelian genetics and the theory of natural selection. It can reasonably be said that Darwin's theory *predicts* some such set of mechanisms. If the theory is true, those mechanisms must exist, and the weight of all the *other* evidence that Darwin himself marshaled, in all the fields that he brought into play, should have given presumptive credibility to that prediction. Darwin was rationally confident of this outcome, but generations of biologists who came after him became absorbed in the details, the puzzles of inheritance, and because of these puzzles, they lost confidence in the larger argument that had nonetheless radically and permanently altered their convictions about the reality of descent with modification.

In this respect, the history of evolutionary theory from about 1860 to about 1920 turns Kuhn's model on its head. Darwin's theory did not provide a paradigm within which scientists busily and almost mechanically went about solving technical "puzzles"—relatively trivial details entailed by the theory. Instead, many of them abandoned the theory and became absorbed in working out the details as an empirical and technical enterprise. The situation in this case bears a fairly close parallel to the developments in geology from the last decade of the eighteenth century through the third decade of the nineteenth. Geologists had become disgusted with the grand theoretical debates between the Wernerian Neptunists—proponents of a universal flood that precipitated the continents—and the Huttonian Vulcanists who identified cataclysmic volcanic activity as the main constructive force in geology. Turning away from large-scale theories, geologists preoccupied themselves instead with the practical work of identifying the total stratigraphic column, an immense and absorbing empirical enterprise. Lyell's *Principles of Geology* constituted something very like a "modern synthesis" between Hutton's large-scale theory of a homeostatic equilibrium and this relatively theory-free empirical research.

Looked at in a negative light, one might say that between the time of Darwin and the Modern Synthesis, geneticists were like the Israelites who had left Egypt and had not yet entered the promised land—they spent decades wandering lost in the wilderness. In a more positive light, one can say that the geneticists constructed their discipline from the ground up, working out the technical structure in empirical, experimental research, and then discovered, to their own surprise, that the structures they had defined fit neatly as mechanisms within a larger logic that united their own discipline once again with all the interconnected fields of evolutionary biology.

CHAPTER 11

The Science Wars in a Long View

Putting the Human in Its Place

Imagine you are taking a quiz in the history of modern critical theory. One of the questions asks you to identify the period in which "the science wars" took place. You know that in 1994 Gross and Levitt published *Higher Superstition: The Academic Left and Its Quarrels with Science.* You probably recall that Alan Sokal's celebrated hoax was published in a special issue of *Social Text* designed specifically to answer Gross and Levitt, and you might also remember that Sokal was himself first alerted to the ideas of postmodern science theory by reading Gross and Levitt. So, the middle of the 1990s seems a likely starting point. You are aware, of course, of the debate between C. P. Snow and F. R. Leavis nearly half a century before, but that took place in an entirely different theoretical and cultural context. Leavis was something of a cross between a modernist and an old-fashioned Victorian humanist, and the last feeble remnants of his generation, like the last veterans of World War Two, are now rapidly dying off. Snow's phrase "the two cultures" has entered into the common parlance, but poststructuralism has radically altered the theoretical character of the conflict between the sciences and the humanities. The poststructuralist revolution overwhelmed the old-fashioned humanists some thirty years ago, and the leaders of that revolution have long been firmly established—to use another of Snow's famous phrases—in the corridors of power. So, say we locate the onset of "the science wars" in the mid-nineties. Can we identify a point of conclusion? Some of the contributors on the *Social Text* side of the conflict have declared that there was no war to lose. For instance, as Barbara Herrnstein Smith explains the situation, the cultural constructivists did not mean what Sokal and his cohorts thought they meant. The scientists were just too little versed in rhetoric to grasp the finer shades of ambiguity in postmodern accounts of science.[1] In any case, whether the science wars were worth fighting or not, are they now over? Could either side stand on a carrier and declare, "Mission accomplished"? Having explained that the Sokal brigades were using high explosives to blow up straw men, could the postmodern science theorists declare that they had already explained quite enough, that they find the whole affair tiresome, and that they have more important issues with which to occupy their attention? Or could Sokal and his cohorts affirm that the postmodern theorists, whether or not they used to say the things the scientists thought they were saying, have stopped saying them? Some of the participants in this conflict would no doubt like to declare victory and go home, but off in the distance one often still hears explosions and sees dark plumes of smoke, followed by the wailing of sirens.

For more than thirty years now, beginning with *Sociobiology: The New Synthesis*, E. O. Wilson has displayed a remarkable capacity to incite explosive responses to the claims of science. In *Consilience: The Unity of Knowledge* (1998), Wilson proposed terms for a peace treaty between the two cultures. The terms were simple—*Anschluss*. The humanities would be enfolded within the larger explanatory contexts of evolutionary social science and evolutionary biology. The two cultures would thus become one. And the response from the humanities? Jubilant crowds from the convention hotels at the MLA conference pouring out into the streets to celebrate the New World Order based on the hegemony of science? Hardly. The lessons of appeasement had not been lost on the members of the MLA. The humanities have their own distinct provinces, their sacred soil. This they must not give up. Quite the contrary. What they must do instead is "hunt down those disciplines whose subject matter they covet and bring them into their own realm."[2] *Lebensraum.* Eastward lies the course of empire. The vast plains of Russia lie open for the taking. One must only sweep the ground clear of its present unworthy inhabitants—illiterate peasants with names like Gross, Levitt, Sokal, and Wilson. And thus the cycle of violence continues.

As a battle over curricular turf, the science wars began in 1880, with the exchange between T. H. Huxley and Matthew Arnold. As a battle over what science can tell us about the meaning of life, the conflict goes back further still, but Arnold and Huxley give classic formulations to basic terms in the debate, and they are the first prominent essayists to link issues of metaphysics and cultural history with propositions about the relative standing of science and humane letters within the university. Huxley gives powerful expression to a materialist metaphysic concordant, as he believes, with the revelations of modern science, and he assesses the main phases of Western cultural history from that metaphysical perspective. Huxley's formulations have had little positive influence on the humanities, but among scientists of broad general culture, Huxley remains a living voice. He is cited with respect by E. O. Wilson, and his central contentions have been taken up and reformulated by Steven Weinberg, a Nobel prize–winning physicist, essayist, and prominent participant in the science wars. In responding to Huxley, Arnold affirms an alternative metaphysic and an alternative conception of the human. For nearly a century, Arnold's humanist idealism provided a central guiding light for literary scholars. In some respects, the poststructuralist revolution produced a radical disjunction with the old Arnoldian episteme, but the deepest underlying impulses in Arnold's defense of "humane letters" are still active in postmodern accounts of science and culture. I admire Arnold enough to have written a book on him, but my own metaphysical and epistemological views are more closely aligned with those of Huxley, Weinberg, and Wilson than with those of Arnold and his descendants in the humanities. For the past fifteen years or so, I have been working to establish linkages between literary study and the evolutionary social sciences. In assessing the possibilities for integrating science and the humanities within a single culture, I am thus of the devil's party.

The occasion for Huxley's essay "Science and Culture" was a celebratory address on the founding of a technical college. Huxley approved the provision for a specifically scientific education for a given set of students, and he defended it as an alternative to the emphasis on Greek and Roman literature that prevailed in most higher education at the time. In support of the pedagogical mission of scientific training, Huxley makes three main points about the cultural significance of modern science: (a) science has fundamentally changed our worldview—our vision of nature and the place of humankind in nature; (b) nature forms a unitary causal order that can be most adequately accessed using scientific methods; and (c), adopting the ethos of science is an ethical imperative for reasons of both intellectual and social responsibility. In supporting these contentions, Huxley gives a synoptic historical account of four phases of Western civilization, the ancient, the medieval, the Renaissance, and the modern. He attributes to the Renaissance the historical mission of recuperating the culture of ancient Greece and Rome, but he segregates the modern world from all preceding epochs. He argues that modern science enforces a worldview that separates the modern period from the Renaissance more widely than the Renaissance was separated from the Middle Ages.[3] As he explains more fully in his essay "On the Advisableness of Improving Natural Knowledge" (1866), before the advent of modern science, humankind had always taken itself "as the standard of comparison, as the centre and measure of the world."[4] He speaks of the animistic fantasies of all primitive peoples and of the ancient Greeks and Romans, and of course he speaks of the spiritualist notions of Christianity and other religions. For Huxley, science in general enforces a materialistic vision of the natural order. Astronomy, first of all, "has filled men's minds with general ideas of a character most foreign to their daily experience." It tells them "that this so vast and seemingly solid earth is but an atom among atoms, whirling no man knows whither, through illimitable space." It "opens up infinite regions where nothing is known . . . but matter and force, operating according to rigid rules." Extending this vision into the range of biology, he affirms that "as the astronomers discover in the earth no centre of the universe, but an eccentric speck, so the naturalists find man to be no centre of the living world, but one amidst endless modifications of life."[5]

In delineating an epochal shift in metaphysical vision, Huxley also identifies a primary source of conflict in the struggle between the humanities and the sciences. In one way or another over the past century, proponents of the humanities have continued to seek to envision humans as the center and measure of the world. Over against that humanistic impulse, the sciences have posed a vision of the world as a vast network of material or physical forces in which the human is but one further link in an unbroken chain of physical causes.

In the terms and concepts Huxley uses for describing cosmology and physiology, he was of course limited to the science available in his own day. The conceptual transformations in these disciplines since Huxley's time have been immense, but Huxley himself would have assimilated these developments with enthusiasm, and they would not have undermined his larger metaphysical vision. We can assess the enduring power of Huxley's vision by comparing it with that of Steven Weinberg, who has been at the forefront of advances in modern particle physics. In *Dreams of a Final Theory* and *Facing Up,* Weinberg outlines the developments in physical knowledge up to the present time, but he also reaffirms the basic principles enunciated by Huxley. He uses one of Huxley's own essays, "On a Piece of Chalk," as the starting point for tracing a sequence of physical causes that leads from common visual perception to the limits of current knowledge in physics. While taking full account of the way in which "principles of symmetry" have replaced older conceptions of "matter," he follows Huxley in affirming that science "gives us access to the logical order built into nature itself."[6] Like Huxley, too, he generalizes from specific discoveries to a larger historical shift in metaphysical vision. He speaks of "the profound cultural effect of the discovery, going back to the work of Newton, that nature is strictly governed by impersonal, mathematical laws."[7] Most importantly, like Huxley, he insists on a certain metaphysical bleakness—a universe of insentient force in which the human occupies a trivial and marginal position:

> Nothing in the last five hundred years has had so great an effect on the human spirit as the discoveries of modern science.... We find that the earth on which we live is a speck of matter revolving around a commonplace star, one of billions in a galaxy of stars, which itself is only one of trillions of galaxies. Even more chilling, we ourselves are the end result of a vast sequence of breedings and eatings, the same process that has also produced the clam and the cactus.... Some of the old magic has gone out of our view of the role of humanity in the universe, its place being taken by what Matthew Arnold called the "note of sadness."...
>
> The human race has had to grow up a good deal in the last five hundred years to confront the fact that we just don't count for much in the grand scheme of things, and the teaching of science as a liberal art helps each of us to grow up as an individual.[8]

Weinberg's quotation on the "note of sadness" is from Arnold's poem "Dover Beach," in which he describes "The Sea of Faith" as a "melancholy, long, withdrawing roar."[9] In his poems, Arnold spoke for a whole phase of Western culture. He gave touchstone expression to an epochal sense of dismay at the dissolution of a religious vision of the cosmos.

Most of Arnold's poems were written in his younger days. In the last thirty years of his career, he chiefly wrote essays in which he offered remedies for the metaphysical sadness articulated in his poems. It is in this later, consolatory phase of Arnold's thinking that we can locate his essay "Literature and Science." Responding directly to Huxley, Arnold poses the question as to whether the predominance of letters in education ought now to pass to science. To answer this question in favor of literary education, he appeals to Plato's argument that an intelligent man "'will prize those studies that result in his soul getting soberness, righteousness, and wisdom, and will less value the others.'"[10] Again invoking Plato, he maintains that the "'fundamental desire'" in "human nature" is the desire for "'good'" ("LS," 63). In Arnold's thinking, the "good" consists ultimately in the harmonious integration of all the human faculties. He develops a teleological scheme of cultural history, quasi-Hegelian, in which a transcendent force works through history toward a culminating realization of a perfected human condition. In Arnold's cultural theory, literature is the most important medium through which we can achieve this "full humanity."[11]

Superficially, Arnold seems to assimilate Darwinian naturalism. He says that in looking through the findings of modern science, "at last we come to propositions so interesting as Mr. Darwin's famous proposition that 'our ancestor was a hairy quadruped furnished with a tail and pointed ears, probably arboreal in his habits'" ("LS," 64). With the deft and good-humored wit so characteristic of his essays, Arnold turns this proposition to the advantage of literary studies. On the grounds that mankind has an innate need for the "good," he concludes that our primate ancestor "carried hidden in his nature, apparently, something destined to develop into a necessity for humane letters" ("LS," 72). By identifying ancient Greek literature as a model for a grand and noble unity of aesthetic perception, he enables himself to draw the still more dramatic inference "that our hairy ancestor carried in his nature, also, a necessity for Greek."

Arnold's manner is charming, and his logic is beguiling, but his facile blending of Plato and Darwin is deceptive. In Darwin's theory of natural selection, there is no transcendent teleological force driving toward some culminating historical realization of psychological and cultural harmony. The driving force in human evolution is the mechanical process of natural selection. The regulative principle that has shaped human nature, as it has shaped the nature of every other species, is inclusive fitness—the transmission of genes. Inclusive fitness has designed human nature in such a way that conflicts of interest are integral and ineradicable. Relations between men and women, parents and children, siblings and other kin, and individuals within a social group—all these relations involve tensions between reciprocal benefits and competing interests. Humans have evolved distinctive capacities for cooperation within groups, but the larger context for the evolution of cooperation within groups is the conflict between competing groups. Within the Darwinian conception of human evolution, there is no transcendent teleological process guiding human cultural history, and there is no transcendent aesthetic and ethical order to which the human mind, through culture, can gain access.

Huxley and Arnold were friends, and the tone of their references to one another is genial. They nonetheless differ profoundly in metaphysical and historical vision, and that difference plays itself out in the subsequent history of the debate over the sciences and the humanities. Huxley announces a radical break in the modern worldview. Arnold, in contrast, gives a strong emphasis to continuity in the cultural imagination of European civilization. Huxley had spoken with scathing contempt of medieval superstition. In reply, Arnold grants that the medieval cosmology is obsolete, but he defends the quality of medieval education on ethical and aesthetic grounds. He concedes that we can no longer uphold traditional Christian beliefs, but he argues that we have no need of those beliefs. We can, he says, look instead to poetry and the other arts to satisfy the emotional and imaginative needs that religion once satisfied. This is a radical proposition. Arnold argues that the real and effective part of religion has always been its unconscious poetry. In the future, he thinks, poetry itself, detached from religious belief, will fulfill all the psychological and moral needs once fulfilled by religion. "The strongest part of our religion to-day is its unconscious poetry. The future of poetry is immense, because in conscious poetry, where it is worthy of its high destiny, our race, as time goes on, will find an ever surer and surer stay."[12]

From our current vantage point, Arnold's proclamation on the future of poetry seems quaint. Almost no one at the present time would invest poetry with this weighty mission. For nearly a century, though, Arnold's belief in the mission of literature had an immense influence in academic literary study and in the wider culture. Until about 1970, the majority of literary scholars could reasonably have been described as "humanists" who shared some important part of Arnold's literary idealism. A humanist in this sense is a scholar who invests literary subjects with an almost sacred value. Such scholars believe that literary works give human beings access to a spiritual realm in which some ultimate harmony or resolution can be glimpsed. Arnold's own humanism mingles Platonic transcendence and Wordsworthian piety with a frank admiration for creative literary genius. He describes great writers as "gifted men, alive and active with extraordinary power at an unusual number of points," and he says that their works "have a fortifying, and elevating, and quickening, and suggestive power" ("LS," 68). The commonly accepted views on such matters have now changed so dramatically that younger scholars might find it hard to credit the fervor with which they were once held. In the mid-century period, Arnoldian idealism informed the New Critics' belief in the poem as a verbal icon and Northrop Frye's Romantic and mystical belief in an ultimate order of literary words equivalent to the mind of God—"the anagogic phase." Arnoldian idealism entered in an attenuated form into Lionel Trilling's concept of literary culture as an apex of "the liberal imagination," and it worked its way also into F. R. Leavis' defense of literary culture over against the claims of science.[13]

The exchange between C. P. Snow and F. R. Leavis is in some ways more important for its symptomatic value than for its substantive intellectual content. Snow speaks of physical science as "the most beautiful and wonderful collective work of the mind of man," but he says nothing of its metaphysical character and very little of its epistemological character.[14] He feels that humanists should know more about physical science and that scientists should read more novels, but he does not seem to register that there are any ultimate questions of meaning at stake. Snow's vision is essentially utilitarian. He advocates more science education chiefly on the grounds that it will produce greater physical comforts for the mass of society. He is sympathetic to the Soviet model of education, and he seems to feel that shared norms of material comfort will eventually heal the ideological wounds of the modern world. In his essay responding to Snow, Leavis gave expression to an aesthetic and moral revulsion that shocked a good many people. Arnold wished to embody in his own prose the genial urbanity of "culture." Leavis aims instead at evoking and denouncing the spiritual emptiness he detects in Snow's techno-managerial perspective on society.

Leavis is no Platonist. His ideal of poetry filters itself through D. H. Lawrence's passionate individualism, but he nonetheless invests that ideal with the power to replace religion and provide spiritual meaning in the modern world. In opposition to Snow's vision of a social collective managed by a bureaucratic elite, Leavis appeals to the Laurentian maxim that "'nothing matters but life.'"[15] He concurs with Lawrence's belief that "only in living individuals is life there, and individual lives cannot be aggregated or equated or dealt with quantitatively in any way" ("Significance," 53–54). Leavis' formulation is typical of a view very wide spread in humanistic thinking—a kind of dualism that separates the world into two parts: a physical natural order that can be known by science and a purely qualitative, subjective human realm that can be accessed only through discursive modes. Leavis presents this subjective realm as an answer to questions about ultimate meaning. "In coming to terms with great literature we discover what at bottom we really believe. What for—what ultimately for? What do men live by?—the questions work and tell at what I can only call a religious depth of thought and feeling" ("Significance," 56). Leavis regards the challenges of rapid technological change chiefly as a threat, and he argues that literary education is a necessary means for meeting that threat. Echoing one of Arnold's phrases, he speaks of having recourse to "our full humanity" ("Significance," 60).

I've said that the essays by both Snow and Leavis seem more important for their symptomatic value than for their actual intellectual content. With respect to Leavis, what I have in mind is the vacuity of the rhetoric through which he seeks to evoke the "full humanity" supposedly to be found in literary training:

[Mankind will need] a basic living deference towards that to which, opening as it does into the unknown and itself unmeasurable, we know we belong. . . . What we need, and shall

> continue to need not less, is something with the livingness of the deepest vital instinct; as intelligence, a power—rooted, strong in experience, and supremely human—of creative response to the new challenges of time. ("Significance," 60–61)

This is a desperate sort of rhetoric—straining for intensity of effect, hyperbolic, disjointed, and vacant of substantive propositions. At the distance of half a century, we can reasonably suggest that Leavis' proclamations represent something like a spasmodic last gasp for old-fashioned literary humanism. The strained and vacant intensity of the rhetoric by which he claims a central place for a literary education goes a long way toward explaining the ultimate collapse of the Arnoldian rationale that sustained the humanities through the first half of the twentieth century. It simply would not wash.

Over the past thirty years or so, two major revolutions have taken place across the disciplines—deconstruction, with its radiations into Foucauldian discourse theory, and sociobiology, with its radiations into evolutionary psychology and behavioral ecology.[16] The deconstructive revolution jettisoned Arnoldian humanism and took on a completely new set of authorities, mostly French. In some ways, the introduction of poststructuralist or postmodernist thinking in the humanities represents a simple reversal in attitudes and concepts that had previously characterized literary training. In place of Platonic idealism and the appeal to some ultimate harmony in culture, deconstruction and Foucauldian discourse theory tend toward metaphysical nihilism and subversive ideology. In place of the appeal to the creative power of gifted individuals, postmodernism transforms the individual into a passive vessel for the circulation of cultural energies. Nonetheless, in two crucial respects, the deconstructive revolution retains continuity with Arnoldian humanism—in its emphasis on verbal culture, and in placing the claims of ethical values over the claims of objective knowledge. Huxley affirms that science bids us "seek for truth not among words, but among things."[17] Postmodern thinking, in contrast, locates ultimate epistemic authority in words—in discourse, in language or semiosis. It systematically deprecates the possibility of objective, empirical knowledge of a real, physical world that exists independently of any human discourse, and it is, in this respect, still qualitative and verbal in orientation. The appeal to values in the deconstructive dispensation is an appeal against established structures of social power. In its programmatic repudiation of existing power structures, postmodern thinking in the humanities seems virtually to invert the conservative tendencies in Arnoldian cultural idealism, but it is continuous with Arnoldian cultural idealism in basing its claim to cultural authority on its claim for wisdom and justice—for representing an enlightened ideological consciousness.

Since the time of Huxley and Arnold, seemingly disparate ideological values in the humanities have converged in a common desire to maintain human experience as "the centre and measure of the world." From the perspective shared by Huxley, Weinberg, and Wilson, that desire entails a false notion of where humans stand in the general scheme of things—a false cosmology, and a false understanding of human evolution. The idealizing sentiments of Arnoldian humanism have now largely faded from sight, and they are not likely to return. The more recent strategy in the humanities has been not to exalt the human but rather to delegitimize the idea of objective scientific knowledge while simultaneously elevating "discourse" to an ultimate ontological category. The cultural study of science from a postmodern perspective deprecates the idea of human nature, but discourse is itself a specifically and distinctively human function. No other species has developed a language sufficiently complex to articulate propositions about the ontological primacy of language. By subsuming the knowledge of nature within the philosophy of "discourse," postmodern science theory indirectly, from the back door, reaffirms the centrality of the human as the measure of all things.

From a consilient perspective, arts and letters—the subjects of the humanities—are encompassed within more elementary domains of knowledge, within psychology, anthropology, and evolutionary biology. The human sciences and life sciences are themselves encompassed within the still more elementary causal domains of chemistry and physics. Literary scholars who accept this consilient conception of their field would not claim that the theory of discourse has an ultimate epistemic authority. They could, however, identify the concepts and concerns that are particular to literature and other humanistic subjects; they could integrate those concepts with broader, deeper causal principles from other domains; and they could in this way gain for their field an empirical validity and a power of progressive development greater than it has ever had before. Scholars who adopt these strategies would be joining a collective scientific effort to put the human in its place. That effort need involve no repudiation of their own humanity, but it would probably suggest new ways to envision that humanity. Humans are imaginative animals. New ways of seeing are new ways of being. In adapting to the changing environment of knowledge, we shall probably discover new forms for what we regard as our "full" humanity.

CHAPTER 12

A Darwinian Revolution in the Humanities

Darwin's *Descent of Man* fed into a larger stream of "naturalistic" thinking in the philosophy and literature of his time. In contrast to the naturalistic visions of philosophers such as Herbert Spencer and Friedrich Nietzsche, Darwin's vision was grounded in careful reasoning about scientific evidence. He linked us with the other animals as no one had ever done before—logically, scientifically, in a cool and methodical spirit of disinterested inquiry. Though he included passages of grand rhetoric, his vision was not at heart rhetorical. Nor was it deeply inflected with any ideological animus. Over the period of a century and a half, these differences of intellectual quality have made a decisive difference in the magnitude and character of Darwin's influence. Nietzsche, violent, ferocious, and never quite sane, has had his day. Spencer grows dusty on the shelves of antiquarian intellectual history. In our thinking on man's place in nature, Darwin is closer to us now than he has ever been before.

On the Origin of Species had an almost immediate impact on biological science—on the recognition that species had evolved and had not just been "created" by divine fiat. Darwin's theory about *how* species had evolved—by means of natural selection, through a process of adaptation—was suspended in controversy for another half century. The Modern Synthesis, integrating genetics with the theory of natural selection, settled that controversy. Though scientific judgment on Darwin's explanation for the mechanism of evolution remained in suspense for decades, the idea of evolution itself—the idea of "descent with modification"—has shed a continuous light on our understanding of other species. The social sciences followed a very different trajectory. For them, the Darwinian dawn was like the light of a day in the far North, when the dawn and dusk have scarcely any time between them. Around the turn of the century, three great minds, those of John Dewey, William James, and Thorstein Veblen, caught something of Darwin's illumination. In the second decade of the twentieth century, though, founding figures in the social sciences turned resolutely away from Darwin's naturalistic vision of man's place in nature. This is a story that has now often been told—how Durkheim, Kroeber, Lowie, and others built the cultural box outside of which no one could think. Humanity produces culture, they declared, and culture produces humanity. Until the latter part of the twentieth century, this vicious conceptual circle formed the boundary for most thinking in the social sciences.[1]

In the magnificent conclusion to *The Descent of Man*, Darwin evoked and affirmed the nobility of the human mind—the "god-like intellect" that has "penetrated into the movements and constitution of the solar system." Darwin would perhaps have been surprised at the extent to which this god-like mind bears within itself the power to be vastly clever in supporting the flimsiest possible ideas. Why did humans—so far along the way in their descent from their "lowly origins"—descend to folly like that of the culturalist circle? How could intelligent people ever have convinced themselves that humans hold themselves up in mid-air, creating cultures out of nothing? Pride, for one thing. If we create culture, and culture creates us, then we create ourselves. Milton's Satan would have understood something of the psychological impulse behind the culturalist theory, and all the more once he discovered, as Nietzsche would have explained to him, that God was dead. With God out of the picture, humans had no choice but to take responsibility for making their own world.

Pride and a sense of ethical responsibility are both real motives, but to make a theory plausible, one needs more than motive. A theory is plausible, on some level, because it appeals to our sense of reality, however fanciful that sense might be. One reality supporting the notion that culture makes human nature is that we do, in fact, live in the imagination. "A fictive covering / Weaves always glistening from the heart and mind."[2] That's a poet talking, Wallace Stevens, and of course poets have a vested interest in the imagination, but then, in that respect, they are only human. Humans are very strange and unusual animals. Like other animals, they are driven by their passions, prompted by their instincts, goaded by their physical needs. Unlike other animals, though, they create imagined worlds and live in them. We know the world is real, and physical, and yet the real physical world for us is always mediated by images and beliefs, dreams and fantasies, ghosts and demons. We have believed in some very strange things—for instance, in the immortality of the soul, the geocentric universe, and the freedom of the will. Is it any wonder, then, that we should look to culture, the fabrications of our minds, and believe, in our simplicity, that culture contains nature?

The culturalist beliefs that ruled the social sciences through most of the twentieth century were not, in the first place, convictions founded on reason and evidence. They were part of an ideology. It is the nature of an ideology fundamentally to subordinate truth to value. Religions are in this respect ideologies, also. Marxism, with all its panoply of science and its plausible appeal to socioeconomic causality, is still an ideology. Veblen saw into the quasi-religious character of the Marxist historical vision.[3] He saw that the Marxist vision is teleological. It is an imaginative, emotional belief in a transcendent force of progress driving toward an ultimate ideal condition, a consummation of history, the final harmonious concord. That ultimate ideal condition consists in brotherhood and cooperation, a social order based on justice and equity. The Marxist state would be a world constructed in concord with our own purposes and ideals.

We can regard the twentieth century as an empirical test for the hypothesis that we could construct a world on this plan alone—posing an ideal social order and building social structures that reflect that ideal. It was an experiment, and the experiment failed. Ideals alone are not a sufficient basis on which to construct a social order. We also have to take account of human nature. What Darwin knew, and what we have now once again begun to realize, is that human nature makes culture. We can still erect ideals and live by them. We can construct social policies that reflect our sense of justice and decency. But we can't do it effectively unless we take account of the materials with which we have to work. Social institutions are made out of people. People are made out of human nature. Understanding human nature—really getting down to the details in neurology, anatomy, physiology, hormones, and behavioral dispositions encoded in genes—that is the only chance we have of constructing social systems that do not blow up in our faces.

Over the past thirty years or so, we have finally started to come to terms with human nature. Edward O. Wilson's *Sociobiology: The New Synthesis*, published in 1975, is a historical landmark. An imaginative arc reaches from the final paragraph of *The Descent of Man* to the final chapter of *Sociobiology*—the chapter on human nature. Both Darwin and Wilson have the larger vision of man's place in nature. More than any other single work, the final chapter of Wilson's book set off the sociobiological revolution in the social sciences. That revolution is now entering a mature phase. All its subsidiary disciplines and schools—behavioral ecology, human ethology, evolutionary psychology, Darwinian anthropology, behavioral genetics, cognitive neuroscience, and the rest—form part of a new paradigm that is becoming ever more firmly established. If it is true, as Dobzhansky famously said, that nothing in biology makes sense except in the light of evolution, it is equally true that nothing in human behavior makes sense except in the light of sociobiology.[4] That is the larger vision and the larger logic. For the details, one can look readily to excellent popular accounts, now multiplying on an almost daily basis, to books by David Buss, Richard Dawkins, Daniel Goleman, Daniel Nettle, Steven Pinker, Matt Ridley, David Sloan Wilson, Frans de Waal, Nicholas Wade, and many others. For slower going, but more massive confirmation, one can look at handbooks such as *The Handbook of Evolutionary Psychology*, edited by David Buss, and *The Oxford Handbook of Evolutionary Psychology*, edited by Robin Dunbar and Louise Barrett.

For the past thirty years or so, while the social sciences were going through a Darwinian revolution, the humanities have been running in an almost exactly opposite direction. While scientists concerned with human behavior have been recognizing that human culture is shaped and constrained by an evolved and adapted human nature, the humanities have been proclaiming, flamboyantly but with a virtuoso skill in sophistical equivocation, that the world is made of words—"discourse," "rhetoric." This too was a revolution—a breaking free from nature and reality, a last euphoric fling into the vanities of imagination. "There is no outside the text."[5] So Derrida told us. Humans did not exist either as individuals or as a species before we thought of them in that way. So Barthes and Foucault told us. Sex is purely a social construct. So a whole generation has told us. None of it was true. Such things are still often said, in a tired and routine way, but deep down, nobody has ever thoroughly believed them. We all wake up at some point and feel the massive, overwhelming reality of our own biological existence in a physical world. Just step off a curb, in a moment of distraction, get brushed by two tons of metal moving at high speed, and you will have an instantaneous, spontaneous conviction that there is indeed a world outside the text.

God died a lingering death in the nineteenth century. The fundamentalists will tell us that reports of His death, like that of Mark Twain, have been greatly exaggerated. But really, there has been no exaggeration. Three or four centuries ago, the most serious thinkers could still easily envision their conceptual constructs as emanations within a divine creation. Not now. Theology is a sideshow at best, and the main intellectual show goes on without any reference at all to transcendent powers. Even the Marxist sublimations of the transcendental spirit in History have now ceased to sway the minds of most serious thinkers. Looked at on an evolutionary scale, the disappearance of divinity from the world has been instantaneous. Looked at on the scale of cultural history, the transition has been more gradual, with many an eddy in intellectual backwaters. During the later part of the nineteenth century and the first half of the twentieth, the humanities have been one such backwater. Matthew Arnold, one of the last great Victorian Men of Letters, saw clearly the fading of the divine light. For him, it was a sad change, a disenchantment. In compassion to himself and his fellows, he suggested a substitute for the romance of religion. He said that the most active parts of religion were morality and poetry, morality lit up by the enchantments of poetry. In the future, he said, poetry would be our new religion. It would be the channel of the transcendent human spirit.[6] Hard to believe now. I mean, it is hard now to believe that anyone ever believed that. But Arnold's essays on religion sold phenomenally well on both sides of the Atlantic, and the Arnoldian religion of poetry and culture were central animating forces in the humanities well into the third quarter of the twentieth century. The New Critics, as they were called in the middle decades of the century, were for the most part both Christians and adherents of the Arnoldian religion of culture. The greatest theoretical mind in literary study in the middle of the twentieth century was that of Northrop Frye, and Frye was both a Christian minister and a Romantic mystic. Most of all, he was a proponent of Culture, in the Arnoldian sense. He believed that the total order of literary words represents an embodiment of the mind of God.[7]

For the first six decades of the twentieth century, the humanities were the chief refuge of mystical fervor in the world of intellect. Then, a revolution took place. If Marx turned Hegel on his head, Derrida turned Frye on his. Frye looked to literature for a spiritual plenitude, and Derrida flipped that vision over into nihilistic vacancy. Endless "deferral" took the place of an ultimate consummation. Derrida often proclaimed the world-historical, apocalyptic character of his vision, and many literary theorists shared in this giddy delusion. Looking back now, both of these visionary phases seem outlandish and a little absurd. The mystical illuminations of Arnoldian humanism were afterglows of a lost cause, and the epochal inversions of deconstruction were baubles of a metaphysical rhetoric more suitable to the thirteenth century than to the twentieth.

For the past fifteen years or so, a counterrevolution has been taking place in the humanities, and especially in literary studies. The literary Darwinists took to heart the vision of *The Descent of Man* and *Sociobiology: The New Synthesis*. Following Darwin, they saw that "man still bears in his bodily frame the indelible stamp of his lowly origin."[8] Like Darwin, they recognized that stamp not only in the human body but also in the human mind. They felt the charm in the very title of the seminal volume in evolutionary psychology, *The Adapted Mind*, and they rallied to the cry for intellectual unification in Wilson's *Consilience: The Unity of Knowledge*.

I think it fairly safe to predict that the profession of literary scholarship will eventually, necessarily, be encompassed within the wider world of naturalistic knowledge. The humanities will not be able to sustain much longer the idea of a world made out of words, either in the mystical version represented by Frye or in the nihilistic version represented by Derrida. The heyday of deconstruction was astonishingly brief—a delirium that swept through English departments, infected almost everyone, and then suddenly departed, supplanted by the political criticism of Foucault and company. Literary study has to have substance. It has to deal with human realities, with psychological impulses and social forces. Derridean wordplay offered too thin an atmosphere in which to breathe. Deconstruction left behind merely a spirit of subversion and a mystified belief in the transcendent reality of "discourse." The substance of discourse was filled in by Althusserian Marxism, Lacanian psychoanalytic theory, and the brooding Foucauldian preoccupation with social "power." For the past two or three decades, that theoretical swirl has been the medium of mainstream thought in the humanities. It cannot last. The Marxists are social theorists, and the Lacanians are psychologists. The forms of psychology and social theory now propounded in the humanities cannot compete effectively with the forms available in the evolutionary human sciences.

In their dependence on jargonized speculative fantasies, the humanities have drifted off into an intellectual third world. That will have to change, and is already changing. The humanities are in crisis and know it. The titles of edited volumes and special issues of journals tell the tale. People in the humanities are not unintelligent. They have simply been trapped in local currents of intellectual history. At some point in the not too distant future, the sheer embarrassment of being unable to contribute in any useful way to the serious world of adult knowledge will, I think, have a decisive effect in reorienting the discipline. At the end of *Evolution and Literary Theory*, I said, "whatever happens within the critical institution as a whole, the pursuit of positive knowledge is available to anyone who desires it. Within this pursuit, the opportunities for real and substantial development in our scientific understanding of culture and literature are now greater than they have ever been before." That was nearly fifteen years ago. Since then, the opportunities have only increased.

Notes

Introduction

1. Joseph Carroll, *The Cultural Theory of Matthew Arnold* (Berkeley: U of California P, 1982); Joseph Carroll, *Wallace Stevens' Supreme Fiction: A New Romanticism* (Baton Rouge: Louisiana State UP, 1987); Joseph Carroll, "Pater's Figures of Perplexity," *Modern Language Quarterly* 52 (1991): 319–40.
2. Joseph Carroll, *Evolution and Literary Theory* (Columbia: U of Missouri P, 1995).
3. D. T. Max, "The Literary Darwinists," *The New York Times Magazine* (6 November 2005): 77.
4. Charles Darwin, *On the Origin of Species by Means of Natural Selection*, ed. Joseph Carroll (1859; Peterborough, Ontario: Broadview, 2003); Joseph Carroll, *Literary Darwinism: Evolution, Human Nature, and Literature* (New York: Routledge, 2004).
5. Frederick Crews, "Apriorism for Empiricists," *Style* 42 (2008): 155–60; William Deresiewicz, "Adaptation: On Literary Darwinism," *The Nation*, 8 June 2009, 26–31; Eugene Goodheart, *Darwinian Misadventures in the Humanities* (New Brunswick, NJ: Transaction, 2007); Eugene Goodheart, "Do We Need Literary Darwinism?" *Style* 42 (2008): 181–85; James M. Mellard, "'No ideas but in things': Fiction, Criticism, and the New Darwinism," *Style* 41 (2007): 1–29; Steven Pinker, "Toward a Consilient Study of Literature," *Philosophy and Literature* 31 (2007): 162–78; Roger Seamon, "Literary Darwinism as Science and Myth," *Style* 42 (2008): 261–65; Edward Slingerland, "Good and Bad Reductionism: Acknowledging the Power of Culture," *Style* 42 (2008): 266–71; Sebastian Smee, "Natural-Born Thrillers," *The Australian Literary Review* (6 May 2009): 17; Ellen Spolsky, "The Centrality of the Exceptional in Literary Study," *Style* 42 (2008): 285–89; Blakey Vermeule, "Response to Joseph Carroll," *Style* 42 (2008): 302–08.

Chapter 1. An Evolutionary Paradigm for Literary Study, with Two Sequels

1. Joseph Carroll, "Adaptationist Literary Study: An Emerging Research Program," *Style* 36 (2003): 596–617; Joseph Carroll, "Literature and

Evolutionary Psychology," in *The Handbook of Evolutionary Psychology*, ed. David M. Buss (Hoboken, NJ: Wiley, 2005), 931–52; Joseph Carroll, "Adaptationist Literary Study: An Introductory Guide," *Ometeca* 10 (2006): 18–31; Joseph Carroll, "Evolutionary Approaches to Literature and Drama," in *The Oxford Handbook of Evolutionary Psychology*, ed. Robin Dunbar and Louise Barrett (Oxford: Oxford UP, 2007), 637–48.

2. Carroll, "Emerging Research Program," 611.
3. Frederick Crews, *Postmodern Pooh* (2001; Evanston, IL: Northwestern UP, 2007).
4. Mellard, "'No ideas but in things,'" 1.
5. Edward O. Wilson, *Consilience: The Unity of Knowledge* (New York: Alfred A. Knopf, 1998).
6. Ellen Spolsky, preface to *The Work of Fiction: Cognition, Culture, and Complexity*, ed. Alan Richardson and Ellen Spolsky (Burlington, VT: Ashgate, 2004), viii, x, x.
7. Alan Richardson, "Studies in Literature and Cognition: A Field Map," in *The Work of Fiction: Cognition, Culture, and Complexity*, ed. Alan Richardson and Ellen Spolsky (Burlington, VT: Ashgate, 2004), 3, 19, 20, 21. For similar appraisals of the disciplinary alignments of the contributors to cognitive poetics, see Tony Jackson, "Issues and Problems in the Blending of Cognitive Science, Evolutionary Psychology, and Literary Study," *Poetics Today* 23 (2002): 161–79; Tony Jackson, "Questioning Interdisciplinarity," *Poetics Today* 21 (2000): 319–47.
8. See Louise Barrett, Robin Dunbar, and John Lycett, *Human Evolutionary Psychology* (Princeton: Princeton UP, 2002), 8–21; David M. Buss, *Evolutionary Psychology: The New Science of the Mind*, 3rd ed. (Boston: Allyn and Bacon, 2007); Robin Dunbar and Louise Barrett, "Evolutionary Psychology in the Round," in *The Oxford Handbook of Evolutionary Psychology*, ed. Robin Dunbar and Louise Barrett (Oxford: Oxford UP, 2007), 3–9; Steven W. Gangestad and Jeffry A. Simpson, "An Introduction to *The Evolution of Mind*: Why We Developed This Book," in *The Evolution of Mind: Fundamental Questions and Controversies*, ed. Steven W. Gangestad and Jeffry A. Simpson (New York: Guilford, 2007), 1–21; Edward H. Hagen and Donald Symons, "Natural Psychology: The Environment of Evolutionary Adaptedness and the Structure of Cognition," in *The Evolution of Mind: Fundamental Questions and Controversies*, ed. Steven W. Gangestad and Jeffry A. Simpson (New York: Guilford, 2007), 38–44; Kevin N. Laland, "Niche Construction, Human Behavioural Ecology and Evolutionary Psychology," in *The Oxford Handbook of Evolutionary Psychology*, ed. Robin Dunbar and Louise Barrett (Oxford: Oxford UP, 2007), 35–47; Kevin N. Laland and Gilian R. Brown, *Sense and Nonsense: Evolutionary Perspectives on Human Behaviour* (Oxford: Oxford UP, 2002); Matteo Mamelli, "Evolution and Psychology in Philosophical Perspective," in *The Oxford Handbook of Evolutionary Psychology*, ed. Robin Dunbar and

Louise Barrett (Oxford: Oxford UP, 2007), 21–34; Steven Pinker, foreword to *The Handbook of Evolutionary Psychology*, ed. David M. Buss (Hoboken, NJ: Wiley, 2005), xi–xvi; Kim Sterelny, *Thought in a Hostile World: The Evolution of Human Cognition* (Oxford: Oxford UP, 2003), 234–35; Edward O. Wilson, "Sociobiology at Century's End," in *Sociobiology: The New Synthesis, Twenty-Fifth Anniversary Edition* (Cambridge: Harvard UP, 2000), v–viii.

9. Jerome Barkow, Leda Cosmides, and John Tooby, eds., *The Adapted Mind: Evolutionary Psychology and the Generation of Culture* (New York: Oxford UP, 1992).

10. See for instance Simon Baron-Cohen, "The Empathizing System: A Revision of the 1994 Model of the Mindreading System," in *Origins of the Social Mind: Evolutionary Psychology and Child Development*, ed. Bruce J. Ellis and David F. Bjorklund (New York: Guilford, 2005), 468–92; Barrett, Dunbar, and Lycett, *Human Evolutionary Psychology*, 295–350; Robert A. Barton, "Evolution of the Social Brain as a Distributed Neural System," in *The Oxford Handbook of Evolutionary Psychology*, ed. Robin Dunbar and Louise Barrett (Oxford: Oxford UP, 2007), 129–44; Derek Bickerton, "From Protolanguage to Language," in *The Speciation of Modern Homo Sapiens*, ed. T. J. Crow (Oxford: Oxford UP, 2002), 103–20; Dan Chiappe and Kevin B. MacDonald, "Metaphor, Modularity, and the Evolution of Conceptual Integration," *Metaphor and Symbol* 15 (2000): 137–58; Antonio R. Damasio, *Descartes' Error: Emotion, Reason, and the Human Brain* (New York: Putnam's, 1994); Terrence W. Deacon, *The Symbolic Species: The Co-Evolution of Language and the Brain* (New York: Norton, 1997); David Geary, *The Origin of Mind: Evolution of Brain, Cognition, and General Intelligence* (Washington, DC: American Psychological Association, 2005); Tjeerd Jellema and David I. Perrett, "Neural Pathways of Social Cognition," in *The Oxford Handbook of Evolutionary Psychology*, ed. Robin Dunbar and Louise Barrett (Oxford: Oxford UP, 2007), 163–77; Simon Kirby, "The Evolution of Language," in *The Oxford Handbook of Evolutionary Psychology*, ed. Robin Dunbar and Louise Barrett (Oxford: Oxford UP, 2007), 669–81; Kevin B. MacDonald and Scott Hershberger, "Theoretical Issues in the Study of Evolution and Development," in *Evolutionary Perspectives on Human Development*, 2nd ed., ed. Robert Burgess and Kevin B. MacDonald (Thousand Oaks: Sage, 2004), 21–72; Steven Mithen, *The Prehistory of the Mind: The Cognitive Origins of Art, Religion, and Science* (London: Thames and Hudson, 1996); Jaak Panksepp, "The Neuroevolutionary and Neuroaffective Psychobiology of the Prosocial Brain," in *The Oxford Handbook of Evolutionary Psychology*, ed. Robin Dunbar and Louise Barrett (Oxford: Oxford UP, 2007), 145–62; Steven Pinker, *How the Mind Works* (New York: Norton, 1977); Steven Pinker, *The Language Instinct: How the Mind Creates Language* (New York: William Morrow, 1994); Steven Pinker, *The Stuff of Thought: Language as a Window into Human Nature* (New York: Viking, 2007); Steven Pinker and Ray Jackendoff, "The Faculty of

Language: What's Special about It?" *Cognition* 95 (2004): 201–36; Henry Plotkin, "The Power of Culture," in *The Oxford Handbook of Evolutionary Psychology*, ed. Robin Dunbar and Louise Barrett (Oxford: Oxford UP, 2007), 11–19; David Premack and Ann James Premack, "Origins of Human Social Competence," in *The Cognitive Neurosciences*, ed. Michael S. Gazzaniga (Cambridge, MA: MIT P, 1995), 205–18; Giacomo Rizzolatti and Leonardo Fogassi, "Mirror Neurons and Social Cognition," in *The Oxford Handbook of Evolutionary Psychology*, ed. Robin Dunbar and Louise Barrett (Oxford: Oxford UP, 2007), 179–95; Michael Tomasello et al., "Understanding and Sharing Intentions: The Origins of Cultural Cognition," *Behavioral and Brain Sciences* 28 (2005): 675–735; Emily Wyman and Michael Tomasello, "The Ontogenetic Origins of Human Cooperation," in *The Oxford Handbook of Evolutionary Psychology*, ed. Robin Dunbar and Louise Barrett (Oxford: Oxford UP, 2007), 227–36.

11. Ellen Spolsky, *Gaps in Nature: Literary Interpretation and the Modular Mind* (Albany: State U of New York P, 1993), 12.

12. Lisa Zunshine, *Why We Read Fiction: Theory of Mind and the Novel* (Columbus: Ohio State UP, 2006).

13. See David Bordwell, *Poetics of Cinema* (New York: Routledge, 2008); David Bordwell and Noël Carroll, eds., *Post-Theory: Reconstructing Film Studies* (Madison: U of Wisconsin P, 1996). Also see Joseph Anderson, *The Reality of Illusion: An Ecological Approach to Cognitive Film Theory* (Carbondale: Southern Illinois UP, 1996); Joseph D. Anderson and Barbara Fisher Anderson, eds., *Moving Image Theory: Ecological Considerations* (Carbondale: Southern Illinois UP, 2005); Torben Grodal, *Embodied Visions: Evolution, Emotion, Culture, and Film* (Oxford: Oxford UP, 2009); Carl Plantinga and Greg M. Smith, eds., *Passionate Views: Film, Cognition, and Emotion* (Baltimore: Johns Hopkins UP, 1999); Murray Smith, "Darwin and the Directors: Film, Emotion, and the Face in the Age of Evolution," *TLS*, 7 February 2003, 13–15; Ed S. Tan, *Emotion and the Structure of Narrative Film: Film as an Emotion Machine*, trans. Barbara Fasting (Mahwah, NJ: Erlbaum, 1996).

14. Brian Boyd, "On the Origin of Comics: New York Double-Take," *The Evolutionary Review: Art, Science, Culture* 1 (2010): 97–111; Brian Boyd, *On the Origin of Stories: Evolution, Cognition, and Fiction* (Cambridge: Harvard UP, 2009); Brian Boyd, "The Art of Literature and the Science of Literature," *The American Scholar* 77.2 (2008): 118–27 (on *Lolita*); Brian Boyd, "Art and Evolution: Spiegelman in *The Narrative Corpse*," *Philosophy and Literature* 32 (2008): 31–57.

15. Frederick Turner, *Natural Classicism: Essays on Literature and Science* (1985; Charlottesville: UP of Virginia, 1992); Michael Winkelman, "Sighs and Tears: Biological Signals and Donne's 'Whining Poetry,'" *Philosophy and Literature* 33 (2009): 329–44.

16. Boyd, *Origin of Stories*; Michelle Scalise-Sugiyama, "Reverse-Engineering Narrative: Evidence of Special Design," in *The Literary Animal: Evolution and the Nature of Narrative*, ed. Jonathan Gottschall

and David Sloan Wilson (Evanston, IL: Northwestern UP, 2005), 177–96; Francis Steen, "The Paradox of Narrative Thinking," *Journal of Cultural and Evolutionary Psychology* 3 (2005): 87–105; Daniel Nettle, "The Wheel of Fire and the Mating Game: Explaining the Origins of Tragedy and Comedy," *Journal of Cultural and Evolutionary Psychology* 3 (2005): 39–56.
17. Darwin, *Origin of Species*, 398.
18. Jonathan Gottschall and David Sloan Wilson, eds., *The Literary Animal: Evolution and the Nature of Narrative* (Evanston, IL: Northwestern UP, 2005); Robin Headlam Wells and Johnjoe McFadden, eds., *Human Nature: Fact and Fiction* (London: Continuum, 2006); Brian Boyd, Joseph Carroll, and Jonathan Gottschall, eds., *Evolution, Literature, and Film: A Reader* (New York: Columbia UP, 2010); Alice Andrews and Joseph Carroll, eds., *The Evolutionary Review: Art, Science, Culture* 1 (2010). For a collection of essays focusing specifically on Hispanic literature, see Jerry Hoeg and Kevin S. Larsen, eds., *Interdisciplinary Essays on Darwinism in Hispanic Literature and Film: The Intersection of Science and the Humanities* (New York: Edwin Mellen, 2009).
19. Robert Storey, *Mimesis and the Human Animal: On the Biogenetic Foundations of Literary Representation* (Evanston, IL: Northwestern UP, 1996).
20. Ellen Dissanayake, *Homo Aestheticus: Where Art Comes from and Why* (1992; Seattle: U of Washington P, 1995); Ellen Dissanayake, *Art and Intimacy: How the Arts Began* (Seattle: U of Washington P, 2000). On evolutionary aesthetics, also see Denis Dutton, *The Art Instinct* (New York: Bloomsbury, 2009). Chapter 3 of part 1 in the present volume contains a commentary on *The Art Instinct*.
21. Jonathan Gottschall, *Literature, Science, and a New Humanities* (New York: Palgrave Macmillan, 2008).
22. For a polemical essay by Boyd, see "Getting It All Wrong," *The American Scholar* 75.4 (2006): 18–30.
23. Harold Fromm, *The Nature of Being Human: From Environmentalism to Consciousness* (Baltimore: Johns Hopkins UP, 2009); Harold Fromm, *Academic Capitalism and Literary Value* (Athens: U of Georgia P, 1991). For other studies combining evolutionary and ecological thinking, see Carroll, *Literary Darwinism*, 16–22, 85–100, 147–85; Nancy Easterlin, "'Loving Ourselves Best of All': Ecocriticism and the Adapted Mind," *Mosaic* 37 (2004): 1–18; Glen Love, *Practical Ecocriticism: Literature, Biology, and the Environment* (Charlottesville: U of Virginia P, 2003). Love also has two important theoretical essays: Glen A. Love, "Ecocriticism and Science: Toward Consilience?" *New Literary History* 30 (1999): 561–76; Glen A. Love, "Science, Anti-Science, and Ecocriticism," *Interdisciplinary Studies in Literature and the Environment* 6 (1999): 65–81.
24. Brett Cooke, *Human Nature in Utopia: Zamyatin's We* (Evanston, IL: Northwestern UP, 2002). Other chief "biopoetical" works by Cooke include "The Promise of a Biothematics," in *Sociobiology and the Arts*, ed. Jean Baptiste Bedaux and Brett Cooke (Amsterdam: Editions Rodopi, 1999), 43–62; and "Sexual Property in Pushkin's

'The Snowstorm': A Darwinist Perspective," in *Biopoetics: Evolutionary Explorations in the Arts*, ed. Brett Cooke and Frederick Turner (Lexington, KY: ICUS, 1999), 175–204.
25. Marcus Nordlund, *Shakespeare and the Nature of Love: Literature, Culture, Evolution* (Evanston, IL: Northwestern UP, 2007). Nordlund also has an important theoretical paper, "Consilient Literary Interpretation," *Philosophy and Literature* 26 (2002): 312–33.
26. Robin Headlam Wells, *Shakespeare's Humanism* (Cambridge: Cambridge UP, 2005).
27. Jonathan Gottschall, *The Rape of Troy: Evolution, Violence, and the World of Homer* (Cambridge: Cambridge UP, 2008).
28. Judith P. Saunders, *Reading Edith Wharton through a Darwinian Lens: Evolutionary Biological Issues in Her Fiction* (Jefferson, NC: McFarland, 2009). Also see Judith Saunders, "Male Reproductive Strategies in Sherwood Anderson's 'The Untold Lie,'" *Philosophy and Literature* 31 (2007): 311–22; Judith Saunders, "Paternal Confidence in Hurston's 'The Gilded Six-Bits,'" in *Evolution, Literature, and Film*, ed. Boyd, Carroll, and Gottschall. David Barash and Nanelle Barash, *Madame Bovary's Ovaries: A Darwinian Look at Literature* (New York: Delacorte, 2005).
29. John A. Johnson, Joseph Carroll, Jonathan Gottschall, and Daniel J. Kruger, "Hierarchy in the Library: Egalitarian Dynamics in Victorian Novels," *Evolutionary Psychology* 6 (2008): 715–38. Also see part 3 in the present volume.
30. Michelle Scalise-Sugiyama, "On the Origins of Narrative: Storyteller Bias as a Fitness-Enhancing Strategy," *Human Nature* 7 (1996): 403–25; Michelle Scalise-Sugiyama, "Cultural Relativism in the Bush: Toward a Theory of Narrative Universals," *Human Nature* 14 (2003): 383–96; Storey, *Mimesis and the Human Animal*; Carroll, *Literary Darwinism*, 129–45, 163–85, 206–16. And see parts 2 and 3 in the present volume.
31. William Flesch, *Comeuppance: Costly Signaling, Altruistic Punishment, and Other Biological Components of Fiction* (Cambridge: Harvard UP, 2008); Michael Austin, *Useful Fictions: Evolution, Anxiety, and the Origins of Literature* (Lincoln: U of Nebraska P, 2011).
32. John Dryden, *Selected Criticism*, ed. James Kinsley and George Parfitt (Oxford: Oxford UP, 1970), 25. For references to other such examples, see Carroll, *Evolution*, 170; Steven Pinker, *The Blank Slate: The Modern Denial of Human Nature* (New York: Viking, 2002), 404–20.
33. Pascal Boyer, "Specialised Inference Engines as Precursors of Creative Imagination?" *Proceedings of the British Academy* 147 (2007): 239–58; Robin Dunbar, "Why Are Good Writers So Rare? An Evolutionary Perspective on Literature," *Journal of Evolutionary and Cultural Psychology* 3 (2005): 7–22; Mithen, *Prehistory of the Mind*; Geary, *Origin of Mind*; Sterelny, *Thought in a Hostile World*.
34. Goodheart, *Darwinian Misadventures*, 18.
35. Richardson, "Studies," 14.
36. Pinker, *How the Mind Works*; Donald Symons, *The Evolution of Human*

Sexuality (New York: Oxford UP, 1979); Donald Symons, "On the Use and Misuse of Darwinism in the Study of Human Behavior," in *The Adapted Mind: Evolutionary Psychology and the Generation of Culture*, ed. Jerome H. Barkow, Leda Cosmides, and John Tooby (Oxford: Oxford UP, 1992), 137–62; John Tooby and Leda Cosmides, "The Psychological Foundations of Culture," in *The Adapted Mind: Evolutionary Psychology and the Generation of Culture*, ed. Jerome Barkow, Leda Cosmides, and John Tooby (New York: Oxford UP, 1992), 19–136.

37. On human life history theory, see Kim Hill, "Evolutionary Biology, Cognitive Adaptations, and Human Culture," in *The Evolution of Mind: Fundamental Questions and Controversies*, ed. Steven W. Gangestad and Jeffry A. Simpson (New York: Guilford, 2007), 348–56; Hillard S. Kaplan and Steven W. Gangestad, "Life History Theory and Evolutionary Psychology," in *The Handbook of Evolutionary Psychology*, ed. David M. Buss (Hoboken, NJ: Wiley, 2005); Hillard S. Kaplan and Steven W. Gangestad, "Optimality Approaches and Evolutionary Psychology: A Call for Synthesis," in *The Evolution of Mind: Fundamental Questions and Controversies*, ed. Steven W. Gangestad and Jeffry A. Simpson (New York: Guilford, 2007), 121–29; Hillard S. Kaplan, Kim Hill, Jane Lancaster, and A. Magdalena Hurtado, "A Theory of Human Life History Evolution: Diet, Intelligence, and Longevity," *Evolutionary Anthropology* 9 (2000): 156–85; Jane B. Lancaster and Hillard S. Kaplan, "Chimpanzee and Human Intelligence: Life History, Diet, and the Mind," in *The Evolution of Mind: Fundamental Questions and Controversies*, ed. Steven W. Gangestad and Jeffry A. Simpson (New York: Guilford, 2007), 111–18; Bobbi S. Low, *Why Sex Matters: A Darwinian Look at Human Behavior* (Princeton: Princeton UP, 2000); Virpi Lummaa, "Life-History Theory, Reproduction, and Longevity in Humans," in *The Oxford Handbook of Evolutionary Psychology*, ed. Robin Dunbar and Louise Barrett (Oxford: Oxford UP, 2007), 397–413; Kevin B. MacDonald, "Life History Theory and Human Reproductive Behavior: Environmental/Contextual Influences and Heritable Variation," *Human Behavior* 8 (1997): 327–59.

38. Christopher Boehm, *Hierarchy in the Forest: The Evolution of Egalitarian Behavior* (Cambridge: Harvard UP, 1999); Arnold Buss, "Evolutionary Perspectives on Personality Traits," in *Handbook of Personality Psychology*, ed. Robert Hogan, John Johnson, and Stephen Briggs (San Diego: Academic P, 1997), 346–66; Denise Cummins, "Dominance, Status, and Social Hierarchies, in *The Handbook of Evolutionary Psychology*, ed. David M. Buss (Hoboken, NJ: Wiley, 2005), 676–97; Jeffrey A. Gray, "The Neuropsychology of Temperament," in *Explorations in Temperament*, ed. Jan Strelau and Alois Angleitner (New York: Plenum, 1991), 105–28; Kevin B. MacDonald, "Evolution, Culture, and the Five-Factor Model," *Journal of Cross-Cultural Psychology* 29 (1998): 119–49; Kevin B. MacDonald, "Evolution, The Five-Factor Model, and Levels of Personality," *Journal of Personality* 63 (1995): 525–67; Robert Plutchik, *Emotions and Life:*

Perspectives from Psychology, Biology, and Evolution (Washington, DC: American Psychological Association, 2003).

39. David F. Bjorklund and Anthony D. Pellegrini, *The Origins of Human Nature: Evolutionary Developmental Psychology* (Washington, DC: American Psychological Association, 2002); David M. Buss, *The Evolution of Desire: Strategies of Human Mating*, rev. ed. (New York: Basic Books, 2003); David M. Buss, *The Dangerous Passion: Why Jealousy Is as Necessary as Love and Sex* (New York: Free P, 2000); Deacon, *The Symbolic Species*; Mark V. Flinn, David C. Geary, and Carol V. Ward, "Ecological Dominance, Social Competition, and Coalitionary Arms Races: Why Humans Evolved Extraordinary Intelligence," *Evolution and Human Behavior* 26 (2005): 10–46; Steven W. Gangestad, "Reproductive Strategies and Tactics," in *The Oxford Handbook of Evolutionary Psychology*, ed. Robin Dunbar and Louise Barrett (Oxford: Oxford UP, 2007), 321–32; David C. Geary, "Evolution of Paternal Investment," in *The Handbook of Evolutionary Psychology*, ed. David M. Buss (Hoboken, NJ: Wiley, 2005), 483–505; David C. Geary, *Male, Female: The Evolution of Human Sex Differences* (Washington, DC: American Psychological Association, 1998); David C. Geary and Mark V. Flinn, "Evolution of Human Parental Behavior and the Human Family," *Parenting: Science and Practice* 1 (2001): 5–61; Hill, "Evolutionary Biology"; Low, *Why Sex Matters*; Catherine A. Salmon and Todd K. Shackelford, eds., *Family Relationships: An Evolutionary Perspective* (Oxford: Oxford UP, 2008); David P. Schmitt, "Fundamentals of Human Mating Strategies," in *The Handbook of Evolutionary Psychology*, ed. David M. Buss (Hoboken, NJ: Wiley, 2005), 258–91; Symons, *Human Sexuality*.

40. Bjorklund and Pellegrini, *Evolutionary Developmental Psychology*; D. Buss, *Evolution of Desire*; Martin Daly and Margo Wilson, *Homicide* (New York: Aldine, 1988); Mark V. Flinn and Carol V. Ward, "Ontogeny and Evolution of the Social Child," in *Origins of the Social Mind*, ed. Bruce J. Ellis and David F. Bjorklund (New York: Guilford, 2005), 19–44; Geary, *Male, Female*; Geary, *Origin of Mind*; Geary and Flinn, "Human Parental Behavior"; William D. Hamilton, "The Genetical Evolution of Social Behavior, I and II," in *Narrow Roads of Gene Land: The Collected Papers of W. D. Hamilton*, vol. 1 (Oxford: W. H. Freeman, 1996), 11–82; Catherine Salmon, "Parental Investment and Parent-Offspring Conflict," in *The Handbook of Evolutionary Psychology*, ed. David M. Buss (Hoboken, NJ: Wiley, 2005), 506–27; Schmitt, "Human Mating Strategies"; Symons, *Human Sexuality*; Robert Trivers, *Social Evolution* (Menlo Park, CA: Benjamin/Cummins, 1985); Robert Trivers, "Parental Investment and Sexual Selection," in *Sexual Selection and the Descent of Man 1871–1971*, ed. Bernard Campbell (Chicago: Aldine, 1972), 136–79.

41. See Richard D. Alexander, *The Biology of Moral Systems* (Hawthorne, NY: Aldine de Gruyter, 1987); Boehm, *Hierarchy in the Forest*; Cummins, "Dominance"; Flinn and Ward, "Ontogeny"; Geary and Flinn, "Human Parental Behavior"; Jeffrey A. Kurland and Steven J. C. Gaulin, "Cooperation and Conflict among Kin," in *The Handbook*

of *Evolutionary Psychology*, ed. David M. Buss (Hoboken, NJ: Wiley, 2005), 447–82; Pinker, *Stuff of Thought*, 380, 401–09; Premack and Premack, "Human Social Competence"; Elliott Sober and David Sloan Wilson, *Unto Others: The Evolution and Psychology of Unselfish Behavior* (Cambridge: Harvard UP, 1998).

42. See Stephen Budiansky, *If a Lion Could Talk: How Animals Think* (London: Weidenfeld and Nicolson, 1998); A. Buss, "Personality Traits"; Darwin, *Descent of Man*; Farah Focquaert and Steven M. Platek, "Social Cognition and the Evolution of Self-Awareness," in *Evolutionary Cognitive Neuroscience*, ed. Steven M. Platek, Julian Paul Keenan, and Todd K. Shackelford (Cambridge, MA: MIT P, 2007), 457–97; Marc Hauser, *Wild Minds: What Animals Really Think* (New York: Henry Holt, 2000); Michael Lewis, "Self-Conscious Emotions: Embarrassment, Pride, Shame, and Guilt," in *Handbook of Emotions*, 2nd ed., ed. Michael Lewis and Jeannette M. Haviland-Jones (New York: Guilford, 2000), 137–56; Delroy L. Paulhus and Oliver P. John, "Egoistic and Moralistic Biases in Self-Perception: The Interplay of Self-Deceptive Styles with Basic Traits and Motives," *Journal of Personality* 66 (1998): 1025–60; Premack and Premack, "Human Social Competence"; Tomasello et al., "Understanding and Sharing Intentions."

43. See Baron-Cohen, "The Empathizing System"; Barrett, Dunbar, and Lycett, *Human Evolutionary Psychology*, 295–321; Premack and Premack, "Human Social Competence"; Rizzolatti and Fogassi, "Mirror Neurons"; Valerie E. Stone, "Theory of Mind and the Evolution of Social Intelligence," in *Social Neuroscience: People Thinking about People*, ed. John T. Cacioppo, Penny S. Visser, and Cynthia L. Pickett (Cambridge, MA: Bradford-MIT P, 2006), 103–30; Tomasello et al., "Understanding and Sharing Intentions"; and Wyman and Tomasello, "Ontogenetic Origins."

44. For formulations on the relations between human nature and culture, see Roy F. Baumeister, *The Cultural Animal: Human Nature, Meaning, and Social Life* (Oxford: Oxford UP, 2005); Boyd, "Art and Evolution"; Brown, "Human Universals"; Dissanayake, *Art and Intimacy*; Nancy Easterlin, "Hans Christian Andersen's Fish Out of Water," *Philosophy and Literature* 25 (2001): 251–77; Jonathan Gottschall, "The Tree of Knowledge and Darwinian Literary Study," *Philosophy and Literature* 27 (2003): 255–68; Headlam Wells, *Shakespeare's Humanism*; Headlam Wells and McFadden, *Human Nature*; Kevin B. MacDonald, "Evolution, Psychology, and a Conflict Theory of Culture," *Evolutionary Psychology* 7 (2009): 208–33; Mithen, *Prehistory of the Mind*; Nordlund, *Shakespeare*; Peter J. Richerson and Robert Boyd, *Not by Genes Alone: How Culture Transformed Human Evolution* (Chicago: U of Chicago P, 2005); Nicholas Wade, *Before the Dawn: Recovering the Lost History of Our Ancestors* (New York: Penguin, 2006); E. Wilson, *Consilience*, 210–37. And see the references later in this chapter in "The Adaptive Function of Literature: A Controversy."

45. Baron-Cohen, "The Empathizing System"; Boyer, "Specialized

Inference Engines"; Robin Dunbar, *The Human Story: A New History of Mankind's Evolution* (London: Faber and Faber, 2004); Geary, *Origin of Mind*, 131–39, 330; Mithen, *Prehistory of the Mind*.

46. Fredric Jameson, *Postmodernism, or, the Cultural Logic of Late Capitalism* (Durham, NC: Duke UP, 1991), ix.
47. Konrad Lorenz, *Behind the Mirror: A Search for a Natural History of Human Knowledge*, trans. Ronald Taylor (1973; New York: Harcourt, 1978), 1–19.
48. David Sloan Wilson, "Evolutionary Social Constructivism," in *The Literary Animal: Evolution and the Nature of Narrative*, ed. Jonathan Gottschall and David Sloan Wilson (Evanston, IL: Northwestern UP, 2005), 20–37.
49. Tooby and Cosmides, "Psychological Foundations," 113.
50. Geary, *Origins of Mind*; Sterelny, *Thought in a Hostile World*.
51. Brian Boyd, "Evolutionary Theories of Art," in *The Literary Animal: Evolution and the Nature of Narrative*, ed. Jonathan Gottschall and David Sloan Wilson (Evanston, IL: Northwestern UP, 2005), 147–76; Carroll, *Literary Darwinism*, 65–69; Carroll, "Literature and Evolutionary Psychology"; Ellen Dissanayake, "What Art Is and What Art Does: An Overview of Contemporary Evolutionary Hypotheses," in *Evolutionary and Neurocognitive Approaches to Aesthetics, Creativity, and the Arts*, ed. Colin Martindale, Paul Locher, and Vladimir M. Petrov (Amityville, NY: Baywood, 2007), 1–14; Pinker, "Consilient Study"; Catherine Salmon and Donald Symons, "Slash Fiction and Human Mating Psychology," *Journal of Sex Research* 41 (2004): 94–100; John Tooby and Leda Cosmides, "Does Beauty Build Adapted Minds? Toward an Evolutionary Theory of Aesthetics, Fiction, and the Arts," *SubStance* 30 (2001): 6–27.
52. Pinker, *How the Mind Works*, 524–43.
53. Geoffrey Miller, *The Mating Mind: How Sexual Choice Shaped the Evolution of Human Nature* (New York: Doubleday, 2000), 281.
54. Pinker, "Consilient Study," 169–70.
55. E. Wilson, *Consilience*, 224–25. Also see Peter Swirski, *Of Literature and Knowledge: Explorations in Narrative Thought Experiments, Evolution, and Game Theory* (London: Routledge, 2007), 68–95.
56. For other arguments on the evolution of human cognitive and behavioral flexibility, see Joseph Carroll, "The Human Revolution and the Adaptive Function of Literature," *Philosophy and Literature* 30 (2006): 33–49; Robert Foley, "The Adaptive Legacy of Human Evolution: A Search for the Environment of Evolutionary Adaptedness," *Evolutionary Anthropology* 4 (1996): 194–203; William Irons, "Adaptively Relevant Environments versus the Environment of Evolutionary Adaptedness," *Evolutionary Anthropology* 6 (1998): 194–204; Mithen, *Prehistory of the Mind*; Rick Potts, "Variability Selection in Hominid Evolution," *Evolutionary Anthropology* 8 (1998): 81–96; Sterelny, *Thought in a Hostile World*; Wade, *Before the Dawn*.
57. Jaak Panksepp and Jules B. Panksepp, "The Seven Sins of Evolutionary Psychology," *Evolution and Cognition* 6 (2000): 126–27.
58. Deacon, *The Symbolic Species*, 22.

59. Gregory Cochran and Henry Harpending, *The 10,000 Year Explosion: How Civilization Accelerated Human Evolution* (New York: Basic Books, 2009); Richerson and Boyd, *Not by Genes Alone*; Wade, *Before the Dawn*. For other contributions to the theory of gene-culture coevolution, see Barrett, Dunbar, and Lycett, *Human Evolutionary Psychology*, 351–83; Robert Boyd and Peter Richerson, "Cultural Adaptation and Maladaptation: Of Kayaks and Commissars," in *The Evolution of Mind: Fundamental Questions and Controversies*, ed. Steven W. Gangestad and Jeffry A. Simpson (New York: Guilford, 2007), 327–31; Joseph Heinrich and Richard McElreath, "Dual-inheritance Theory: The Evolution of Human Cultural Capacities and Cultural Evolution," in *The Oxford Handbook of Evolutionary Psychology*, ed. Robin Dunbar and Louise Barrett (Oxford: Oxford UP, 2007), 555–70; Hill, "Evolutionary Biology"; Laland, "Niche Construction"; Charles J. Lumsden and Edward O. Wilson, *Promethean Fire: Reflections on the Origin of Mind* (Cambridge: Harvard UP, 1983); Richard McElreath and Joseph Henrich, "Modeling Cultural Evolution," in *The Oxford Handbook of Evolutionary Psychology*, ed. Robin Dunbar and Louise Barrett (Oxford: Oxford UP, 2007), 571–85; Plotkin, "Power of Culture"; Stephen Sheenan, "Evolutionary Perspectives in Archaeology: From Culture History to Cultural Evolution," in *The Oxford Handbook of Evolutionary Psychology*, ed. Robin Dunbar and Louise Barrett (Oxford: Oxford UP, 2007), 587–97; Daniel Lord Smail, *On Deep History and the Brain* (Berkeley: U of California P, 2008); Sterelny, *Thought in a Hostile World*; Tomasello et al., "Understanding and Sharing Intentions"; David S. Wilson, *Evolution for Everyone: How Darwin's Theory Can Change the Way We Think about Our Lives* (New York: Delacorte, 2007); E. Wilson, *Consilience*; Richard Wrangham, *Catching Fire: How Cooking Made Us Human* (New York: Basic Books, 2009).
60. Derek Bickerton, *Adam's Tongue: How Humans Made Language, How Language Made Humans* (New York: Hill and Wang, 2009); Derek Bickerton, "Foraging versus Social Intelligence in the Evolution of Protolanguage," in *The Transition to Language*, ed. Alison Wray (Oxford: Oxford UP, 2000), 207–25; Bickerton, "From Protolanguage to Language"; Derek Bickerton, *Language and Species* (Chicago: U of Chicago P, 1990); Wolfgang Enard et al., "Molecular Evolution of FOXP2, a Gene Involved in Speech and Language," *Nature* 418 (2002): 869–72.
61. Boyd, *Origin of Stories*; Brown, *Human Universals*; Dissanayake, *Art and Intimacy*; Dutton, *The Art Instinct*; Tooby and Cosmides, "Does Beauty Build Adapted Minds?"
62. Carroll, "The Human Revolution"; Cochran and Harpending, *The 10,000 Year Explosion*; Gregory Currie, *Arts and Minds* (Oxford: Clarendon, 2004); Richard Klein, with Blake Edgar, *The Dawn of Human Culture* (New York: John Wiley and Sons, 2002); Wade, *Before the Dawn*; Paul Mellars, Katie Boyle, Ofer Bar-Yosef, and Chris Stringer, eds., *Rethinking the Human Revolution: New Behavioural and Biological Perspectives on the Origin and Dispersal of Modern Humans*

(Exeter, UK: MacDonald Institute, 2007); Paul Mellars and Chris Stringer, eds., *The Human Revolution: Behavioural and Biological Perspectives on the Origins of Modern Humans* (Princeton: Princeton UP, 1989); Steven Mithen, "Mind, Brain, and Material Culture: An Archaeological Perspective," in *Evolution and the Human Mind: Modularity, Language, and Meta-Cognition*, ed. Peter Carruthers and Andrew Chamberlain (Cambridge: Cambridge UP, 2000), 207–17; Mithen, *Prehistory of the Mind*; Christopher Stringer and Clive Gamble, *In Search of the Neanderthals* (New York: Thames and Hudson, 1993).

63. Deacon, *The Symbolic Species*, 374–75; Christopher S. Henshilwood and Curtis W. Marean, "The Origin of Modern Human Behavior: Critique of the Models and Their Test Implications," *Current Anthropology* 44 (2003): 627–51; Sally McBrearty and Alison S. Brooks, "The Revolution That Wasn't: A New Interpretation of the Origin of Modern Human Behavior," *Journal of Human Evolution* 39 (2000): 453–563; Smail, *On Deep History*, 195; Sterelny, *Thought in a Dangerous World*, 115–16.
64. Pinker, "Consilient Study."
65. See Jackson, "Questioning Interdisciplinarity," 319, 322, 328; Pinker, "Consilient Study," 175.
66. Gottschall, *Rape of Troy*, 4–5.
67. Scalise-Sugiyama, "Cultural Relativism in the Bush"; Storey, *Mimesis and the Human Animal*, 131–35.
68. Daniel Nettle, "What Happens in Hamlet? Exploring the Psychological Foundations of Drama," in *The Literary Animal: Evolution and the Nature of Narrative*, ed. Jonathan Gottschall and David Sloan Wilson (Evanston, IL: Northwestern UP, 2005), 56–75; Nettle, "The Wheel of Fire."
69. Boyd, "Art and Evolution" (on Spiegelman); Boyd, "The Art of Literature" (on *Lolita*); Brian Boyd, "Literature and Evolution: A Bio-Cultural Approach" (on *Hamlet*), *Philosophy and Literature* 29 (2005): 1–23; Boyd, *Origin of Stories* (on *The Odyssey* and *Horton Hears a Who*). On the use of adaptationist categories in the analysis of postmodern fiction, also see Nancy Easterlin, "Do Cognitive Predispositions Predict or Determine Literary Value Judgments? Narrativity, Plot, and Aesthetics," in *Biopoetics: Evolutionary Explorations in the Arts*, ed. Brett Cooke and Frederick Turner (Lexington, KY: ICUS, 1999), 241–62.
70. Daly and Wilson, *Homicide*, 107–21; Carl Degler, *In Search of Human Nature: The Decline and Revival of Darwinism in American Social Thought* (New York: Oxford UP, 1991), 245–69.
71. Michelle Scalise-Sugiyama, "New Science, Old Myth: An Evolutionary Critique of the Oedipal Paradigm," *Mosaic* 34 (2001): 121–36; Nancy Easterlin, "Psychoanalysis and the 'Discipline of Love,'" *Philosophy and Literature* 24 (2000): 261–79; Dylan Evans, "From Lacan to Darwin," in *The Literary Animal: Evolution and the Nature of Narrative*, ed. Jonathan Gottschall and David Sloan Wilson (Evanston, IL: Northwestern UP, 2005), 38–55.

72. Ian Jobling, "Personal Justice and Homicide in Scott's *Ivanhoe*: An Evolutionary Psychological Perspective," *Interdisciplinary Literary Studies* 2 (2001): 31.
73. Cooke, *Human Nature in Utopia*; Saunders, *Reading Edith Wharton*.
74. Carroll, *Literary Darwinism*, 129–45; 163–85; 206–16.
75. Joseph Carroll, Jonathan Gottschall, John Johnson, and Daniel Kruger, *Graphing Jane Austen: Paleolithic Politics in British Novels of the Nineteenth Century* (under submission).
76. See Abrams, "How to Do Things with Texts"; Bordwell, *Poetics of Cinema*; Bordwell and Carroll, *Post-Theory*; Boyd, "Art and Evolution" and "Getting It All Wrong"; Carroll, *Evolution and Literary Darwinism*, 15–27; Cooke, *Human Nature* and "Toward"; Dissanayake, *Homo Aestheticus*; Crews, *Postmodern Pooh*; Fromm, *Academic Capitalism*; Gottschall, *Literature* and "The Tree"; Headlam Wells, *Shakespeare's Humanism*; Headlam Wells and McFadden, eds., *Human Nature*; Love, "Science"; Ian McEwan, "Literature, Science, and Human Nature," in *The Literary Animal: Evolution and the Nature of Narrative*, ed. Jonathan Gottschall and David Sloan Wilson (Evanston, IL: Northwestern UP, 2005), 5–19; Edward Slingerland, *What Science Offers the Humanities: Integrating Body and Culture* (Cambridge: Cambridge UP, 2008); Storey, *Mimesis*; E. O. Wilson, *Consilience*.
77. Jackson, "Issues and Problems," 177, 178.
78. Goodheart, *Darwinian Misadventures*, 23, 20, 37.
79. Northrop Frye, *The Anatomy of Criticism: Four Essays* (Princeton: Princeton UP, 1957); Meyer H. Abrams, *The Mirror and the Lamp: Romantic Theory and the Critical Tradition* (New York: Oxford UP, 1953); Ian Watt, *The Rise of the Novel* (Berkeley: U of California P, 1957).
80. For instances and further references, see Colin Martindale, *The Clockwork Muse: The Predictability of Artistic Change* (New York: Basic Books, 1990); David S. Miall, *Literary Reading: Empirical and Theoretical Studies* (New York: Peter Lang, 2006); Franco Moretti, *Graphs, Maps, Trees: Abstract Models for a Literary History* (London: Verso, 2005); Willie van Peer, *Muses and Measures: Empirical Research Methods for the Humanities* (Newcastle upon Tyne: Cambridge Scholars, 2007). Also see the website for the International Society for the Empirical Study of Literature and Media: http://www.psych.ualberta.ca/IGEL/, accessed 5 August 2009.
81. Gottschall, *Literature*; Daniel Kruger, Maryanne Fisher, and Ian Jobling, "Proper and Dark Heroes as Dads and Cads: Alternative Mating Strategies in British and Romantic Literature," *Human Nature* 14 (2003): 305–17; Paul Matthews and Louise Barrett, "Small-Screen Social Groups: Soap Operas and Social Networks," *Journal of Cultural and Evolutionary Psychology* 3 (2005): 75–86; James Stiller and Matthew Hudson, "Weak Links and Scene Cliques within the Small World of Shakespeare," *Journal of Cultural and Evolutionary Psychology* 3 (2005): 57–73; James Stiller, Daniel Nettle, and Robin Dunbar, "The Small World of Shakespeare's Plays," *Human Nature* 14

(2003): 397–408; David Miall and Ellen Dissanayake, "The Poetics of Babytalk," *Human Nature* 14 (2003): 337–64.
82. Carroll, Gottschall, Johnson, and Kruger, *Graphing Jane Austen*.
83. Models for scientifically oriented interdisciplinary study can be found in a consortium of evolutionary studies programs (EvoS): http://www.evostudies.org/scholarls.html. The first EvoS program was established at SUNY Binghamton by the evolutionary biologist D. S. Wilson.
84. The whole exchange—target article, responses, and rejoinder—are in *Style* 42 (2008): 103–412.
85. Brett Cooke, "Compliments and Complements," *Style* 42 (2008): 150.
86. Darwin, *Origin of Species*, 398.
87. Claude C. Albritton, Jr., *The Abyss of Time: Changing Conceptions of the Earth's Antiquity after the Sixteenth Century* (San Francisco, CA: Freeman, Cooper, 1980); Anthony Hallam, *Great Geological Controversies* (Oxford: Oxford UP, 1983); Simon Winchester, *The Map That Changed the World: William Smith and the Birth of Modern Geology* (New York: HarperCollins, 2001).
88. To gain a synoptic overview of the current state of thinking on these various topics, see the short essays by numerous leading researchers in Steven W. Gangestad and Jeffry A. Simpson, eds., *The Evolution of Mind: Fundamental Questions and Controversies* (New York: Guilford, 2007). For more detailed accounts of research on specific topics, see David M. Buss, ed., *The Handbook of Evolutionary Psychology* (Hoboken, NJ: Wiley, 2005); Robin Dunbar and Louise Barrett, eds., *The Oxford Handbook of Evolutionary Psychology* (Oxford: Oxford UP, 2007).
89. Bickerton, "From Protolanguage to Language"; Enard et al., "FOXP2"; Mithen, *Prehistory of the Mind*; Wade, *Before the Dawn*.
90. For examples of sociobiological thinking that emphasizes fitness maximization as a direct motive, see Alexander, *Biology of Moral Systems*; Laura L. Betzig, *Despotism and Differential Reproduction: A Darwinian View of History* (Hawthorne, NY: Aldine de Gruyter, 1986); Napoleon A. Chagnon, *Yanomamö: The Fierce People*, 3rd ed. (New York: Holt, Rinehart, and Winston, 1979); Richard Dawkins, *The Selfish Gene* (New York: Oxford UP, 1976); Gottschall, *Rape of Troy*. On relations between sociobiology, behavioral ecology, and evolutionary psychology, see Barrett, Dunbar, and Lycett, *Human Evolutionary Psychology*, 8–21; D. Buss, *Evolutionary Psychology*, 13–18; Dunbar and Barrett, "Evolutionary Psychology"; Gangestad and Simpson, "An Introduction"; Hagen and Symons, "Natural Psychology"; Hill, "Evolutionary Biology"; Laland and Brown, *Sense and Nonsense*; Mamelli, "Evolution and Psychology"; Pinker, foreword to *The Handbook of Evolutionary Psychology*; Sterelny, *Thought in a Hostile World*, 234–35; E. Wilson, *Sociobiology*, v–viii.
91. Boehm, *Hierarchy in the Forest*; Irenäus Eibl-Eibesfeldt, "Us and the Others: The Familial Roots of Ethnonationalism," in *Ethnic Conflict and Indoctrination: Altruism and Identity in Evolutionary Perspective*, ed.

Irenäus Eibl-Eibesfeldt and Frank K. Salter (New York: Berghahn, 1998), 21–53; Hill, "Evolutionary Biology"; Richerson and Boyd, *Not by Genes Alone*; Sterelny, *Thought in a Hostile World*; Sober and Wilson, *Unto Others*; D. Wilson, *Evolution for Everyone*; David Sloan Wilson, "Group-Level Evolutionary Processes," in *The Oxford Handbook of Evolutionary Psychology*, ed. Robin Dunbar and Louise Barrett (Oxford: Oxford UP, 2007), 49–55; David Sloan Wilson and Edward O. Wilson, "Rethinking the Theoretical Foundation of Sociobiology," *The Quarterly Review of Biology* 82 (2007): 327–48.
92. Hill, "Evolutionary Biology," 351.
93. Sterelny, *Thought in a Hostile World*; Tomasello et al., "Understanding and Sharing Intentions."
94. Hill, "Evolutionary Biology," 353.
95. Hill, "Evolutionary Biology," 351, 352–53.
96. Richerson and Boyd, *Not by Genes Alone*, 180.
97. For a taxonomic analysis of the theoretical options available within gene-culture coevolution and cultural evolution, see Geoffrey M. Hodgson, "Taxonomizing the Relationship between Biology and Economics: A Very Long Engagement," *Journal of Bioeconomics* 9 (2007): 169–85.
98. Frederick Crews, "Apriorism for Empiricists," *Style* 42 (2008): 157; Jacov Rofé, "Does Repression Exist? Memory, Pathogenic Unconscious and Clinical Evidence," *Review of General Psychology* 12 (2008): 63.
99. Timothy D. Wilson, *Strangers to Ourselves: Discovering the Adaptive Unconscious* (Cambridge: Harvard UP, 2002), 39. Also see Rofé, "Does Repression Exist?" 67–68; Ran R. Hassin, James S. Uleman, and John A. Bargh, eds., *The New Unconscious* (New York: Oxford UP, 2005).
100. Pinker, "Consilient Study."
101. See Ernst Mayr, "How to Carry Out the Adaptationist Program?" *The American Naturalist* 121 (1983): 326–28; Leda Cosmides, John Tooby, and Jerome Barkow, "Introduction: Evolutionary Psychology and Conceptual Integration," in *The Adapted Mind: Evolutionary Psychology and the Generation of Culture*, ed. Jerome Barkow, Leda Cosmides, John Tooby (New York: Oxford UP, 1992), 9–10.
102. Gottschall, *Literature*.

Chapter 3. A Meta-Review of *The Art Instinct*

1. Dutton, *The Art Instinct* (cited hereafter, in the text, as *AI*).
2. Gottschall and Wilson, eds., *The Literary Animal*; Boyd, *Origin of Stories*.
3. Pinker, *How the Mind Works*; Pinker, *Language Instinct*; E. Wilson, *Consilience*.
4. Ellen Dissanayake, *What Is Art For?* (Seattle: U of Washington P, 1988); Dissanayake, *Homo Aestheticus*; Dissanayake, *Art and Intimacy*.

5. Maureen Mullarkey, "Wired for Art," *The Weekly Standard*, 2 March 2009, 30.
6. Darwin, *Origin of Species*, 379.
7. Brian Boyd, "Art and Selection," *Philosophy and Literature* 33 (2009): 204–20; Alexander Nehamas, "What Does Evolution Say about Why We Make Art?" *The American Scholar* 78.2 (2009): 118–25; John Onians, "Evolved Tastes," *The Wilson Quarterly* 33.2 (2009): 109–10; James Q. Wilson, "The Evolution of Art," *Newsweek*, International Edition, 2 March 2009.
8. Miller, *Mating Mind*.
9. Smee, "Natural-Born Thrillers," 17.
10. Deresiewicz, "Adaptation," 30.
11. Slingerland, *What Science Offers*.

Chapter 4. Three Scenarios for Literary Darwinism

1. M. H. Abrams, "The Transformation of English Studies: 1930–1995," *Daedalus* 126 (1997): 105–32; Carroll, *Literary Darwinism*, 3–40; Northrop Frye, *The Critical Path: An Essay on the Social Context of Literary Criticism* (Bloomington: Indiana UP, 1971).
2. Louis Menand, "Dangers Within and Without," *Profession 2005*, ed. Rosemary G. Feal (New York: Modern Language Association of America, 2005), 14.
3. Paul Boghossian, *Fear of Knowledge: Against Relativism and Constructivism* (Oxford: Oxford UP, 2006).
4. M. H. Abrams, "How to Do Things with Texts," *Partisan Review* 46 (1979): 566–88; William E. Cain, *The Crisis in Criticism: Theory, Literature, and Reform in English Studies* (Baltimore: Johns Hopkins UP, 1984), chap. 2; Carroll, *Evolution*, 56–68; John R. Searle, "Literary Theory and Its Discontents," *New Literary History* 25 (1994): 637–67; John R. Searle, "The Word Turned Upside Down," *New York Review of Books* 30.16 (27 October 1983): 74–79.
5. Michel Foucault, *Language, Counter-Memory, Practice: Selected Essays and Interviews*, trans. D. F. Bouchard and S. Simon, ed. D. F. Bouchard (Ithaca: Cornell UP, 1977).
6. Gillian Beer, *Darwin's Plots: Evolutionary Narrative in Darwin, George Eliot, and Nineteenth-Century Fiction* (London: Routledge, 1983); George Levine, *Darwin and the Novelists: Patterns of Science in Victorian Fiction* (Chicago: U of Chicago P, 1988); Spolsky, *Gaps in Nature*.
7. Degler, *In Search of Human Nature*; Pinker, *Blank Slate*.
8. Larry Arnhart, *Darwinian Natural Right: The Biological Ethics of Human Nature* (Albany: State U of New York P, 1998); Darwin, *Descent of Man*; L. D. Katz, ed., *Evolutionary Origins of Morality: Cross-Disciplinary Perspectives* (Bowling Green, OH: Imprint Academic, 2000); Peter Singer, *A Darwinian Left* (New Haven: Yale UP, 2000).
9. Goodheart, *Darwinian Misadventures*.
10. Smail, *On Deep History*.
11. E. Wilson, *Consilience*.

12. Smail, *On Deep History*.
13. Peter Turchin, *War and Peace and War: The Rise and Fall of Empires* (New York: Plume, 2007).
14. Cooke, *Human Nature*; Gottschall, *Rape of Troy*.
15. Gottschall, *Literature*.

Chapter 5. Aestheticism, Homoeroticism, and Christian Guilt in *The Picture of Dorian Gray*

1. See Edward S. Brinkley, "Homosexuality as (Anti)Illness: Oscar Wilde's *The Picture of Dorian Gray* and Gabriele d'Annunzio's *Il Piacere*," *Studies in Twentieth-Century Literature* 22 (1998): 61–82; Joseph Bristow, "'A Complex Multiform Creature': Wilde's Sexual Identities," in *The Cambridge Companion to Oscar Wilde*, ed. Peter Raby (Cambridge: Cambridge UP, 1997), 195–218; Ed Cohen, "Writing Gone Wilde: Homoerotic Desire in the Closet of Representation," *PMLA* 102 (1987): 801–13; Liang-ya Liou, "The Politics of a Transgressive Desire: Oscar Wilde's *The Picture of Dorian Gray*," *Studies in Language and Literature* 6 (1994): 101–25; Jeff Nunokawa, "Homosexual Desire and the Effacement of the Self in *The Picture of Dorian Gray*," *American Imago: Studies in Psychoanalysis and Culture* 49 (1992): 311–21; Eve Kosofsky Sedgwick, *Epistemology of the Closet* (Berkeley: U of California P, 1990), chap. 3; Ian Small, *Oscar Wilde: Recent Research, a Supplement to "Oscar Wilde Revalued"* (Greensboro, NC: ELT P, 2000), chap. 3.
2. Philip Cohen, *The Moral Vision of Oscar Wilde* (Cranbury, NJ: Associated University Presses, 1978), chap. 4; Jeffrey Meyers, *Homosexuality and Literature 1890–1930* (Montreal: McGill-Queens UP, 1977), chap. 2; Claude J. Summers, *Gay Fictions Wilde to Stonewall: Studies in a Male Homosexual Literary Tradition* (New York: Frederick Ungar, 1990), chap. 2.
3. Jonathan Dollimore, *Sexual Dissidence: Augustine to Wilde, Freud to Foucault* (Oxford: Oxford UP, 1991), 64.
4. Barbara Charlesworth, *Dark Passages: The Decadent Consciousness in Victorian Literature* (Madison: U of Wisconsin P, 1965), 54, 79.
5. Oscar Wilde, *The Letters of Oscar Wilde*, ed. Rupert Hart-Davis (New York: Harcourt, Brace, 1962), 352; cited hereafter, in the text, as *Letters*.
6. Textualist readings include those of Nils Clausson, "Culture and Corruption: Paterian Self-Development versus Gothic Degeneration in Oscar Wilde's *The Picture of Dorian Gray*," *Papers on Language and Literature* 39 (2003): 339–64; Elana Gomel, "Oscar Wilde, *The Picture of Dorian Gray*, and the (Un)Death of the Author," *Narrative* 12 (2001): 74–92; Michael Patrick Gillespie, "Picturing Dorian Gray: Resistant Readings in Wilde's Novel," *English Literature in Transition 1880–1920* 35 (1992): 7–25.
7. Oscar Wilde, *Complete Shorter Fiction*, ed. Isobel Murray (Oxford: Oxford UP, 1979), 108.

8. Oscar Wilde, *The Picture of Dorian Gray*, ed. Donald L. Lawler (New York: W. W. Norton, 1988), 64; cited hereafter, in the text, as *DG*.
9. Instances include Julia Prewitt Brown, *Cosmopolitan Criticism: Oscar Wilde's Philosophy of Art* (Charlottesville: UP of Virginia, 1997); Richard Ellmann, "Overtures to Salome," in *Oscar Wilde: A Collection of Critical Essays*, ed. Richard Ellmann (Englewood Cliffs, NJ: Prentice-Hall, 1969), 87–91; Robert K. Martin, "Parody and Homage: The Presence of Pater in *Dorian Gray*," *The Victorian Newsletter* 69 (Spring 1983): 15–18; John Paul Riquelme, "Oscar Wilde's Aesthetic Gothic: Walter Pater, Dark Enlightenment, and *The Picture of Dorian Gray*," *Modern Fiction Studies* 46 (2000): 609–31.
10. Walter Pater, *The Renaissance: Studies in Art and Poetry*, ed. Donald Hall (Berkeley: U of California P, 1980), 187–88; cited hereafter, in the text, as *Renaissance*.
11. Darwin, *Descent of Man*, 1:88–91.
12. Symons, *Human Sexuality*, chap. 9, "Test Cases: Hormones and Homosexuals."
13. See Joseph Bristow, *Effeminate England: Homoerotic Writing after 1885* (New York: Columbia UP, 1995); Ed Cohen, *Talk on the Wilde Side: Toward a Genealogy of a Discourse on Male Sexualities* (New York: Routledge, 1993); Thais Morgan, "Victorian Effeminacies," in *Victorian Sexual Dissidence*, ed. Richard Dellamora (Chicago: U of Chicago P, 1999), 109–25; Alan Sinfield, *The Wilde Century: Effeminacy, Oscar Wilde and the Queer Movement* (London: Cassell, 1994).
14. Elaine Showalter, *Sexual Anarchy: Gender and Culture at the Fin de Siècle* (New York: Penguin, 1990), 175–78.
15. Joseph Conrad, *Heart of Darkness*, ed. Robert Hampson (London: Penguin, 1995), 80.

Chapter 6. The Cuckoo's History: Human Nature in *Wuthering Heights*

1. Miriam Allott, introduction to *Emily Brontë: "Wuthering Heights": A Casebook*, ed. Miriam Allott (Houndmills: Macmillan, 1970), 12; Harold Fromm, *Academic Capitalism and Literary Value* (Athens, GA: U of Georgia P, 1991), 128. For more recent surveys of postmodern criticism on *Wuthering Heights*, see *Emily Brontë: Wuthering Heights*, ed. Patsy Stoneman (Cambridge: Icon, 2000); Gillian Frith, "Decoding Wuthering Heights," in *Critical Essays on Emily Brontë*, ed. Thomas John Winnifrith (New York: Hall-Simon and Schuster, 1997), 243–61. For prominent instances of humanist criticism of *Wuthering Heights*, see Miriam Allott, "The Rejection of Heathcliff?" *Essays in Criticism* 8 (1958): 27–47; Lord David Cecil, *Early Victorian Novelists: Essays in Revaluation* (Indianapolis: Bobbs-Merrill, 1935); Q. D. Leavis, "A Fresh Approach to Wuthering Heights," in *Lectures in America*, by F. R. Leavis and Q. D. Leavis (New York: Pantheon-Random, 1969), 85–138; John K. Mathison, "Nelly Dean and the Power of *Wuthering*

Heights," *Nineteenth-Century Fiction* 11 (1956): 106–29; Martha Nussbaum, "*Wuthering Heights*: The Romantic Ascent," *Philosophy and Literature* 20 (1996): 362–82; Dorothy Van Ghent, *The English Novel: Form and Function* (1953; New York: Harper Torchbooks-Harper, 1961). For prominent instances of postmodern criticism, see Nancy Armstrong, "Emily Brontë In and Out of Her Time," *Genre* 15 (1982): 243–64; Terry Eagleton, *Myths of Power: A Marxist Study of the Brontës* (1975; Houndmills: Palgrave, 2005); Sandra M. Gilbert and Susan Gubar, *The Madwoman in the Attic: The Woman Writer and the Nineteenth-Century Literary Imagination* (New Haven: Yale UP, 1979); Margaret Homans, "The Name of the Mother in *Wuthering Heights*," in *Wuthering Heights: Complete, Authoritative Text with Biographical and Historical Contexts, Critical History, and Essays from Five Contemporary Critical Perspectives*, ed. Linda H. Peterson (Boston: Bedford-St. Martin's, 1992), 341–58; Carol Jacobs, *Uncontainable Romanticism: Shelley, Brontë, Kleist* (Baltimore: Johns Hopkins UP, 1989); J. Hillis Miller, *Fiction and Repetition: Seven English Novels* (Cambridge, MA: Harvard UP, 1982).
2. Frye, *Anatomy of Criticism.*
3. On Brontë's familiarity with pre-Darwinian knowledge of animal husbandry and natural history, see Barbara Munson Goff, "Between Natural Theology and Natural Selection: Breeding the Human Animal in Wuthering Heights," *Victorian Studies* 27 (1984): 477–508.
4. Emily Brontë, *Wuthering Heights: The 1847 Text, Backgrounds and Criticism*, ed. Richard J. Dunn, 4th ed. (New York: Norton, 2003), 77; cited hereafter, in the text, as *WH.*
5. On the scope and significance of parental investment theory, see Aurelio José Figueredo et al., "The K-Factor, Covitality, and Personality: A Psychometric Test of Life History Theory," *Human Nature* 18 (2007): 47–73; Trivers, "Parental Investment."
6. In one of the best and most influential humanist interpretations of the novel, Cecil identifies the imagery of storm and calm as symbols for an ultimate metaphysical equilibrium—"a cosmic harmony" (*Early Victorian Novelists*, 174).
7. Van Ghent, *Form and Function*, 158.
8. On the juvenile character of the adult romantic relation between Catherine and Heathcliff, see Cecil, *Early Victorian Novelists*, 167; Edward Mendelson, *The Things That Matter: What Seven Classic Novels Have to Say about the Stages of Life* (New York: Pantheon, 2006), 47–55; Patricia Meyer Spacks, *The Female Imagination* (New York: Knopf, 1975), 138; Patsy Stoneman, "The Brontë Myth," in *The Cambridge Companion to the Brontës*, ed. Heather Glen (Cambridge: Cambridge UP, 2002), 234–35; Van Ghent, *Form and Function*, 158–59, 169. Gilbert and Gubar observe that "all the Brontë novels betray intense feelings of motherlessness, orphanhood, destitution" (*Madwoman*, 251). In *A Future for Astyanax: Character and Desire in Literature* (Boston: Little, Brown, 1969), Leo Bersani notes that "the emotional register of the novel is that of hysterical children"

(203). For a Darwinian perspective on the lasting traumatic effects of maternal separation, see John Bowlby, *Attachment and Loss*, 3 vols., 2nd ed. (London: Hogarth, 1982). On Brontë's own traumatized response to motherlessness, see Edward Chitham, *A Life of Emily Brontë* (Oxford: Blackwell, 1987), 205, 210, 213–14. In *Narcissism and the Novel* (New York: New York UP, 1990), Jeffrey Berman reads the novel psycho-biographically in the light of Bowlby's concepts (78–112). Wion, using a psychoanalytic framework, characterizes Heathcliff as a maternal surrogate for Catherine Earnshaw. See Philip K. Wion, "The Absent Mother in Emily Brontë's *Wuthering Heights*," *American Imago* 42 (1985): 146. Massé gives a Freudian account of Catherine's narcissism. See Michelle A. Massé, "'He's More Myself Than I Am': Narcissism and Gender in *Wuthering Heights*," in *Psychoanalyses/Feminisms*, ed. Peter L. Rudnytsky and Andrew M. Gordon (Albany: State U of New York P, 2000), 135–53. Moglen and Schapiro both contrast the narcissistic disorders of the first generation with the norm of maturity in the second. See Helene Moglen, "The Double Vision of *Wuthering Heights*: A Clarifying View of Female Development," *Centennial Review* 15 (1971): 398; and Barbara Ann Schapiro, *Literature and the Relational Self* (New York: New York UP, 1994), 49. Bersani, in contrast, though using a similar Freudian vocabulary, valorizes the disintegrative emotional violence of the older generation (*Astyanax*, 214–15, 221–22).

9. On the supernatural and animistic aspects of Brontë's imagination, see Cecil, *Early Victorian Novelists*, 162–70; Ingrid Geerken, "'The Dead Are Not Annihilated': Mortal Regret in *Wuthering Heights*," *Journal of Narrative Theory* 34 (2004): 374–76, 385–86; John Maynard, "The Brontës and Religion," in *The Cambridge Companion to the Brontës*, ed. Heather Glen (Cambridge: Cambridge UP, 2002), 204–09; Derek Traversi, "*Wuthering Heights* after a Hundred Years," in *Emily Brontë: Wuthering Heights: A Casebook*, ed. Miriam Allott (Houndmills: Macmillan, 1970), 157–76; Van Ghent, *Form and Function*, 164–65. On the problematic divisions between the natural and social worlds in *Wuthering Heights*, see Eagleton, *Myths of Power*, 97–111. Despite seeking resolution in a utopian social norm alien to Brontë's own perspective, Eagleton's critique powerfully registers psychosocial stress in the novel.

Chapter 7. Intentional Meaning in *Hamlet*: An Evolutionary Perspective

1. Ernest Jones, *Hamlet and Oedipus* (1949; New York: Norton, 1976); Jacques Lacan, "Desire and the Interpretation of Desire in *Hamlet*," *Yale French Studies* 55/56 (1977): 11–52; Janet Adelman, *Suffocating Mothers: Fantasies of Maternal Origin in Shakespeare's Plays, "Hamlet" to "The Tempest"* (New York: Routledge, 1992); Marjorie Garber, *Shakespeare's Ghost Writers: Literature as Uncanny Causality* (New York: Methuen, 1987); Michael Bristol, "'Funeral-Bak'd Meats':

Carnival and the Carnivalesque in *Hamlet*," in *William Shakespeare: "Hamlet,"* ed. Susanne L. Wofford (Boston: St. Martin's P, 1994), 348–67; Michael Neill, "*Hamlet*: A Modern Perspective," in *William Shakespeare: The Tragedy of Hamlet, Prince of Denmark*, ed. Barbara A. Mowat and Paul Werstine (New York: Washington Square P, 1992), 311–12.
2. Headlam Wells, *Shakespeare's Humanism*; Nordlund, *Shakespeare*.
3. Storey, *Mimesis*, 131–35; Scalise-Sugiyama, "Cultural Relativism"; Boyd, "Literature and Evolution," 18–19; Laura Bohannan, "Shakespeare in the Bush," *Natural History* 75 (August–September 1966): 28–33.
4. Boyd, *Origin of Stories*, 2.
5. J. Dover Wilson, *What Happens in Hamlet?* (Cambridge: Cambridge UP, 1935).
6. Nettle, "What Happens in *Hamlet*?" 71–72.
7. Nettle, "What Happens in *Hamlet*?" 71.
8. Nettle, "What Happens in *Hamlet*?" 73.
9. For instance Franco Zeffirelli's with Mel Gibson as Hamlet and Glenn Close as Gertrude. John Knapp discusses Derek Jacobi's similar performance. See John V. Knapp, "Family Games and Imbroglio in *Hamlet*," in *Reading the Family Dance: Family Systems Therapy and Literary Study*, ed. John V. Knapp and Kenneth Womack (Newark: U of Delaware P, 2003), 194–95.
10. John V. Knapp, "Family-Systems Psychotherapy and Psychoanalytic Literary Criticism: A Comparative Critique," *Mosaic* 37.1 (2004): 149–66.
11. Boyd, "Literature and Evolution," 9.
12. Nettle, "What Happens in *Hamlet*?" 69.
13. William Hazlitt, *Characters of Shakespeare's Plays* (1817; London: Oxford UP, 1955), 84–85.
14. Tooby and Cosmides, "Does Beauty Build Adapted Minds?" 19.
15. Hazlitt, *Characters*, 81, 80, 80–81.
16. A. C. Bradley, *Shakespearean Tragedy: Lectures on "Hamlet," "Othello," "King Lear," and "Macbeth"* (1904; London: Penguin, 1991), 117–18.
17. Bradley, *Shakespearean Tragedy*, 125, 125, 126.
18. Darwin, *Descent of Man*, 2: 405. For commentary on the web of allusions in the conclusion to *Descent*, see Carroll, *Evolution*, 256–58.
19. Daly and Wilson, *Homicide*, 107–21; Degler, *In Search of Human Nature*, 245–69; Scalise-Sugiyama, "New Science."
20. Darwin, *Descent of Man*, 1: 80.
21. Dissanayake, *Art and Intimacy*; Easterlin, "Psychoanalysis."
22. John Bowlby, *Attachment*, 2nd ed., vol. 1 of *Attachment and Loss* (New York: Basic Books, 1982), 232.
23. Bowlby, *Attachment and Loss*; Peter C. Whybrow, *A Mood Apart: Depression, Mania, and Other Afflictions of the Self* (New York: Basic Books, 1997); Lewis Wolpert, *Malignant Sadness: The Anatomy of Depression*, 3rd ed. (London: Faber and Faber, 2006).

24. Caroline Spurgeon, *Leading Motives in the Imagery of Shakespeare's Tragedies* (London: Oxford UP, 1930), 10–14.
25. Bradley, *Shakespearean Tragedy*, 121.
26. Peter Kramer, *Against Depression* (New York: Viking, 2005); Francis Mark Mondimore, *Depression: The Mood Disease*, 3rd ed. (Baltimore: Johns Hopkins UP, 2006); Jim Phelps, *Why Am I Still Depressed? Recognizing and Managing the Ups and Downs of Bipolar II and Soft Bipolar Disorder* (New York: McGraw-Hill, 2006); Andrew Solomon, *The Noonday Demon: An Atlas of Depression* (New York: Scribner, 2001); Whybrow, *A Mood Apart*; Wolpert, *Malignant Sadness*.
27. Samuel H. Barondes, *Better Than Prozac: Creating the Next Generation of Psychiatric Drugs* (Oxford: Oxford UP, 2003); Nell Casey, ed., *Unholy Ghost: Writers on Depression* (New York: HarperCollins, 2002); Richard J. Davidson, Diego Pizzagalli, Jack B. Nitschke, and Katherine Putnam, "Depression: Perspectives from Affective Neuroscience," *Annual Review of Psychology* 53 (2002): 545–74; Ian H. Gotlib and Constance L. Hammen, eds., *Handbook of Depression*, 2nd ed. (New York: Guilford, 2009); Kramer, *Against Depression*; Raymond Lam and Hiram Mok, *Depression* (Oxford: Oxford UP, 2008); Mondimore, *Depression*; A. A. Nierenberg, D. Doughtery, and J. F. Rosenbaum, "Dopaminergic Agents and Stimulants as Antidepressant Augmentation Strategies," *Journal of Clinical Psychiatry* 59, supplement 5 (1998): 60–63; Phelps, *Why Am I Still Depressed?*; Donald S. Robinson, "The Role of Dopamine and Norepinephrine in Depression," *Primary Psychiatry* 14.5 (2007): 21–23; Solomon, *The Noonday Demon*; William Styron, *Darkness Visible: A Memoir of Madness* (New York: Random House, 1990); Herman M. van Praag, Ron de Kloet, and Jim van Os, *Stress, the Brain, and Depression* (Cambridge: Cambridge UP, 2004); Whybrow, *A Mood Apart*; Wolpert, *Malignant Sadness*.
28. Bradley, *Shakespearean Tragedy*, 120, 119.
29. A. Buss, "Personality Traits"; David M. Buss, "Social Adaptation and Five Major Factors of Personality," in *The Five-Factor Model of Personality: Theoretical Perspectives*, ed. Jerry S. Wiggins (New York: Guilford, 1996), 180–207; Paul T. Costa and Robert R. McCrae, "Personality Trait Structure as a Human Universal," *American Psychologist* 52 (1997): 509–16; Oliver P. John, Alois Angleitner, and Fritz Ostendorf, "The Lexical Approach to Personality: A Historical Review of Trait Taxonomic Research," *European Journal of Personality* 2 (1988): 171–203; MacDonald, "Levels of Personality"; MacDonald, "Life History Theory"; Daniel Nettle, "Individual Differences," in *The Oxford Handbook of Evolutionary Psychology*, ed. Robin Dunbar and Louise Barrett (Oxford: Oxford UP, 2007), 479–90; Daniel Nettle, *Personality: What Makes You the Way You Are* (Oxford: Oxford UP, 2007); Lawrence A. Pervin and Oliver P. John, eds., *Handbook of Personality*, 2nd ed. (New York: Guilford, 1999); Dirk J. M. Smits and Paul de Boeck, "From BIS/BAS to the Big Five," *European Journal of Personality* 20 (2006): 255–70; Jerry S. Wiggins, ed., *The Five-*

Factor Model of Personality: Theoretical Perspectives (New York: Guilford, 1996).
30. Costa and McCrae, "Personality Trait Structure"; Nettle, *Personality*.
31. Darwin, *Descent of Man*, 1: 88–89.
32. Boehm, *Hierarchy in the Forest*; Deacon, *The Symbolic Species*; Dissanayake, *Art and Intimacy*; Dutton, *The Art Instinct*; Hill, "Evolutionary Biology"; MacDonald, "Conflict Theory of Culture"; Mithen, *Prehistory of the Mind*; Panksepp and Panksepp, *Seven Sins*; Tomasello et al., "Understanding and Sharing Attention"; Tooby and Cosmides, "Does Beauty Build Adapted Minds?"; Wade, *Before the Dawn*; E. Wilson, *Consilience*.
33. Johann Wilhelm von Goethe, *Wilhelm Meister's Apprenticeship and Travels*, trans. Thomas Carlyle, vol. 23 of *The Works of Thomas Carlyle* (London: Chapman and Hall, 1899), 282. Boyd gives a similarly concise verbal portrait of Hamlet's personality ("Literature and Evolution," 18).
34. Samuel Johnson, *Johnson on Shakespeare*, ed. Arthur Sherbo, vol. 8 of *The Yale Edition of the Works of Samuel Johnson* (New Haven: Yale UP, 1968), 1011.
35. Carroll, *Literary Darwinism*, 190–91, 200, 206; Kevin B. MacDonald, "A Perspective on Darwinian Psychology: The Importance of Domain-General Mechanisms, Plasticity, and Individual Differences," *Ethology and Sociobiology* 12 (1991): 449–80; Nettle, "Individual Differences."
36. Deresiewicz, "Adaptation"; Smee, "Natural-Born Thrillers."
37. Judith Rich Harris, *No Two Alike: Human Nature and Human Individuality* (New York: Norton, 2007); Nettle, *Personality*; Frank J. Sulloway, *Born to Rebel: Birth Order, Family Dynamics, and Creative Lives* (New York: Pantheon, 1996).
38. Carroll, *Matthew Arnold*, 1–37.
39. Matthew Arnold, *On the Classical Tradition*, vol. 1 of *The Complete Prose Works of Matthew Arnold*, ed. R. H. Super (Ann Arbor: U of Michigan P, 1960), 1.

Chapter 8. Agonistic Structure in Victorian Novels: Doing the Math

1. On the evolution of cooperative social groups, see Alexander, *Biology of Moral Systems*; Darwin, *Descent of Man*; D. Wilson, *Evolution for Everyone*. On egalitarianism in hunter-gatherers, see Boehm, *Hierarchy in the Forest*.
2. On literature as a form of "simulation," see Keith Oatley, "Why Fiction May Be Twice as True as Fact: Fiction as Cognitive and Emotional Simulation," *Review of General Psychology* 3 (1999): 101–17; Keith Oatley, "Emotions and the Story Worlds of Fiction," in *Narrative Impact: Social and Cognitive Foundations*, ed. M. C. Green, J. J. Strange, and T. C. Brock (Mahwah, NJ: Erlbaum, 2002), 36–69; Ed S. Tan, "Emotions, Art, and the Humanities," in *Handbook of Emotions*, 2nd ed., ed. M. Lewis and J. M. Haviland-Jones (New

York: Guilford, 2000), 116–34. On the parallel responses to "real" and "fictive" people, see Gordon H. Bower and Daniel G. Morrow, "Mental Models in Narrative Comprehension," *Science*, New Series, 247 (1990): 44–48; Herbert Grabes, "Turning Words on the Page into 'Real' People," *Style* 38 (2004): 221–35.
3. George Eliot, *Middlemarch: An Authoritative Text, Backgrounds, Reviews, and Criticism*, ed. Bert G. Hornback. 2nd ed. (1871–1872; New York: W. W. Norton, 2000), 185. For a recent, biologically informed version of the idea of the "unconscious" mind, see T. Wilson, *Strangers to Ourselves*. Also see Hassin, Uleman, and Bargh, *The New Unconscious*; Jiro Tanaka, "What Is Copernican? A Few Common Barriers to Darwinian Thinking about the Mind," *The Evolutionary Review* 1 (2010): 6–12.
4. Johnson, Carroll, Gottschall, and Kruger, "Hierarchy in the Library."
5. On the conserved features of human nature, see A. Buss, "Personality Traits"; Lancaster and Kaplan, "Chimpanzee and Human Intelligence." And see "A Model of Human Nature" in the first chapter of this book. On reproduction and the family, see Bjorklund and Pellegrini, *Evolutionary Developmental Psychology*; D. Buss, *Evolution of Desire*; Flinn and Ward, "Ontogeny"; Geary and Flinn, "Human Parental Behavior." On social organization, see Cummins, "Dominance"; Kurzban and Neuberg, "Ingroup and Outgroup Relations." On evolved dispositions for culture, see Hill, "Evolutionary Biology."
6. D. Buss, *Evolution of Desire*; Gangestad, "Reproductive Strategies"; Geary, *Male, Female*; Jonathan Gottschall, "Greater Emphasis on Female Attractiveness in Homo sapiens: A Revised Solution to an Old Evolutionary Riddle," *Evolutionary Psychology* 5 (2007): 347–58; Schmitt, "Human Mating Strategies."
7. McCrae and Costa, "Personality Trait Structure"; Nettle, *Personality*.
8. Paul Ekman, *Emotions Revealed: Recognizing Faces and Feelings to Improve Communication and Emotional Life* (New York: Henry Holt, 2003); Plutchik, *Emotions and Life*.
9. Boehm, *Hierarchy in the Forest*; Richerson and Boyd, *Not by Genes Alone*; Sterelny, *Thought in a Hostile World*.

Chapter 9. Quantifying Agonistic Structure in *The Mayor of Casterbridge*

1. John Paterson, "*The Mayor of Casterbridge* as Tragedy," *Victorian Studies* 3 (1959): 151, 152, 156, 154. Other critiques invoking the model of retributive justice include Jean R. Brooks, *Thomas Hardy: The Poetic Structure* (Ithaca, NY: Cornell UP, 1971); Pamela Dalziel, introduction to *The Mayor of Casterbridge*, by Thomas Hardy (Oxford: Oxford UP, 2004); Eugene W. Davis, "Comparatively Modern Skeletons in the Garden: A Reconsideration of *The Mayor of Casterbridge*," *English*

Literature in Transition: 1880–1920, special series, 3 (1985): 108–20; D. A. Dike, "A Modern Oedipus: *The Mayor of Casterbridge*," *Essays in Criticism* 2 (1952): 169–79; Albert J. Guerard, *Thomas Hardy: The Novels and Stories* (Cambridge: Harvard UP, 1949); Robert B. Heilman, "Hardy's *Mayor* and the Problem of Intention," in *The Workings of Fiction: Essays by Robert Bechtold Heilman* (Columbia: U of Missouri P, 1991), 280–89; Bruce Johnson, *True Correspondence: A Phenomenology of Thomas Hardy's Novels* (Tallahassee: UP of Florida, 1983); Frederick Karl, "*The Mayor of Casterbridge*: A New Fiction Defined," *Modern Fiction Studies* 6 (1960): 195–213; Jeannette King, *Tragedy in the Victorian Novel: Theory and Practice in the Novels of George Eliot, Thomas Hardy, and Henry James* (Cambridge: Cambridge UP, 1978); Christopher Lane, *The Burdens of Intimacy: Psychoanalysis and Victorian Masculinity* (Chicago: U of Chicago P, 1999); Laurence Lerner, *Thomas Hardy's Mayor of Casterbridge: Tragedy or Social History?* (London: Sussex UP, 1975); J. Hillis Miller, *Thomas Hardy: Distance and Desire* (Cambridge: Harvard UP, 1970); Kevin Z. Moore, "Death against Life: Hardy's Mortified and Mortifying 'Man of Character' in *The Mayor of Casterbridge*," *Ball State U Forum* 24 (1983): 13–25; Kevin Z. Moore, *The Descent of the Imagination: Postromantic Culture in the Later Novels of Thomas Hardy* (New York: New York UP, 1990); Craig Raine, "Conscious Artistry in *The Mayor of Casterbridge*," in *New Perspectives on Thomas Hardy*, ed. Charles Pettit (New York: St. Martin's, 1994), 156–71; Annie Ramel, "The Crevice in the Canvas: A Study of *The Mayor of Casterbridge*," *Victorian Literature and Culture* 26 (1998): 259–72.

2. George Levine, *The Realistic Imagination: English Fiction from Frankenstein to Lady Chatterley* (Chicago: U of Chicago P, 1981), 232, 232, 244. Other critiques invoking the model of Promethean Romantic heroism include Simon Gatrell, *Thomas Hardy and the Proper Study of Mankind* (Charlottesville: U of Virginia P, 1993); Frank R. Giordano, Jr., *"I'd Have My Life Unbe": Thomas Hardy's Self-Destructive Characters* (Tuscaloosa: U of Alabama P, 1984); Guerard, *Thomas Hardy*; Bert G. Hornback, *The Metaphor of Chance: Vision and Technique in the Works of Thomas Hardy* (Athens: Ohio UP, 1971); Irving Howe, *Thomas Hardy* (New York: Macmillan, 1967); Karl, "A New Fiction Defined"; Robert Langbaum, *Thomas Hardy in Our Time* (New York: St. Martin's, 1995); Lerner, *Tragedy or Social History?*; Michael Millgate, *Thomas Hardy: His Career as a Novelist* (1971; New York: St. Martin's, 1994); Michael Valdez Moses, "Agon in the Marketplace: *The Mayor of Casterbridge* as Bourgeois Tragedy," *South Atlantic Quarterly* 87 (1988): 219–51; Ted R. Spivey, "Thomas Hardy's Tragic Hero," *Nineteenth-Century Fiction* 9 (1954): 179–91; Keith Wilson, introduction to *The Mayor of Casterbridge*, by Thomas Hardy (London: Penguin, 1997), xxi–xli; Virginia Woolf, "The Novels of Thomas Hardy," *The Second Common Reader* (1932; New York: Harcourt, 1960), 222–33.

3. Richard Holt Hutton, review of *The Mayor of Casterbridge*, by Thomas

Hardy, in *Thomas Hardy: The Critical Heritage*, ed. R. G. Cox (1979; London: Routledge, 1995), 138, 138–39.
4. Elaine Showalter, "The Unmanning of the Mayor of Casterbridge," in *Critical Approaches to the Fiction of Thomas Hardy*, ed. Dale Kramer (London: Macmillan, 1979), 103, 104. Other critiques invoking the model of redemptive change include Dalziel, introduction; Gatrell, *Proper Study*; Ian Gregor, *The Great Web: The Form of Hardy's Major Fiction* (Totowa, NJ: Rowman, 1974); Langbaum, *Thomas Hardy*; Paterson, "Tragedy"; Spivey, "Tragic Hero"; T. R. Wright, *Hardy and the Erotic* (New York: St. Martin's, 1989).
5. Thomas Hardy, *The Life and Work of Thomas Hardy*, ed. Michael Millgate (London: Macmillan, 1984), 182.
6. Lord David Cecil, *Hardy the Novelist: An Essay in Criticism* (Indianapolis: Bobbs, 1943), 222.
7. Thomas Hardy, *The Mayor of Casterbridge: An Authoritative Text, Backgrounds and Contexts, Criticism*, ed. Phillip Mallett, 2nd ed. (New York: Norton, 2001), 252.
8. On Elizabeth-Jane's role as observer and reflective consciousness, and on Hardy's identification with her, see Brooks, *Poetic Structure*, 212; J. B. Bullen, *The Expressive Eye: Fiction and Perception in the Work of Thomas Hardy* (Oxford: Oxford UP, 1986), 157–59; John Goode, *Thomas Hardy: The Offensive Truth* (Oxford: Blackwell, 1988), 78–94; Gregor, *The Great Web*, 388; Julie Grossman, "Thomas Hardy and the Role of Observer," *ELH* 56 (1989): 619, 633–36; Lars Hartveit, *The Art of Persuasion: A Study of Six Novels* (Bergen: Universitetsforlaget, 1977), 50–70; Pamela Jekel, *Thomas Hardy's Heroines: A Chorus of Priorities* (Troy, NY: Whitson, 1986), 131–43; Langbaum, *Thomas Hardy*, 129; Millgate, *Career as a Novelist*, 228–29; Penelope Vigar, *The Novels of Thomas Hardy: Illusion and Reality* (London: Athlone, 1978), 164–65.
9. Review of *The Mayor of Casterbridge*, by Thomas Hardy, in *Thomas Hardy: The Critical Heritage*, ed. R. G. Cox. (1979; London: Routledge, 1995), 136.
10. Phillip Mallett, a note on the text, in *The Mayor of Casterbridge: An Authoritative Text, Backgrounds and Contexts, Criticism*, by Thomas Hardy, ed. Phillip Mallett, 2nd. ed. (New York: Norton, 2001), xiv–xv.

Chapter 10. The Power of Darwin's Vision

1. Ernst Mayr, *One Long Argument: Charles Darwin and the Genesis of Modern Evolutionary Thought* (Cambridge: Harvard UP, 1991), 164, vii, ix.
2. Mark Ridley, *Evolution* (Boston: Blackwell, 1993), 3, 59.
3. Michael Ghiselin, *The Triumph of the Darwinian Method* (1969; Chicago: U of Chicago P, 1984), 232.
4. Darwin, *Origin of Species*, 97; hereafter, in this chapter, cited in the text as *OS*.

5. Thomas Henry Huxley, "On the Reception of the *Origin of Species*," in *The Life and Letters of Charles Darwin, Including an Autobiographical Chapter*, ed. Francis Darwin, foreword by George Gaylord Simpson, 2 vols. (1887; New York: Basic Books, 1959), 1: 551.
6. Charles Darwin, *The Autobiography of Charles Darwin, 1809–1882, with Original Omissions Restored*, ed. Nora Barlow (London: Collins, 1958), 138–39.
7. Ghiselin, *Darwinian Method*, 83.
8. Darwin, *Descent of Man*, 1: 14–16.
9. John A. Moore, *Science as a Way of Knowing: The Foundations of Modern Biology* (Cambridge: Harvard UP, 1993), 407, 409–10.
10. Stephen Toulmin and June Goodfield, *The Discovery of Time* (Chicago: U of Chicago P, 1965), 169. Also see Albritton, *Abyss of Time*; Hallam, *Great Geological Controversies*; David Oldroyd, *Thinking about the Earth: A History of Ideas in Geology* (Cambridge: Harvard UP, 1996); Martin S. J. Rudwick, *The Meaning of Fossils: Episodes in the History of Paleontology* (Chicago: U of Chicago P, 1972); Winchester, *Map That Changed the World*.
11. Ghiselin, *Darwinian Method*, 232.
12. Thomas Henry Huxley, *Darwiniana*, vol. 2 of *Collected Essays* (New York: Macmillan, 1893), 246.
13. Joseph Hooker, review of *On the Origin of Species*, in *Darwin and His Critics: The Reception of Darwin's Theory of Evolution by the Scientific Community*, ed. David L. Hull (Chicago: U of Chicago P, 1973), 83.
14. Darwin, *Autobiography*, 145, 140, 140, 140, 141, 141, 141.
15. Darwin, *Autobiography*, 141.
16. Darwin, *Descent of Man*, 1: 169.
17. Darwin, *Autobiography*, 43, 85, and see 138.
18. Ghiselin, *Darwinian Method*, 237, 237, 238.
19. Ghiselin, *Darwinian Method*, 243.
20. Mayr, *One Long Argument*, 43; George Gaylord Simpson, *The Meaning of Evolution: A Study of the History of Life and of Its Significance for Man*, 2nd ed. (New Haven: Yale UP, 1967), 266.
21. See for instance two letters to Hooker (11 January 1844 and 10 November 1844) and two to Lyell (12 March 1863 and 17 March 1863), in Darwin, *Life and Letters*, 1: 383, 1: 390, 2: 198–99, 2: 201.
22. Carroll, *Evolution*, 21–24, 184–91, 200–01.
23. John Maynard Smith, ed., *Evolution Now: A Century after Darwin* (San Francisco: W. H. Freeman, 1982), 91–92.
24. Herbert Spencer, *An Autobiography*, 2 vols. (London: Williams and Norgate, 1904), 1: 176–77.
25. Darwin, *Descent of Man*, 1: 100–01.
26. Darwin, *Autobiography*, 109.
27. Herbert Spencer, *First Principles* (London: Williams and Norgate, 1862), 490.
28. See Herbert Spencer, *The Principles of Biology*, 2 vols. (London:

Williams and Norgate, 1864–1867), 1: 444–46; Spencer, *Autobiography*, 2: 50, 2: 100.
29. Darwin, *Autobiography*, 109.
30. Darwin, *Autobiography*, 120.
31. Darwin, *Autobiography*, 122.
32. Darwin, *Autobiography*, 122–21.
33. Charles Darwin and Alfred Russel Wallace, *Evolution by Natural Selection*, with a foreword by Gavin De Beer (Cambridge: Cambridge UP, 1958), 33–34, 215n. This volume contains the 1842 sketch, the 1844 manuscript, and the 1858 Linnean Society papers by Darwin and Wallace.
34. The Linnean Society paper by Darwin and Wallace is included in Darwin and Wallace, *Evolution by Natural Selection*. For other references that support this interpretation of divergence, see the commentary by Glick and Kohn in Charles Darwin, *Darwin on Evolution: The Development of the Theory of Natural Selection*, ed. Thomas F. Glick and David Kohn (Indianapolis: Hackett, 1996), 127–30; and see Dov Ospovat, *The Development of Darwin's Theory: Natural History, Natural Theology, and Natural Selection, 1838–1859* (Cambridge: Cambridge UP, 1981), chap. 7.
35. Charles Darwin, *Journal of Researches into the Geology and Natural History of the Various Countries Visited by H.M.S. Beagle, Under the Command of Captain Fitzroy from 1832 to 1836*, 2nd ed. (London: Murray, 1845), 346.
36. Darwin and Wallace, *Evolution by Natural Selection*, 114, 115, 115.
37. Darwin, *Darwin on Evolution*, 130; Darwin and Wallace, *Evolution by Natural Selection*, 45.
38. Darwin and Wallace, *Evolution by Natural Selection*, 135.
39. Adrian Desmond and James Moore, *Darwin* (New York: Warner, 1991); Michael Ruse, *The Darwinian Revolution: Science Red in Tooth and Claw*, 2nd ed. (Chicago: U of Chicago P, 1999).
40. Cited in David Oldroyd, *Darwinian Impacts: An Introduction to the Darwinian Revolution*, 2nd ed. (Atlantic Highlands, NJ: Humanities P, 1983), 84.
41. Alfred Russel Wallace, in Charles Darwin, *More Letters of Charles Darwin: A Record of His Work in a Series of Hitherto Unpublished Letters*, 2 vols., ed. Francis Darwin and A. C. Seward (London: Murray, 1903), 2: 36; David L. Hull, *Science as a Process: An Evolutionary Account of the Social and Conceptual Development of Science* (Chicago: U of Chicago P, 1988), 279.
42. Ghiselin, *Darwinian Method*, 77.
43. John Maynard Smith, *The Theory of Evolution* (Cambridge: Cambridge UP, 1993), 43.
44. Mayr, *One Long Argument*, 132.
45. Huxley, "On the Reception."
46. Mayr, *One Long Argument*, 132.
47. Ridley, *Evolution*, 18.
48. Karl Popper, "Normal Science and Its Dangers," in *Criticism and*

the Growth of Knowledge, ed. Imre Lakatos and Alan Musgrave (Cambridge: Cambridge UP, 1970), 56.
49. Ghiselin, Darwinian Method, 45.
50. Maynard Smith, Theory of Evolution, 43.
51. Darwin, Autobiography, 140.
52. Mayr, One Long Argument, 37, 95.

Chapter 11. The Science Wars in a Long View: Putting the Human in Its Place

1. Barbara Herrnstein Smith, Scandalous Knowledge: Science, Truth, and the Human (Edinburgh: Edinburgh UP, 2006).
2. Menand, "Dangers Within and Without," 14.
3. Thomas Henry Huxley, Science and Education, vol. 3 of Collected Essays (New York: D. Appleton, 1898), 234–59.
4. Thomas Henry Huxley, Method and Results, vol. 1 of Collected Essays (New York: D. Appleton, 1899), 34.
5. Huxley, Method and Results, 35, 37.
6. Steven Weinberg, Dreams of a Final Theory: The Search for the Fundamental Laws of Nature (New York: Pantheon, 1992), 3, 46.
7. Steven Weinberg, Facing Up: Science and Its Cultural Adversaries (Cambridge: Harvard UP, 2001), 146.
8. Facing Up, 4–5.
9. Matthew Arnold, Poetical Works, ed. C. B. Tinker and H. F. Lowry (London: Oxford UP, 1950), 211.
10. Matthew Arnold, "Literature and Science," in Philistinism in England and America, vol. 10 of The Complete Prose Works of Matthew Arnold, ed. R. H. Super (Ann Arbor: U of Michigan P, 1974), 55; cited hereafter, in the text, as "LS."
11. Matthew Arnold, Culture and Anarchy, vol. 5 of The Complete Prose Works of Matthew Arnold, ed. R. H. Super (Ann Arbor: U of Michigan P, 1965), 236. For a more extensive account of Arnold's teleological scheme of cultural history, see Carroll, Matthew Arnold, 69–84.
12. Matthew Arnold, "On Poetry," in English Literature and Irish Politics, vol. 9 in The Complete Prose Works of Matthew Arnold, ed. R. H. Super (Ann Arbor: U of Michigan P, 1973), 63. On the development of Arnold's thinking about the relations between poetry and religion, see Carroll, Cultural Theory, 85–123.
13. For Frye's account of the anagogic phase, see Anatomy of Criticism, 115–28. For a commentary on Frye's literary idealism from a Darwinian perspective, see Carroll, Evolution, 383–90.
14. C. P. Snow, The Two Cultures, with an introduction by Stefan Collini (Cambridge: Cambridge UP, 1993), 14.
15. F. R. Leavis, "Two Cultures? The Significance of Lord Snow," in Nor Shall My Sword: Discourses on Pluralism, Compassion, and Social Hope (London: Chatto and Windus, 1972), 53; hereafter cited as "Significance."

16. On the radiations of sociobiology, see Barrett, Dunbar, and Lycett, *Human Evolutionary Psychology*, 8–21.
17. Huxley, *Science and Education*, 150.

Chapter 12. A Darwinian Revolution in the Humanities

1. Brown, *Human Universals*, 1–38; Degler, *In Search of Human Nature*; Robin Fox, *The Search for Society: Quest for a Biosocial Science and Morality* (New Brunswick: Rutgers UP, 1989), 53–105; Derek Freeman, *The Fateful Hoaxing of Margaret Mead: A Historical Analysis of Her Samoan Research* (Boulder, CO: Westview, 1999), 17–27; Pinker, *The Blank Slate*; Henry Plotkin, *Evolutionary Thought in Psychology: A Brief History* (Malden, MA: Blackwell, 2004); Tooby and Cosmides, "Psychological Foundations," 28.
2. Wallace Stevens, "Notes Toward a Supreme Fiction," in *Collected Poetry and Prose*, ed. Frank Kermode and Joan Richardson (New York: Library of America, 1997), 342.
3. Thorstein Veblen, *The Place of Science in Modern Civilization*, with an introduction by Warren J. Samuels (1919; New Brunswick, NJ: Transaction, 1990), 409–30.
4. Theodosius Dobzhansky, "Nothing in Biology Makes Sense Except in the Light of Evolution," *American Biology Teacher* 35 (1973): 125–29.
5. Jacques Derrida, *Of Grammatology*, trans. Gayatri Chakravorty Spivak (Baltimore: Johns Hopkins UP, 1976), 158.
6. Matthew Arnold, *English Literature and Irish Politics*, 63.
7. Frye, *Anatomy of Criticism*, 115–28.
8. Darwin, *Descent of Man*, 2: 405.

References

Abrams, Meyer H. "How to Do Things with Texts." *Partisan Review* 46 (1979): 566–88.

———. *The Mirror and the Lamp: Romantic Theory and the Critical Tradition.* New York: Oxford UP, 1953.

———. "The Transformation of English Studies: 1930–1995." *Daedalus* 126 (1997): 105–32.

Adelman, Janet. *Suffocating Mothers: Fantasies of Maternal Origin in Shakespeare's Plays, "Hamlet" to "The Tempest."* New York: Routledge, 1992.

Albritton, Claude C., Jr. *The Abyss of Time: Changing Conceptions of the Earth's Antiquity after the Sixteenth Century.* San Francisco, CA: Freeman, Cooper, 1980.

Alexander, Richard D. *The Biology of Moral Systems.* Hawthorne, NY: Aldine de Gruyter, 1987.

Allott, Miriam. Introduction to *Emily Brontë: "Wuthering Heights": A Casebook*, ed. Miriam Allott, 11–36. Houndmills: Macmillan, 1970.

———. "The Rejection of Heathcliff?" *Essays in Criticism* 8 (1958): 27–47.

Anderson, Joseph. *The Reality of Illusion: An Ecological Approach to Cognitive Film Theory.* Carbondale: Southern Illinois UP, 1996.

———, and Barbara Fisher Anderson, eds. *Moving Image Theory: Ecological Considerations.* Carbondale: Southern Illinois UP, 2005.

Andrews, Alice, and Joseph Carroll, eds. *The Evolutionary Review: Art, Science, Culture* 1 (2010).

Armstrong, Nancy. "Emily Brontë In and Out of Her Time." *Genre* 15 (1982): 243–64.

Arnhart, Larry. *Darwinian Natural Right: The Biological Ethics of Human Nature.* Albany: State U of New York P, 1998.

Arnold, Matthew. *Culture and Anarchy.* Vol. 5 of *The Complete Prose Works of Matthew Arnold.* Ed. R. H. Super. Ann Arbor: U of Michigan P, 1965.

———. "Literature and Science." In *Philistinism in England and America.* Vol. 10 of *The Complete Prose Works of Matthew Arnold*, ed. R. H. Super, 53–73. Ann Arbor: U of Michigan P, 1974.

———. *On the Classical Tradition*. Vol. 1 of *The Complete Prose Works of Matthew Arnold*, ed. R. H. Super. Ann Arbor: U of Michigan P, 1960.

Austin, Michael. *Useful Fictions: Evolution, Anxiety, and the Origins of Literature*. Lincoln: U of Nebraska P, 2011.

Barash, David, and Nanelle Barash. *Madame Bovary's Ovaries: A Darwinian Look at Literature*. New York: Delacorte, 2005.

Barkow, Jerome, Leda Cosmides, and John Tooby, eds. *The Adapted Mind: Evolutionary Psychology and the Generation of Culture*. New York: Oxford UP, 1992.

Baron-Cohen, Simon. "The Empathizing System: A Revision of the 1994 Model of the Mindreading System." In *Origins of the Social Mind: Evolutionary Psychology and Child Development*, ed. Bruce J. Ellis and David F. Bjorklund, 468–92. New York: Guilford, 2005.

Barondes, Samuel H. *Better Than Prozac: Creating the Next Generation of Psychiatric Drugs*. Oxford: Oxford UP, 2003.

Barrett, Louise, Robin Dunbar, and John Lycett. *Human Evolutionary Psychology*. Princeton: Princeton UP, 2002.

Barton, Robert A. "Evolution of the Social Brain as a Distributed Neural System." In *The Oxford Handbook of Evolutionary Psychology*, ed. Robin Dunbar and Louise Barrett, 129–44. Oxford: Oxford UP, 2007.

Baumeister, Roy F. *The Cultural Animal: Human Nature, Meaning, and Social Life*. Oxford: Oxford UP, 2005.

Beer, Gillian. *Darwin's Plots: Evolutionary Narrative in Darwin, George Eliot, and Nineteenth-Century Fiction*. London: Routledge, 1983.

Berman, Jeffrey. *Narcissism and the Novel*. New York: New York UP, 1990.

Bersani, Leo. *A Future for Astyanax: Character and Desire in Literature*. Boston: Little, Brown, 1969.

Betzig, Laura L. *Despotism and Differential Reproduction: A Darwinian View of History*. Hawthorne, NY: Aldine de Gruyter, 1986.

Bickerton, Derek. *Adam's Tongue: How Humans Made Language, How Language Made Humans*. New York: Hill and Wang, 2009.

———. "Foraging versus Social Intelligence in the Evolution of Protolanguage." In *The Transition to Language*, ed. Alison Wray, 207–25. Oxford: Oxford UP, 2000.

———. "From Protolanguage to Language." In *The Speciation of Modern Homo Sapiens*, ed. T. J. Crow, 103–20. Oxford: Oxford UP, 2002.

———. *Language and Species*. Chicago: U of Chicago P, 1990.

Bjorklund, David F., and Anthony D. Pellegrini. *The Origins of Human Nature: Evolutionary Developmental Psychology*. Washington, DC: American Psychological Association, 2002.

Boehm, Christopher. *Hierarchy in the Forest: The Evolution of Egalitarian Behavior*. Cambridge: Harvard UP, 1999.

Boghossian, Paul. *Fear of Knowledge: Against Relativism and Constructivism*. Oxford: Oxford UP, 2006.

Bohannan, Laura. "Shakespeare in the Bush." *Natural History* 75 (August–September 1966): 28–33.

Bordwell, David. *Poetics of Cinema*. New York: Routledge, 2008.

———, and Noël Carroll, eds. *Post-Theory: Reconstructing Film Studies*. Madison: U of Wisconsin P, 1996.
Bower, Gordon H., and Daniel G. Morrow. "Mental Models in Narrative Comprehension." *Science*, New Series, 247(1990): 44–48.
Bowlby, John. *Attachment*, 2nd ed. Vol. 1 of *Attachment and Loss*. New York: Basic Books, 1982.
———. *Attachment and Loss*, 2nd ed. 3 Vols. London: Hogarth, 1982.
Boyd, Brian. "Art and Evolution: Spiegelman in *The Narrative Corpse*." *Philosophy and Literature* 32 (2008): 31–57.
———. "The Art of Literature and the Science of Literature." *The American Scholar* 77.2 (2008): 118–27.
———. "Evolutionary Theories of Art." In *The Literary Animal: Evolution and the Nature of Narrative*, ed. Jonathan Gottschall and David Sloan Wilson, 147–76. Evanston, IL: Northwestern UP, 2005.
———. "Getting It All Wrong." *The American Scholar* 75.4 (2006): 18–30.
———. "Literature and Evolution: A Bio-Cultural Approach." *Philosophy and Literature* 29 (2005): 1–23.
———. "On the Origin of Comics: New York Double-Take." *The Evolutionary Review: Art, Science, Culture* 1 (2010): 97–111.
———. *On the Origin of Stories: Evolution, Cognition, and Fiction*. Cambridge: Harvard UP, 2009.
———, Joseph Carroll, and Jonathan Gottschall, eds. *Evolution, Literature, and Film: A Reader*. New York: Columbia UP, 2010.
Boyd, Robert, and Peter Richerson. "Cultural Adaptation and Maladaptation: Of Kayaks and Commissars." In *The Evolution of Mind: Fundamental Questions and Controversies*, ed. Steven W. Gangestad and Jeffry A. Simpson, 327–31. New York: Guilford, 2007.
Boyer, Pascal. "Specialised Inference Engines as Precursors of Creative Imagination?" *Proceedings of the British Academy* 147 (2007): 239–58.
Bradley, A. C. *Shakespearean Tragedy: Lectures on "Hamlet," "Othello," "King Lear," and "Macbeth."* 1904; London: Penguin, 1991.
Brinkley, Edward S. "Homosexuality as (Anti)Illness: Oscar Wilde's *The Picture of Dorian Gray* and Gabriele d'Annunzio's *Il Piacere*." *Studies in Twentieth-Century Literature* 22 (1998): 61–82.
Bristol, Michael. "'Funeral-Bak'd Meats': Carnival and the Carnivalesque in *Hamlet*." In *William Shakespeare: "Hamlet,"* ed. Susanne L. Wofford, 348–67. Boston: St. Martin's P, 1994.
Bristow, Joseph. "'A Complex Multiform Creature': Wilde's Sexual Identities." In *The Cambridge Companion to Oscar Wilde*, ed. Peter Raby, 195–218. Cambridge: Cambridge UP, 1997.
———. *Effeminate England: Homoerotic Writing after 1885*. New York: Columbia UP, 1995.
Brontë, Emily. *"Wuthering Heights": The 1847 Text, Backgrounds and Criticism*, ed. Richard J. Dunn, 4th ed. New York: Norton, 2003.
Brooks, Jean R. *Thomas Hardy: The Poetic Structure*. Ithaca, NY: Cornell UP, 1971.
Brown, Donald. *Human Universals*. Philadelphia: Temple UP, 1991.

———. "Human Universals and Their Implications." In *Being Human: Anthropological Universality and Particularity in Transdisciplinary Perspectives*, ed. Neil Roughley, 156–74. New York: Walter de Gruyter, 2000.

Budiansky, Stephen. *If a Lion Could Talk: How Animals Think*. London: Weidenfeld and Nicolson, 1998.

Bullen, J. B. *The Expressive Eye: Fiction and Perception in the Work of Thomas Hardy*. Oxford: Oxford UP, 1986.

Buss, Arnold. "Evolutionary Perspectives on Personality Traits." In *Handbook of Personality Psychology*, ed. Robert Hogan, John Johnson, and Stephen Briggs, 346–66. San Diego: Academic P, 1997.

Buss, David M. *The Dangerous Passion: Why Jealousy Is as Necessary as Love and Sex*. New York: Free Press, 2000.

———. *Evolutionary Psychology: The New Science of the Mind*, 3rd ed. Boston: Allyn and Bacon, 2007.

———. *The Evolution of Desire: Strategies of Human Mating*, rev. ed. New York: Basic Books, 2003.

———. *The Handbook of Evolutionary Psychology*. Hoboken, NJ: Wiley, 2005.

———. "Social Adaptation and Five Major Factors of Personality." In *The Five-Factor Model of Personality: Theoretical Perspectives*, ed. Jerry S. Wiggins, 180–207. New York: Guilford, 1996.

Cain, William E. *The Crisis in Criticism: Theory, Literature, and Reform in English Studies*. Baltimore: Johns Hopkins UP, 1984.

Carroll, Joseph. "Adaptationist Literary Study: An Emerging Research Program." *Style* 36 (2003): 596–617.

———. "Adaptationist Literary Study: An Introductory Guide." *Ometeca* 10 (2006): 18–31.

———. *The Cultural Theory of Matthew Arnold*. Berkeley: U of California P, 1982.

———. "Evolutionary Approaches to Literature and Drama." In *The Oxford Handbook of Evolutionary Psychology*, ed. Robin Dunbar and Louise Barrett, 637–48. Oxford: Oxford UP, 2007.

———. "An Evolutionary Paradigm for Literary Study." *Style* 42 (2008): 103–35.

———. *Evolution and Literary Theory*. Columbia: U of Missouri P, 1995.

———. "The Human Revolution and the Adaptive Function of Literature." *Philosophy and Literature* 30 (2006): 33–49.

———. *Literary Darwinism: Evolution, Human Nature, and Literature*. New York: Routledge, 2004.

———. "Literature and Evolutionary Psychology." In *The Handbook of Evolutionary Psychology*, ed. David Buss, 931–52. Hoboken, NJ: Wiley, 2005.

———. "Pater's Figures of Perplexity." *Modern Language Quarterly* 52 (1991): 319–40.

———. *Wallace Stevens' Supreme Fiction: A New Romanticism*. Baton Rouge: Louisiana State UP, 1987.

Casey, Nell, ed. *Unholy Ghost: Writers on Depression*. New York: HarperCollins, 2002.

Cecil, Lord David. *Early Victorian Novelists: Essays in Revaluation.* Indianapolis: Bobbs-Merrill, 1935.

———. *Hardy the Novelist: An Essay in Criticism.* Indianapolis: Bobbs-Merrill, 1943.

Chagnon, Napoleon A. *Yanomamö: The Fierce People,* 3rd ed. New York: Holt, Rinehart, and Winston, 1979.

Charlesworth, Barbara. *Dark Passages: The Decadent Consciousness in Victorian Literature.* Madison: U of Wisconsin P, 1965.

Chiappe, Dan, and Kevin B. MacDonald. "Metaphor, Modularity, and the Evolution of Conceptual Integration." *Metaphor and Symbol* 15 (2000): 137–58.

Chitham, Edward. *A Life of Emily Brontë.* Oxford: Blackwell, 1987.

Clausson, Nils. "Culture and Corruption: Paterian Self-Development *versus* Gothic Degeneration in Oscar Wilde's *The Picture of Dorian Gray.*" *Papers on Language and Literature* 39 (2003): 339–64.

Cochran, Gregory, and Henry Harpending. *The 10,000 Year Explosion: How Civilization Accelerated Human Evolution.* New York: Basic Books, 2009.

Cohen, Ed. *Talk on the Wilde Side: Toward a Genealogy of a Discourse on Male Sexualities.* New York: Routledge, 1993.

———. "Writing Gone Wilde: Homoerotic Desire in the Closet of Representation." *PMLA* 102 (1987): 801–13.

Cohen, Philip. *The Moral Vision of Oscar Wilde.* Cranbury, NJ: Associated University Presses, 1978.

Conrad, Joseph. *Heart of Darkness,* ed. Robert Hampson. London: Penguin, 1995.

Cooke, Brett. "Compliments and Complements." *Style* 42.2/3 (2008): 150–54.

———. *Human Nature in Utopia: Zamyatin's* We. Evanston, IL: Northwestern UP, 2002.

———. "The Promise of a Biothematics." In *Sociobiology and the Arts,* ed. Jean Baptiste Bedaux and Brett Cooke, 43–62. Amsterdam: Editions Rodopi, 1999.

———. "Sexual Property in Pushkin's 'The Snowstorm': A Darwinist Perspective." In *Biopoetics: Evolutionary Explorations in the Arts,* ed. Brett Cooke and Frederick Turner, 175–204. Lexington, KY: ICUS, 1999.

Cosmides, Leda, and John Tooby, and Jerome Barkow. "Introduction: Evolutionary Psychology and Conceptual Integration." In *The Adapted Mind: Evolutionary Psychology and the Generation of Culture,* ed. Jerome Barkow, Leda Cosmides, and John Tooby, 3–15. New York: Oxford UP, 1992.

Costa, Paul T., and Robert R. McCrae. "Personality Trait Structure as a Human Universal." *American Psychologist* 52 (1997): 509–16.

Crews, Frederick. "Apriorism for Empiricists." *Style* 42.2/3 (2008): 155–60.

———. *Postmodern Pooh.* 2001; Evanston, IL: Northwestern UP, 2007.

Cummins, Denise. "Dominance, Status, and Social Hierarchies." In *The Handbook of Evolutionary Psychology*, ed. David M. Buss, 676–97. Hoboken, NJ: Wiley, 2005.

Currie, Gregory. *Arts and Minds.* Oxford: Clarendon, 2004.

Daly, Martin, and Margo Wilson. *Homicide.* New York: Aldine, 1988.

Dalziel, Pamela. Introduction to *The Mayor of Casterbridge*, by Thomas Hardy. Oxford: Oxford UP, 2004.

Damasio, Antonio R. *Descartes' Error: Emotion, Reason, and the Human Brain.* New York: Putnam's, 1994.

Darwin, Charles. *The Autobiography of Charles Darwin, 1809–1882, with Original Omissions Restored*, ed. Nora Barlow. London: Collins, 1958.

———. *Darwin on Evolution: The Development of the Theory of Natural Selection*, ed. Thomas F. Glick and David Kohn. Indianapolis: Hackett, 1996.

———. *The Descent of Man, and Selection in Relation to Sex*, ed. John Tyler Bonner and Robert M. May. 2 vols. in 1. 1871; Princeton: Princeton UP, 1981.

———. *Journal of Researches into the Geology and Natural History of the Various Countries Visited by H.M.S. Beagle, Under the Command of Captain Fitzroy from 1832 to 1836*, 2nd ed. London: Murray, 1845.

———. *More Letters of Charles Darwin. A Record of His Work in a Series of Hitherto Unpublished Letters.* 2 vols., ed. Francis Darwin and A. C. Seward. London: Murray, 1903.

———. *On the Origin of Species by Means of Natural Selection.* Ed. Joseph Carroll. 1859; Peterborough, Ontario: Broadview, 2003.

———, and Alfred Russel Wallace. *Evolution by Natural Selection.* With a foreword by Gavin De Beer. Cambridge: Cambridge UP, 1958. This volume contains the 1842 sketch; the 1844 manuscript; and the 1858 Linnean Society papers by Darwin and Wallace.

Davidson, Richard J., Diego Pizzagalli, Jack B. Nitschke, and Katherine Putnam. "Depression: Perspectives from Affective Neuroscience." *Annual Review of Psychology* 53 (2002): 545–74.

Davis, Eugene W. "Comparatively Modern Skeletons in the Garden: A Reconsideration of *The Mayor of Casterbridge*." *English Literature in Transition: 1880–1920*, special series 3 (1985): 108–20.

Dawkins, Richard. *The Selfish Gene.* New York: Oxford UP, 1976.

Deacon, Terrence W. *The Symbolic Species: The Co-Evolution of Language and the Brain.* New York: Norton, 1997.

Degler, Carl. *In Search of Human Nature: The Decline and Revival of Darwinism in American Social Thought.* New York: Oxford UP, 1991.

Deresiewicz, William. "Adaptation: On Literary Darwinism." *The Nation*, 8 June 2009, 26–31 .

Desmond, Adrian, and James Moore. *Darwin.* New York: Warner, 1991.

Dike, D. A. "A Modern Oedipus: *The Mayor of Casterbridge*." *Essays in Criticism* 2 (1952): 169–79.

Dissanayake, Ellen. *Art and Intimacy: How the Arts Began.* Seattle: U of Washington P, 2000.

———. *Homo Aestheticus: Where Art Comes from and Why.* 1992; Seattle: U of Washington P, 1995.
———. "What Art Is and What Art Does: An Overview of Contemporary Evolutionary Hypotheses." In *Evolutionary and Neurocognitive Approaches to Aesthetics, Creativity, and the Arts*, ed. Colin Martindale, Paul Locher, and Vladimir M. Petrov, 1–14. Amityville, NY: Baywood, 2007.
———. *What Is Art For?* Seattle: U of Washington P, 1988.
Dollimore, Jonathan. *Sexual Dissidence: Augustine to Wilde, Freud to Foucault.* Oxford: Oxford UP, 1991.
Dryden, John. *Selected Criticism.* Ed. James Kinsley and George Parfitt. Oxford: Oxford UP, 1970.
Dunbar, Robin. *The Human Story: A New History of Mankind's Evolution.* London: Faber and Faber, 2004.
———. "Why Are Good Writers So Rare? An Evolutionary Perspective on Literature." *Journal of Evolutionary and Cultural Psychology* 3 (2005): 7–22.
———, and Louise Barrett. "Evolutionary Psychology in the Round." In *The Oxford Handbook of Evolutionary Psychology*, ed. Robin Dunbar and Louise Barrett, 3–9. Oxford: Oxford UP, 2007.
———, eds. *The Oxford Handbook of Evolutionary Psychology.* Oxford: Oxford UP, 2007.
Dutton, Denis. *The Art Instinct.* New York: Bloomsbury, 2009.
Eagleton, Terry. *Myths of Power: A Marxist Study of the Brontës.* 1975; Houndmills: Palgrave, 2005.
Easterlin, Nancy. "Do Cognitive Predispositions Predict or Determine Literary Value Judgments? Narrativity, Plot, and Aesthetics." In *Biopoetics: Evolutionary Explorations in the Arts*, ed. Brett Cooke and Frederick Turner, 241–62. Lexington, KY: ICUS, 1999.
———. "Hans Christian Andersen's Fish Out of Water." *Philosophy and Literature* 25 (2001): 251–77.
———. "'Loving Ourselves Best of All': Ecocriticism and the Adapted Mind." *Mosaic* 37 (2004): 1–18.
———. "Psychoanalysis and the 'Discipline of Love,'" *Philosophy and Literature* 24 (2000): 261–79.
Eibl-Eibesfeldt, Irenäus. "Us and the Others: The Familial Roots of Ethnonationalism." In *Ethnic Conflict and Indoctrination: Altruism and Identity in Evolutionary Perspective*, ed. Irenäus Eibl-Eibesfeldt and Frank K. Salter, 21–53. New York: Berghahn, 1998.
Ekman, Paul. *Emotions Revealed: Recognizing Faces and Feelings to Improve Communication and Emotional Life.* New York: Henry Holt, 2003.
Eliot, George. *Middlemarch: An Authoritative Text, Backgrounds, Reviews, and Criticism*, ed. Bert G. Hornback. 2nd ed. 1871–1872; New York: W. W. Norton, 2000.
Ellmann, Richard. "Overtures to Salome." In *Oscar Wilde: A Collection of Critical Essays*, ed. Richard Ellmann, 87–91. Englewood Cliffs, NJ: Prentice-Hall, 1969.
Enard, Wolfgang, et al. "Molecular Evolution of *FOXP2*, a Gene Involved in Speech and Language." *Nature* (2002): 869–72.

Evans, Dylan. "From Lacan to Darwin." In *The Literary Animal: Evolution and the Nature of Narrative*, ed. Jonathan Gottschall and David Sloan Wilson, 38–55. Evanston, IL: Northwestern UP, 2005.

Figueredo, Aurelio José, et al. "The K-Factor, Covitality, and Personality: A Psychometric Test of Life History Theory." *Human Nature* 18 (2007): 47–73.

Flesch, William. *Comeuppance: Costly Signaling, Altruistic Punishment, and Other Biological Components of Fiction*. Cambridge: Harvard UP, 2008.

Flinn, Mark V., David C. Geary, and Carol V. Ward. "Ecological Dominance, Social Competition, and Coalitionary Arms Races: Why Humans Evolved Extraordinary Intelligence." *Evolution and Human Behavior* 26 (2005): 10–46.

Flinn, Mark V., and Carol V. Ward. "Ontogeny and Evolution of the Social Child." In *Origins of the Social Mind*, ed. Bruce J. Ellis and David F. Bjorklund, 19–44. New York: Guilford, 2005.

Focquaert, Farah, and Steven M. Platek. "Social Cognition and the Evolution of Self–Awareness." In *Evolutionary Cognitive Neuroscience*, ed. Steven M. Platek, Julian Paul Keenan, and Todd K. Shackelford, 457–97. Cambridge: MIT P, 2007.

Foley, Robert. "The Adaptive Legacy of Human Evolution: A Search for the Environment of Evolutionary Adaptedness." *Evolutionary Anthropology* 4 (1996): 194–203.

Foucault, Michel. *Language, Counter–Memory, Practice: Selected Essays and Interviews*, trans. D. F. Bouchard and S. Simon, ed. D. F. Bouchard. Ithaca: Cornell UP, 1977.

Frith, Gillian. "Decoding *Wuthering Heights*." In *Critical Essays on Emily Brontë*, ed. Thomas John Winnifrith, 243–61. New York: Hall-Simon and Schuster, 1997.

Fromm, Harold. *Academic Capitalism and Literary Value*. Athens: U of Georgia P, 1991.

———. *The Nature of Being Human: From Environmentalism to Consciousness*. Baltimore: Johns Hopkins UP, 2009.

Frye, Northrop. *The Anatomy of Criticism: Four Essays*. Princeton: Princeton UP, 1957.

———. *The Critical Path: An Essay on the Social Context of Literary Criticism*. Bloomington: Indiana UP, 1971.

Gangestad, Steven W. "Reproductive Strategies and Tactics." In *The Oxford Handbook of Evolutionary Psychology*, ed. Robin Dunbar and Louise Barrett, 321–32. Oxford: Oxford UP, 2007.

———, and Jeffry A. Simpson, eds. *The Evolution of Mind: Fundamental Questions and Controversies*. New York: Guilford, 2007.

———. "An Introduction to *The Evolution of Mind*: Why We Developed This Book." In *The Evolution of Mind: Fundamental Questions and Controversies*, ed. Steven W. Gangestad and Jeffry A. Simpson, 1–21. New York: Guilford, 2007.

Garber, Marjorie. *Shakespeare's Ghost Writers: Literature as Uncanny Causality*. New York: Methuen, 1987.

Gatrell, Simon. *Thomas Hardy and the Proper Study of Mankind.* Charlottesville: U of Virginia P, 1993.
Geary, David C. "Evolution of Paternal Investment." In *The Handbook of Evolutionary Psychology,* ed. David M. Buss, 483–505. Hoboken, NJ: Wiley, 2005.
———. *Male, Female: The Evolution of Human Sex Differences.* Washington, DC: American Psychological Association, 1998.
———. *The Origin of Mind: Evolution of Brain, Cognition, and General Intelligence.* Washington, DC: American Psychological Association, 2005.
———, and Mark V. Flinn. "Evolution of Human Parental Behavior and the Human Family." *Parenting: Science and Practice* 1 (2001): 5–61.
Geerken, Ingrid. "'The Dead Are Not Annihilated': Mortal Regret in *Wuthering Heights.*" *Journal of Narrative Theory* 34 (2004): 373–406.
Ghiselin, Michael. *The Triumph of the Darwinian Method.* 1969; Chicago: U of Chicago P, 1984.
Gilbert, Sandra M., and Susan Gubar. *The Madwoman in the Attic: The Woman Writer and the Nineteenth-Century Literary Imagination.* New Haven: Yale UP, 1979.
Gillespie, Michael Patrick. "Picturing Dorian Gray: Resistant Readings in Wilde's Novel." *English Literature in Transition 1880–1920* 35 (1992): 7–25.
Giordano, Frank R., Jr. *"I'd Have My Life Unbe": Thomas Hardy's Self-Destructive Characters.* Tuscaloosa: U of Alabama P, 1984.
Gomel, Elana. "Oscar Wilde, *The Picture of Dorian Gray,* and the (Un)Death of the Author." *Narrative* 12 (2001): 74–92.
Goethe, Johann Wilhelm von. *Wilhelm Meister's Apprenticeship and Travels.* Trans. Thomas Carlyle. Vol. 23 of *The Works of Thomas Carlyle.* London: Chapman and Hall, 1899.
Goode, John. *Thomas Hardy: The Offensive Truth.* Oxford: Blackwell, 1988.
Goodheart, Eugene. *Darwinian Misadventures in the Humanities.* New Brunswick, NJ: Transaction, 2007.
———. "Do We Need Literary Darwinism?" *Style* 42 (2008): 181–85.
Gotlib, Ian H., and Constance L. Hammen, eds. *Handbook of Depression,* 2nd ed. New York: Guilford, 2009.
Gottschall, Jonathan. "Greater Emphasis on Female Attractiveness in Homo sapiens: A Revised Solution to an Old Evolutionary Riddle." *Evolutionary Psychology* 5 (2007): 347–58.
———. *Literature, Science, and a New Humanities.* New York: Palgrave Macmillan, 2008.
———. *The Rape of Troy: Evolution, Violence, and the World of Homer.* Cambridge: Cambridge UP, 2008.
———. "The Tree of Knowledge and Darwinian Literary Study." *Philosophy and Literature* 27 (2003): 255–68.
———, and David Sloan Wilson, eds. *The Literary Animal: Evolution and the Nature of Narrative.* Evanston, IL: Northwestern UP, 2005.
Grabes, Herbert. "Turning Words on the Page into 'Real' People." *Style* 38 (2004): 221–35.

Griffiths, Paul. *What Emotions Really Are: The Problem of Psychological Categories.* Chicago: U of Chicago P, 1997.

Grodal, Torben. *Embodied Visions: Evolution, Emotion, Culture, and Film.* Oxford: Oxford UP, 2009.

Gray, Jeffrey A. "The Neuropsychology of Temperament." In *Explorations in Temperament*, ed. Jan Strelau and Alois Angleitner, 105–28. New York: Plenum, 1991.

Gregor, Ian. *The Great Web: The Form of Hardy's Major Fiction.* Totowa, NJ: Rowman, 1974.

Grossman, Julie. "Thomas Hardy and the Role of Observer." *ELH* 56 (1989): 619–38.

Guerard, Albert J. *Thomas Hardy: The Novels and Stories.* Cambridge: Harvard UP, 1949.

Hagen, Edward H., and Donald Symons. "Natural Psychology: The Environment of Evolutionary Adaptedness and the Structure of Cognition." In *The Evolution of Mind: Fundamental Questions and Controversies*, ed. Steven W. Gangestad and Jeffry A. Simpson, 38–44. New York: Guilford, 2007.

Hallam, Anthony. *Great Geological Controversies.* Oxford: Oxford UP, 1983.

Hamilton, William D. "The Genetical Evolution of Social Behavior, I and II." In *Narrow Roads of Gene Land: The Collected Papers of W. D. Hamilton*, Vol. 1. Oxford: W. H. Freeman, 1996.

Hardy, Thomas. *The Life and Work of Thomas Hardy*, ed. Michael Millgate. London: Macmillan, 1984.

———. *The Mayor of Casterbridge: An Authoritative Text, Backgrounds and Contexts, Criticism*, ed. Phillip Mallett, 2nd ed. New York: Norton, 2001.

Harris, Judith Rich. *No Two Alike: Human Nature and Human Individuality.* New York: Norton, 2007.

Hartveit, Lars. *The Art of Persuasion: A Study of Six Novels.* Bergen: Universitetsforlaget, 1977.

Hassin, Ran R., James S. Uleman, and John A. Bargh, eds. *The New Unconscious.* New York: Oxford UP, 2005.

Hauser, Marc. *Wild Minds: What Animals Really Think.* New York: Henry Holt, 2000.

Hazlitt, William. *Characters of Shakespeare's Plays.* 1817; London: Oxford UP, 1955.

Headlam Wells, Robin. *Shakespeare's Humanism.* Cambridge: Cambridge UP, 2005.

———, and Johnjoe McFadden, eds. *Human Nature: Fact and Fiction.* London: Continuum, 2006.

Heilman, Robert B. "Hardy's *Mayor* and the Problem of Intention." In *The Workings of Fiction: Essays by Robert Bechtold Heilman.* Columbia: U of Missouri P, 1991. Originally published in *Criticism* 5 (1963): 199–213.

Henrich, Joseph, and Richard McElreath. "Dual-Inheritance Theory: The Evolution of Human Cultural Capacities and Cultural Evolution." In *The Oxford Handbook of Evolutionary Psychology*, ed. Robin Dunbar and Louise Barrett, 555–70. Oxford: Oxford UP, 2007.

Henshilwood, Christopher S., and Curtis W. Marean. "The Origin of Modern Human Behavior: Critique of the Models and Their Test Implications." *Current Anthropology* 44 (2003): 627–51.

Herrnstein Smith, Barbara. *Scandalous Knowledge: Science, Truth, and the Human.* Edinburgh: Edinburgh UP, 2006.

Hill, Kim. "Evolutionary Biology, Cognitive Adaptations, and Human Culture." In *The Evolution of Mind: Fundamental Questions and Controversies*, ed. Steven W. Gangestad and Jeffry A. Simpson, 348–56. New York: Guilford, 2007.

Hodgson, Geoffrey M. "Taxonomizing the Relationship between Biology and Economics: A Very Long Engagement." *Journal of Bioeconomics* 9 (2007): 169–85.

Hoeg, Jerry, and Kevin S. Larsen, eds. *Interdisciplinary Essays on Darwinism in Hispanic Literature and Film: The Intersection of Science and the Humanities.* New York: Edwin Mellen, 2009.

Homans, Margaret. "The Name of the Mother in *Wuthering Heights*." In *Wuthering Heights: Complete, Authoritative Text with Biographical and Historical Contexts, Critical History, and Essays from Five Contemporary Critical Perspectives*, ed. Linda H. Peterson, 341–58. Boston: Bedford-St. Martin's, 1992.

Hooker, Joseph. Review of *On the Origin of Species*. In *Darwin and His Critics: The Reception of Darwin's Theory of Evolution by the Scientific Community*, ed. David L. Hull, 81–85. Chicago: U of Chicago P, 1973.

Hornback, Bert G. *The Metaphor of Chance: Vision and Technique in the Works of Thomas Hardy.* Athens: Ohio UP, 1971.

Howe, Irving. *Thomas Hardy*. New York: Macmillan, 1967.

Hull, David L. *Science as a Process: An Evolutionary Account of the Social and Conceptual Development of Science.* Chicago: U of Chicago P, 1988.

Hutton, Richard Holt. Review of *The Mayor of Casterbridge*, by Thomas Hardy. In *Thomas Hardy: The Critical Heritage*, ed. R. G. Cox, 136–40. 1979; London: Routledge, 1995. Originally published in *Spectator* 5 (June 1886): 752–53.

Huxley, Thomas Henry. *Darwiniana*. Vol. 2 of *Collected Essays*. New York: Macmillan, 1893.

———. *Method and Results*. Vol. 1 of *Collected Essays*. New York: D. Appleton, 1899.

———. "On the Reception of the *Origin of Species*." In *The Life and Letters of Charles Darwin, Including an Autobiographical Chapter*, ed. Francis Darwin, foreword by George Gaylord Simpson, 2 vols. 1887; New York: Basic Books, 1959.

———. *Science and Education*. Vol. 3 of *Collected Essays*. New York: D. Appleton, 1898.

Irons, William. "Adaptively Relevant Environments versus the Environment of Evolutionary Adaptedness." *Evolutionary Anthropology* 6 (1998): 194–204.
Jackson, Tony. "Issues and Problems in the Blending of Cognitive Science, Evolutionary Psychology, and Literary Study." *Poetics Today* 23 (2002): 161–79.
———. "Questioning Interdisciplinarity." *Poetics Today* 21 (2000): 319–47.
Jacobs, Carol. *Uncontainable Romanticism: Shelley, Brontë, Kleist*. Baltimore: Johns Hopkins UP, 1989.
Jameson, Fredric. *Postmodernism, or, the Cultural Logic of Late Capitalism*. Durham, NC: Duke UP, 1991.
Jekel, Pamela. *Thomas Hardy's Heroines: A Chorus of Priorities*. Troy, NY: Whitson, 1986.
Jellema, Tjeerd, and David I. Perrett. "Neural Pathways of Social Cognition." In *The Oxford Handbook of Evolutionary Psychology*, ed. Robin Dunbar and Louise Barrett, 163–77. Oxford: Oxford UP, 2007.
Jobling, Ian. "Personal Justice and Homicide in Scott's *Ivanhoe*: An Evolutionary Psychological Perspective." *Interdisciplinary Literary Studies* 2 (2001): 29–43.
John, Oliver P., Alois Angleitner, and Fritz Ostendorf. "The Lexical Approach to Personality: A Historical Review of Trait Taxonomic Research." *European Journal of Personality* 2 (1988): 171–203.
Johnson, Bruce. *True Correspondence: A Phenomenology of Thomas Hardy's Novels*. Tallahassee: UP of Florida, 1983.
Johnson, John A., Joseph Carroll, Jonathan Gottschall, and Daniel J. Kruger. "Hierarchy in the Library: Egalitarian Dynamics in Victorian Novels." *Evolutionary Psychology* 6 (2008): 715–38.
Johnson, Samuel. *Johnson on Shakespeare*, ed. Arthur Sherbo. Vol. 8 of *The Yale Edition of the Works of Samuel Johnson*. New Haven: Yale UP, 1968.
Jones, Ernest. *Hamlet and Oedipus*. 1949; New York: Norton, 1976.
Kaplan, Hillard S., and Steven W. Gangestad. "Life History Theory and Evolutionary Psychology." In *The Handbook of Evolutionary Psychology*, ed. David M. Buss, 68–95. Hoboken, NJ: Wiley, 2005.
———. "Optimality Approaches and Evolutionary Psychology: A Call for Synthesis." In *The Evolution of Mind: Fundamental Questions and Controversies*, ed. Steven W. Gangestad and Jeffry A. Simpson, 121–29. New York: Guilford, 2007.
Kaplan, Hillard S., Kim Hill, Jane Lancaster, and A. Magdalena Hurtado. "A Theory of Human Life History Evolution: Diet, Intelligence, and Longevity." *Evolutionary Anthropology* 9 (2000): 156–85.
Karl, Frederick. "*The Mayor of Casterbridge*: A New Fiction Defined." *Modern Fiction Studies* 6 (1960): 195–213.
Katz, L. D., ed. *Evolutionary Origins of Morality: Cross-Disciplinary Perspectives*. Bowling Green, OH: Imprint Academic, 2000.

King, Jeannette. *Tragedy in the Victorian Novel: Theory and Practice in the Novels of George Eliot, Thomas Hardy, and Henry James.* Cambridge: Cambridge UP, 1978.
Kirby, Simon. "The Evolution of Language." In *The Oxford Handbook of Evolutionary Psychology*, ed. Robin Dunbar and Louise Barrett, 669–81. Oxford: Oxford UP, 2007.
Klein, Richard, with Blake Edgar. *The Dawn of Human Culture.* New York: John Wiley and Sons, 2002.
Knapp, John V. "Family Games and Imbroglio in *Hamlet.*" In *Reading the Family Dance: Family Systems Therapy and Literary Study*, ed. John V. Knapp and Kenneth Womack, 194–220. Newark: U of Delaware P, 2003.
———. "Family-Systems Psychotherapy and Psychoanalytic Literary Criticism: A Comparative Critique." *Mosaic* 37 (2004): 149–66.
Kramer, Peter. *Against Depression.* New York: Viking, 2005.
Kruger, Daniel, Maryanne Fisher, and Ian Jobling. "Proper and Dark Heroes as Dads and Cads: Alternative Mating Strategies in British and Romantic Literature." *Human Nature* 14 (2003): 305–17.
Kurland, Jeffrey A., and Steven J. C. Gaulin. "Cooperation and Conflict among Kin." In *The Handbook of Evolutionary Psychology*, ed. David M. Buss, 447–82. Hoboken, NJ: Wiley, 2005.
Kurzban, Robert, and Steven Neuberg. "Managing Ingroup and Outgroup Relations." In *The Handbook of Evolutionary Psychology*, ed. David M. Buss, 653–75. Hoboken, NJ: Wiley, 2005.
Lacan, Jacques. "Desire and the Interpretation of Desire in *Hamlet.*" *Yale French Studies* 55/56 (1977): 11–52.
Laland, Kevin N. "Niche Construction, Human Behavioural Ecology and Evolutionary Psychology." In *The Oxford Handbook of Evolutionary Psychology*, ed. Robin Dunbar and Louise Barrett, 35–47. Oxford: Oxford UP, 2007.
———, and Gilian R. Brown, *Sense and Nonsense: Evolutionary Perspectives on Human Behaviour.* Oxford: Oxford UP, 2002.
Lam, Raymond, and Hiram Mok. *Depression.* Oxford: Oxford UP, 2008.
Lancaster, Jane B., and Hillard S. Kaplan. "Chimpanzee and Human Intelligence: Life History, Diet, and the Mind." In *The Evolution of Mind: Fundamental Questions and Controversies*, ed. Steven W. Gangestad and Jeffry A. Simpson, 111–18. New York: Guilford, 2007.
Lane, Christopher. *The Burdens of Intimacy: Psychoanalysis and Victorian Masculinity.* Chicago: U of Chicago P, 1999.
Langbaum, Robert. *Thomas Hardy in Our Time.* New York: St. Martin's, 1995.
Leavis, F. R. "Two Cultures? The Significance of Lord Snow." In *Nor Shall My Sword: Discourses on Pluralism, Compassion, and Social Hope.* London: Chatto and Windus, 1972.
Leavis, Q. D. "A Fresh Approach to *Wuthering Heights.*" In *Lectures in America*, by F. R. Leavis and Q. D. Leavis, 85–138. New York: Pantheon-Random, 1969.
Lerner, Laurence. *Thomas Hardy's Mayor of Casterbridge: Tragedy or Social History?* London: Sussex UP, 1975.

Levine, George. *Darwin and the Novelists: Patterns of Science in Victorian Fiction*. Chicago: U of Chicago P, 1988.
———. *The Realistic Imagination: English Fiction from Frankenstein to Lady Chatterley*. Chicago: U of Chicago P, 1981.
Lewis, Michael. "Self-Conscious Emotions: Embarrassment, Pride, Shame, and Guilt." In *Handbook of Emotions*, 2nd ed., ed. Michael Lewis and Jeannette M. Haviland-Jones, 137–56. New York: Guilford, 2000.
Liou, Liang-ya. "The Politics of a Transgressive Desire: Oscar Wilde's *The Picture of Dorian Gray*." *Studies in Language and Literature* 6 (1994): 101–25.
Lorenz, Konrad. *Behind the Mirror: A Search for a Natural History of Human Knowledge*, trans. Ronald Taylor. 1973; New York: Harcourt, 1978.
Love, Glen A. "Ecocriticism and Science: Toward Consilience?" *New Literary History* 30 (1999): 561–76.
———. *Practical Ecocriticism: Literature, Biology, and the Environment*. Charlottesville: U of Virginia P, 2003.
———. "Science, Anti-Science, and Ecocriticism." *Interdisciplinary Studies in Literature and the Environment* 6 (1999): 65–81.
Low, Bobbi S. *Why Sex Matters: A Darwinian Look at Human Behavior*. Princeton: Princeton UP, 2000.
Lummaa, Virpi. "Life-History Theory, Reproduction, and Longevity in Humans." In *The Oxford Handbook of Evolutionary Psychology*, ed. Robin Dunbar and Louise Barrett, 397–413. Oxford: Oxford UP, 2007.
Lumsden, Charles J., and Edward O. Wilson. *Promethean Fire: Reflections on the Origin of Mind*. Cambridge: Harvard UP, 1983.
Lyell, Charles. *Principles of Geology*, 3 vols., ed. Martin S. J. Rudwick. 1830–33; Chicago: U of Chicago P, 1990.
McBrearty, Sally, and Alison S. Brooks. "The Revolution That Wasn't: A New Interpretation of the Origin of Modern Human Behavior." *Journal of Human Evolution* 39 (2000): 453–563.
McCrae, Robert R., and Paul T. Costa. "Personality Trait Structure as a Human Universal." *American Psychologist* 52 (1997): 509–16.
MacDonald, Kevin B. "Evolution, Culture, and the Five-Factor Model." *Journal of Cross-Cultural Psychology* 29 (1998): 119–49.
———. "Evolution, Psychology, and a Conflict Theory of Culture." *Evolutionary Psychology* 7 (2009): 208–33.
———. "Evolution, the Five-Factor Model, and Levels of Personality." *Journal of Personality* 63 (1995): 525–67.
———. "Life History Theory and Human Reproductive Behavior: Environmental/Contextual Influences and Heritable Variation." *Human Behavior* 8 (1997): 327–59.
———. "A Perspective on Darwinian Psychology: The Importance of Domain-General Mechanisms, Plasticity, and Individual Differences." *Ethology and Sociobiology* 12 (1991): 449–80.
———, and Scott Hershberger. "Theoretical Issues in the Study of Evolution and Development." In *Evolutionary Perspectives on Human Development*, 2nd ed., ed. Robert Burgess and Kevin B. MacDonald, 21–72. Thousand Oaks: Sage, 2004.

McElreath, Richard, and Joseph Henrich. "Modeling Cultural Evolution." In *The Oxford Handbook of Evolutionary Psychology*, ed. Robin Dunbar and Louise Barrett, 571–85. Oxford: Oxford UP, 2007.

McEwan, Ian. "Literature, Science, and Human Nature." In *The Literary Animal: Evolution and the Nature of Narrative*, ed. Jonathan Gottschall and David Sloan Wilson, 5–19. Evanston, IL: Northwestern UP, 2005.

Mallett, Phillip. A Note on the Text. In *The Mayor of Casterbridge: An Authoritative Text, Backgrounds and Contexts, Criticism*, by Thomas Hardy, ed. Phillip Mallett, 2nd ed., xiii–xvii. New York: Norton, 2001.

Mamelli, Matteo. "Evolution and Psychology in Philosophical Perspective." In *The Oxford Handbook of Evolutionary Psychology*, ed. Robin Dunbar and Louise Barrett, 21–34. Oxford: Oxford UP, 2007.

Martin, Robert K. "Parody and Homage: The Presence of Pater in *Dorian Gray*." *The Victorian Newsletter* 69 (Spring 1983): 15–18.

Martindale, Colin. *The Clockwork Muse: The Predictability of Artistic Change*. New York: Basic Books, 1990.

Massé, Michelle A. "'He's More Myself Than I Am': Narcissism and Gender in *Wuthering Heights*." In *Psychoanalyses/Feminisms*, ed. Peter L. Rudnytsky and Andrew M. Gordon, 135–53. Albany: State U of New York P, 2000.

Mathison, John K. "Nelly Dean and the Power of *Wuthering Heights*." *Nineteenth-Century Fiction* 11 (1956): 106–29.

Matthews, Paul, and Louise Barrett. "Small-Screen Social Groups: Soap Operas and Social Networks." *Journal of Cultural and Evolutionary Psychology* 3 (2005): 75–86.

Max, D. T. "The Literary Darwinists." *The New York Times Magazine* (6 November 2005): 74–79.

Maynard, John. "The Brontës and Religion." In *The Cambridge Companion to the Brontës*, ed. Heather Glen, 204–09. Cambridge: Cambridge UP, 2002.

Maynard Smith, John. *The Theory of Evolution*. Cambridge: Cambridge UP, 1993.

———, ed. *Evolution Now: A Century after Darwin*. San Francisco: W. H. Freeman, 1982.

Review of *The Mayor of Casterbridge*, by Thomas Hardy. In *Thomas Hardy: The Critical Heritage*, ed. R. G. Cox, 134–36. 1979; London: Routledge, 1995. Originally published in *Saturday Review*, May 1886: 757.

Mayr, Ernst. "How to Carry Out the Adaptationist Program?" *The American Naturalist* 121 (1983): 326–28.

———. *One Long Argument: Charles Darwin and the Genesis of Modern Evolutionary Thought*. Cambridge: Harvard UP, 1991.

Mellard, James M. "'No ideas but in things': Fiction, Criticism, and the New Darwinism." *Style* 41 (2007): 1–29.

Mellars, Paul, Katie Boyle, Ofer Bar-Yosef, and Chris Stringer, eds. *Rethinking the Human Revolution: New Behavioural and Biological Perspectives on the Origin and Dispersal of Modern Humans*. Exeter, UK: MacDonald Institute, 2007.

Mellars, Paul, and Chris Stringer, eds. *The Human Revolution: Behavioural and Biological Perspectives on the Origins of Modern Humans.* Princeton: Princeton UP, 1989.

Menand, Louis. "Dangers Within and Without." In *Profession 2005,* ed. Rosemary G. Feal, 10–17. New York: Modern Language Association of America, 2005.

Mendelson, Edward. *The Things That Matter: What Seven Classic Novels Have to Say about the Stages of Life.* New York: Pantheon, 2006.

Meyers, Jeffrey. *Homosexuality and Literature 1890–1930.* Montreal: McGill-Queens UP, 1977.

Miall, David S. *Literary Reading: Empirical and Theoretical Studies.* New York: Peter Lang, 2006.

———, and Ellen Dissanayake. "The Poetics of Babytalk." *Human Nature* 14 (2003): 337–64.

Miller, Geoffrey. *The Mating Mind: How Sexual Choice Shaped the Evolution of Human Nature.* New York: Doubleday, 2000.

Miller, J. Hillis. *Fiction and Repetition: Seven English Novels.* Cambridge: Harvard UP, 1982.

———. *Thomas Hardy: Distance and Desire.* Cambridge: Harvard UP, 1970.

Millgate, Michael. *Thomas Hardy: His Career as a Novelist.* 1971; New York: St. Martin's, 1994.

Mithen, Steven. "Mind, Brain, and Material Culture: An Archaeological Perspective." In *Evolution and the Human Mind: Modularity, Language, and Meta-Cognition,* ed. Peter Carruthers and Andrew Chamberlain, 207–17. Cambridge: Cambridge UP, 2000.

———. *The Prehistory of the Mind: The Cognitive Origins of Art, Religion, and Science.* London: Thames and Hudson, 1996.

Moglen, Helene. "The Double Vision of *Wuthering Heights*: A Clarifying View of Female Development." *Centennial Review* 15 (1971): 391–405.

Mondimore, Francis Mark. *Depression: The Mood Disease,* 3rd ed. Baltimore: Johns Hopkins UP, 2006.

Moore, John A. *Science as a Way of Knowing: The Foundations of Modern Biology.* Cambridge: Harvard UP, 1993.

Moore, Kevin Z. "Death against Life: Hardy's Mortified and Mortifying 'Man of Character' in *The Mayor of Casterbridge.*" *Ball State University Forum* 24 (1983): 13–25.

———. *The Descent of the Imagination: Postromantic Culture in the Later Novels of Thomas Hardy.* New York: New York UP, 1990.

Moretti, Franco. *Graphs, Maps, Trees: Abstract Models for a Literary History.* London: Verso, 2005.

Morgan, Thais. "Victorian Effeminacies." In *Victorian Sexual Dissidence,* ed. Richard Dellamora. Chicago: U of Chicago P, 1999.

Moses, Michael Valdez. "Agon in the Marketplace: *The Mayor of Casterbridge* as Bourgeois Tragedy." *South Atlantic Quarterly* 87 (1988): 219–51.

Munson Goff, Barbara. "Between Natural Theology and Natural Selection: Breeding the Human Animal in *Wuthering Heights.*" *Victorian Studies* 27 (1984): 477–508.

Neill, Michael. "*Hamlet*: A Modern Perspective." In *William Shakespeare: The Tragedy of Hamlet, Prince of Denmark*, ed. Barbara A. Mowat and Paul Werstine, 311–12. New York: Washington Square P, 1992.
Nettle, Daniel. "Individual Differences." In *The Oxford Handbook of Evolutionary Psychology*, ed. Robin Dunbar and Louise Barrett, 479–90. Oxford: Oxford UP, 2007.
———. *Personality: What Makes You the Way You Are*. Oxford: Oxford UP, 2007.
———. "What Happens in *Hamlet*? Exploring the Psychological Foundations of Drama." In *The Literary Animal: Evolution and the Nature of Narrative*, ed. Jonathan Gottschall and David Sloan Wilson, 56–75. Evanston, IL: Northwestern UP, 2005.
———. "The Wheel of Fire and the Mating Game: Explaining the Origins of Tragedy and Comedy." *Journal of Cultural and Evolutionary Psychology* 3 (2005): 39–56.
Nierenberg, A. A., D. Doughtery, and J. F. Rosenbaum. "Dopaminergic Agents and Stimulants as Antidepressant Augmentation Strategies." *Journal of Clinical Psychiatry* 59, supplement 5 (1998): 60–63.
Nordlund, Marcus. *Shakespeare and the Nature of Love: Literature, Culture, Evolution*. Evanston, IL: Northwestern UP, 2007.
———. "Consilient Literary Interpretation." *Philosophy and Literature* 26 (2002): 312–33.
Nunokawa, Jeff. "Homosexual Desire and the Effacement of the Self in *The Picture of Dorian Gray*." *American Imago: Studies in Psychoanalysis and Culture* 49 (1992): 311–21.
Nussbaum, Martha. "*Wuthering Heights*: The Romantic Ascent." *Philosophy and Literature* 20 (1996): 362–82.
Oatley, Keith. "Emotions and the Story Worlds of Fiction." In *Narrative Impact: Social and Cognitive Foundations*, ed. M. C. Green, J. J. Strange, and T. C. Brock, 36–69. Mahwah, NJ: Erlbaum, 2002.
———. "Why Fiction May Be Twice as True as Fact: Fiction as Cognitive and Emotional Simulation." *Review of General Psychology* 3 (1999): 101–17.
Oldroyd, David. *Darwinian Impacts: An Introduction to the Darwinian Revolution*, 2nd ed. Atlantic Highlands, NJ: Humanities P, 1983.
———. *Thinking about the Earth: A History of Ideas in Geology*. Cambridge: Harvard UP, 1996.
Ospovat, Dov. *The Development of Darwin's Theory: Natural History, Natural Theology, and Natural Selection, 1838–1859*. Cambridge: Cambridge UP, 1981.
Panksepp, Jaak. "The Neuroevolutionary and Neuroaffective Psychobiology of the Prosocial Brain." In *The Oxford Handbook of Evolutionary Psychology*, ed. Robin Dunbar and Louise Barrett, 145–62. Oxford: Oxford UP, 2007.
———, and Jules B. Panksepp. "The Seven Sins of Evolutionary Psychology." *Evolution and Cognition* 6 (2000): 108–31.
Pater, Walter. *The Renaissance: Studies in Art and Poetry*, ed. Donald Hall. Berkeley: U of California P, 1980.

Paterson, John. "*The Mayor of Casterbridge* as Tragedy." *Victorian Studies* 3 (1959): 151–72.
Paulhus, Delroy L., and Oliver P. John. "Egoistic and Moralistic Biases in Self-Perception: The Interplay of Self-Deceptive Styles with Basic Traits and Motives." *Journal of Personality* 66 (1998): 1025–60.
Pervin, Lawrence A., and Oliver P. John, eds. *Handbook of Personality*, 2nd ed. New York: Guilford, 1999.
Phelps, Jim. *Why Am I Still Depressed? Recognizing and Managing the Ups and Downs of Bipolar II and Soft Bipolar Disorder.* New York: McGraw-Hill, 2006.
Pinker, Steven. *The Blank Slate: The Modern Denial of Human Nature.* New York: Viking, 2002.
———. Foreword to *The Handbook of Evolutionary Psychology*, ed. David M. Buss, xi–xvi. Hoboken, NJ: Wiley, 2005.
———. *How the Mind Works.* New York: Norton, 1977.
———. *The Language Instinct: How the Mind Creates Language.* New York: William Morrow, 1994.
———. *The Stuff of Thought: Language as a Window into Human Nature.* New York: Viking, 2007.
———. "Toward a Consilient Study of Literature." *Philosophy and Literature* 31 (2007): 162–78.
———, and Ray Jackendoff. "The Faculty of Language: What's Special about It?" *Cognition* 95 (2004): 201–36.
Plantinga, Carl, and Greg M. Smith, eds. *Passionate Views: Film, Cognition, and Emotion.* Baltimore: Johns Hopkins UP, 1999.
Plotkin, Henry. "The Power of Culture." In *The Oxford Handbook of Evolutionary Psychology*, ed. Robin Dunbar and Louise Barrett, 11–19. Oxford: Oxford UP, 2007.
Plutchik, Robert. *Emotions and Life: Perspectives from Psychology, Biology, and Evolution.* Washington, DC: American Psychological Association, 2003.
Popper, Karl. "Normal Science and Its Dangers." In *Criticism and the Growth of Knowledge*, ed. Imre Lakatos and Alan Musgrave. Cambridge: Cambridge UP, 1970.
Potts, Rick. "Variability Selection in Hominid Evolution." *Evolutionary Anthropology* 8 (1998): 81–96.
Premack, David, and Ann James Premack. "Origins of Human Social Competence." In *The Cognitive Neurosciences*, ed. Michael S. Gazzaniga, 205–18. Cambridge: MIT P, 1995.
Prewitt Brown, Julia. *Cosmopolitan Criticism: Oscar Wilde's Philosophy of Art.* Charlottesville: UP of Virginia, 1997.
Raine, Craig. "Conscious Artistry in *The Mayor of Casterbridge*." In *New Perspectives on Thomas Hardy*, ed. Charles Pettit, 156–71. New York: St. Martin's, 1994. Originally published as the introduction to *The Mayor of Casterbridge*, by Thomas Hardy. New York: Knopf, 1993.
Ramel, Annie. "The Crevice in the Canvas: A Study of *The Mayor of Casterbridge*." *Victorian Literature and Culture* 26 (1998): 259–72.

Richardson, Alan. "Studies in Literature and Cognition: A Field Map." In *The Work of Fiction: Cognition, Culture, and Complexity*, ed. Alan Richardson and Ellen Spolsky, 1–29. Burlington, VT: Ashgate, 2004.

Richerson, Peter J., and Robert Boyd. *Not by Genes Alone: How Culture Transformed Human Evolution*. Chicago: U of Chicago P, 2005.

Ridley, Mark. *Evolution*. Boston: Blackwell, 1993.

Riquelme, John Paul. "Oscar Wilde's Aesthetic Gothic: Walter Pater, Dark Enlightenment, and *The Picture of Dorian Gray*." *Modern Fiction Studies* 46 (2000): 609–31.

Rizzolatti, Giacomo, and Leonardo Fogassi. "Mirror Neurons and Social Cognition." In *The Oxford Handbook of Evolutionary Psychology*, ed. Robin Dunbar and Louise Barrett, 179–95. Oxford: Oxford UP, 2007.

Robinson, Donald S. "The Role of Dopamine and Norepinephrine in Depression." *Primary Psychiatry* 14.5 (2007): 21–23.

Rofé, Jacov. "Does Repression Exist? Memory, Pathogenic Unconscious and Clinical Evidence." *Review of General Psychology* 12 (2008): 63–85.

Rudwick, Martin S. J. *The Meaning of Fossils: Episodes in the History of Paleontology*. Chicago: U of Chicago P, 1972.

Ruse, Michael. *The Darwinian Revolution: Science Red in Tooth and Claw*, 2nd ed. Chicago: U of Chicago P, 1999.

Salmon, Catherine. "Parental Investment and Parent-Offspring Conflict." In *The Handbook of Evolutionary Psychology*, ed. David M. Buss, 506–27. Hoboken, NJ: Wiley, 2005.

Salmon, Catherine, and Todd Shackelford, eds. *Family Relationships: An Evolutionary Perspective*. Oxford: Oxford UP, 2008.

Salmon, Catherine, and Donald Symons. "Slash Fiction and Human Mating Psychology." *Journal of Sex Research* 41 (2004): 94–100.

Saunders, Judith. "Male Reproductive Strategies in Sherwood Anderson's 'The Untold Lie.'" *Philosophy and Literature* 31 (2007): 311–22.

———. "Paternal Confidence in Hurston's 'The Gilded Six-Bits.'" In *Evolution, Literature, and Film*, ed. Brian Boyd, Joseph Carroll, and Jonathan Gottschall. New York: Columbia UP, 2010.

———. *Reading Edith Wharton through a Darwinian Lens: Evolutionary Biological Issues in Her Fiction*. Jefferson, NC: McFarland, 2009.

Scalise-Sugiyama, Michelle. "Cultural Relativism in the Bush: Toward a Theory of Narrative Universals." *Human Nature* 14 (2003): 383–96.

———. "New Science, Old Myth: An Evolutionary Critique of the Oedipal Paradigm." *Mosaic* 34 (2001): 121–36.

———. "On the Origins of Narrative: Storyteller Bias as a Fitness-Enhancing Strategy." *Human Nature* 7 (1996): 403–25.

———. "Reverse-Engineering Narrative: Evidence of Special Design." In *The Literary Animal: Evolution and the Nature of Narrative*, ed. Jonathan Gottschall and David Sloan Wilson, 177–96. Evanston, IL: Northwestern UP, 2005.

Schapiro, Barbara Ann. *Literature and the Relational Self*. New York: New York UP, 1994.

Schmitt, David P. "Fundamentals of Human Mating Strategies." In *The Handbook of Evolutionary Psychology*, ed. David M. Buss, 258–91. Hoboken, NJ: Wiley, 2005.

Seamon, Roger. "Literary Darwinism as Science and Myth." *Style* 42 (2008): 261–65.

Searle, John R. "Literary Theory and Its Discontents." *New Literary History* 25 (1994): 637–67.

———. "The Word Turned Upside Down." *New York Review of Books*, 27 October 1983: 74–79.

Sedgwick, Eve Kosofsky. *Epistemology of the Closet*. Berkeley: U of California P, 1990.

Shakespeare, William. *The Tragedy of Hamlet, Prince of Denmark*, ed. Frank Kermode. In *The Riverside Shakespeare*, 2nd ed., ed. G. Blakemore Evans et al., 1182–1245. Boston: Houghton Mifflin, 1997.

Sheenan, Stephen. "Evolutionary Perspectives in Archaeology: From Culture History to Cultural Evolution." In *The Oxford Handbook of Evolutionary Psychology*, ed. Robin Dunbar and Louise Barrett, 587–97. Oxford: Oxford UP, 2007.

Showalter, Elaine. *Sexual Anarchy: Gender and Culture at the Fin de Siècle*. New York: Penguin, 1990.

———. "The Unmanning of the Mayor of Casterbridge." In *Critical Approaches to the Fiction of Thomas Hardy*, ed. Dale Kramer, 99–115. London: Macmillan, 1979.

Simpson, George Gaylord. *The Meaning of Evolution: A Study of the History of Life and of Its Significance for Man*, 2nd ed. New Haven: Yale UP, 1967.

Sinfield, Alan. *The Wilde Century: Effeminacy, Oscar Wilde and the Queer Movement*. London: Cassell, 1994.

Singer, Peter. *A Darwinian Left*. New Haven: Yale UP, 2000.

Slingerland, Edward. *What Science Offers the Humanities: Integrating Body and Culture*. Cambridge: Cambridge UP, 2008.

Smail, Daniel Lord. *On Deep History and the Brain*. Berkeley: U of California P, 2008.

Small, Ian. *Oscar Wilde: Recent Research, a Supplement to "Oscar Wilde Revalued."* Greensboro, NC: ELT Press, 2000.

Smee, Sebastian. "Natural-Born Thrillers." *The Australian Literary Review* (6 May 2009): 17.

Smith, Murray. "Darwin and the Directors: Film, Emotion, and the Face in the Age of Evolution." *TLS* (7 February 2003): 13–15.

Smits, Dirk J. M., and Paul de Boeck. "From BIS/BAS to the Big Five." *European Journal of Personality* 20 (2006): 255–70.

Snow, C. P. *The Two Cultures*, with an introduction by Stefan Collini. 1959; Cambridge: Cambridge UP, 1993.

Sober, Elliott, and David Sloan Wilson. *Unto Others: The Evolution and Psychology of Unselfish Behavior*. Cambridge: Harvard UP, 1998.

Solomon, Andrew. *The Noonday Demon: An Atlas of Depression*. New York: Scribner, 2001.

Spacks, Patricia Meyer. *The Female Imagination*. New York: Knopf, 1975.

Spencer, Herbert. *An Autobiography*. 2 vols. London: Williams and Norgate, 1904.
———. *First Principles*. London: Williams and Norgate, 1862.
———. *The Principles of Biology*. 2 vols. London: Williams and Norgate, 1864–1867.
Spivey, Ted R. "Thomas Hardy's Tragic Hero." *Nineteenth-Century Fiction* 9 (1954): 179–91.
Spolsky, Ellen. "The Centrality of the Exceptional in Literary Study." *Style* 42 (2008): 285–89.
———. *Gaps in Nature: Literary Interpretation and the Modular Mind*. Albany: State U of New York P, 1993.
———. Preface. *The Work of Fiction: Cognition, Culture, and Complexity*, ed. Alan Richardson and Ellen Spolsky, vii–xiii. Burlington, VT: Ashgate, 2004.
Spurgeon, Caroline. *Leading Motives in the Imagery of Shakespeare's Tragedies*. London: Oxford UP, 1930.
Steen, Francis. "The Paradox of Narrative Thinking." *Journal of Cultural and Evolutionary Psychology* 3 (2005): 87–105.
Sterelny, Kim. *Thought in a Hostile World: The Evolution of Human Cognition*. Oxford: Oxford UP, 2003.
Stiller, James, and Matthew Hudson. "Weak Links and Scene Cliques within the Small World of Shakespeare." *Journal of Cultural and Evolutionary Psychology* 3 (2005): 57–73.
Stiller, James, Daniel Nettle, and Robin Dunbar. "The Small World of Shakespeare's Plays." *Human Nature* 14 (2003): 397–408.
Stone, Valerie E. "Theory of Mind and the Evolution of Social Intelligence." In *Social Neuroscience: People Thinking about People*, ed. John T. Cacioppo, Penny S. Visser, and Cynthia L. Pickett, 103–30. Cambridge: Bradford-MIT P, 2006.
Stoneman, Patsy. "The Brontë Myth." In *The Cambridge Companion to the Brontës*, ed. Heather Glen, 214–41. Cambridge: Cambridge UP, 2002.
Storey, Robert. *Mimesis and the Human Animal: On the Biogenetic Foundations of Literary Representation*. Evanston, IL: Northwestern UP, 1996.
Stringer, Christopher, and Clive Gamble. *In Search of the Neanderthals*. New York: Thames and Hudson, 1993.
Style 42.2–3 (2008): 103–411. Special issue on literary Darwinism.
Styron, William. *Darkness Visible: A Memoir of Madness*. New York: Random House, 1990.
Sulloway, Frank J. *Born to Rebel: Birth Order, Family Dynamics, and Creative Lives*. New York: Pantheon, 1996.
Summers, Claude J. *Gay Fictions Wilde to Stonewall: Studies in a Male Homosexual Literary Tradition*. New York: Frederick Ungar, 1990.
Swirski, Peter. *Of Literature and Knowledge: Explorations in Narrative Thought Experiments, Evolution, and Game Theory*. London: Routledge, 2007.
Symons, Donald. *The Evolution of Human Sexuality*. New York: Oxford UP, 1979.

———. "On the Use and Misuse of Darwinism in the Study of Human Behavior." In *The Adapted Mind: Evolutionary Psychology and the Generation of Culture*, ed. Jerome H. Barkow, Leda Cosmides, and John Tooby, 137–162. Oxford: Oxford UP, 1992.

Tan, Ed S. "Emotions, Art, and the Humanities." In *Handbook of Emotions*, 2nd ed., ed. M. Lewis and J. M. Haviland-Jones, 116–34. New York: Guilford, 2000.

———. *Emotion and the Structure of Narrative Film: Film as an Emotion Machine*, trans, Barbara Fasting. Mahwah, NJ: Erlbaum, 1996.

Tanaka, Jiro. "What Is Copernican? A Few Common Barriers to Darwinian Thinking about the Mind." *The Evolutionary Review*, 1 (2010): 6–12.

Tomasello, Michael, et al. "Understanding and Sharing Intentions: The Origins of Cultural Cognition." *Behavioral and Brain Sciences* 28 (2005): 675–735.

Tooby, John, and Leda Cosmides. "Does Beauty Build Adapted Minds? Toward an Evolutionary Theory of Aesthetics, Fiction, and the Arts." *SubStance* 30 (2001): 6–27.

———. "The Psychological Foundations of Culture." In *The Adapted Mind: Evolutionary Psychology and the Generation of Culture*, ed. Jerome Barkow, Leda Cosmides, and John Tooby, 19–136. New York: Oxford UP, 1992.

Toulmin, Stephen, and June Goodfield, *The Discovery of Time*. Chicago: U of Chicago P, 1965.

Traversi, Derek. "*Wuthering Heights* after a Hundred Years." In *Emily Brontë: "Wuthering Heights": A Casebook*, ed. Miriam Allott, 157–76. Houndmills: Macmillan, 1970. Originally published in *Dublin Review* 202 (Spring 1949): 154–68.

Trivers, Robert. "Parental Investment and Sexual Selection." In *Sexual Selection and the Descent of Man 1871–1971*, ed. Bernard Campbell, 136–79. Chicago: Aldine, 1972.

———. *Social Evolution*. Menlo Park, CA: Benjamin/Cummins, 1985.

Turchin, Peter. *War and Peace and War: The Rise and Fall of Empires*. New York: Plume, 2007.

Turner, Frederick. *Natural Classicism: Essays on Literature and Science*. 1985; Charlottesville: UP of Virginia, 1992.

Van Ghent, Dorothy. *The English Novel: Form and Function*. 1953; New York: Harper Torchbooks-Harper, 1961.

van Peer, Willie. *Muses and Measures: Empirical Research Methods for the Humanities*. Newcastle upon Tyne: Cambridge Scholars, 2007.

van Praag, Herman M., Ron de Kloet, and Jim van Os. *Stress, the Brain, and Depression*. Cambridge: Cambridge UP, 2004.

Vermeule, Blakey. "Response to Joseph Carroll." *Style* 42 (2008): 302–08.

Vigar, Penelope. *The Novels of Thomas Hardy: Illusion and Reality*. London: Athlone, 1978.

Wade, Nicholas. *Before the Dawn: Recovering the Lost History of Our Ancestors*. New York: Penguin, 2006.

Watt, Ian. *The Rise of the Novel*. Berkeley: U of California P, 1957.

Weinberg, Steven. *Dreams of a Final Theory: The Search for the Fundamental Laws of Nature*. New York: Pantheon, 1992.
——. *Facing Up: Science and Its Cultural Adversaries*. Cambridge: Harvard UP, 2001.
Whybrow, Peter C. *A Mood Apart: Depression, Mania, and Other Afflictions of the Self*. New York: Basic Books, 1997.
Wiggins, Jerry S., ed. *The Five-Factor Model of Personality: Theoretical Perspectives*. New York: Guilford, 1996.
Wilde, Oscar. *Complete Shorter Fiction*, ed. Isobel Murray. Oxford: Oxford UP, 1979.
——. *The Letters of Oscar Wilde*, ed. Rupert Hart-Davis. New York: Harcourt, Brace, 1962.
——. *The Picture of Dorian Gray*, ed. Donald L. Lawler. 1890; New York: W. W. Norton, 1988.
Wilson, David Sloan. "Evolutionary Social Constructivism." In *The Literary Animal: Evolution and the Nature of Narrative*, ed. Jonathan Gottschall and David Sloan Wilson, 20–37. Evanston, IL: Northwestern UP, 2005.
——. *Evolution for Everyone: How Darwin's Theory Can Change the Way We Think about Our Lives*. New York: Delacorte, 2007.
——. "Group-Level Evolutionary Processes." In *The Oxford Handbook of Evolutionary Psychology*, ed. Robin Dunbar and Louise Barrett, 49–55. Oxford: Oxford UP, 2007.
——, and Edward O. Wilson. "Rethinking the Theoretical Foundation of Sociobiology." *The Quarterly Review of Biology* 82 (2007): 327–48.
Wilson, Edward O. *Consilience: The Unity of Knowledge*. New York: Alfred A. Knopf, 1998.
——. "Sociobiology at Century's End." In *Sociobiology: The New Synthesis, Twenty-Fifth Anniversary Edition*, v–viii. Cambridge: Harvard UP, 2000.
Wilson, J. Dover. *What Happens in Hamlet?*. Cambridge: Cambridge UP, 1935.
Wilson, Keith. Introduction to *The Mayor of Casterbridge*, by Thomas Hardy. London: Penguin, 1997.
Wilson, Timothy D. *Strangers to Ourselves: Discovering the Adaptive Unconscious*. Cambridge: Harvard UP, 2002.
Winchester, Simon. *The Map That Changed the World: William Smith and the Birth of Modern Geology*. New York: HarperCollins, 2001.
Winkelman, Michael. "Sighs and Tears: Biological Signals and Donne's 'Whining Poetry.'" *Philosophy and Literature* 33 (October 2009): 329–44.
Wion, Philip K. "The Absent Mother in Emily Brontë's *Wuthering Heights*." *American Imago* 42 (1985): 143–64.
Wolpert, Lewis. *Malignant Sadness: The Anatomy of Depression*, 3rd ed. London: Faber and Faber, 2006.
Woolf, Virginia. "The Novels of Thomas Hardy." In *The Second Common Reader*, 222–33. 1932; New York: Harcourt, 1960.
Wrangham, Richard. *Catching Fire: How Cooking Made Us Human*. New York: Basic Books, 2009.

Wright, T. R. *Hardy and the Erotic*. New York: St. Martin's, 1989.
Wyman, Emily, and Michael Tomasello. "The Ontogenetic Origins of Human Cooperation." In *The Oxford Handbook of Evolutionary Psychology*, ed. Robin Dunbar and Louise Barrett, 227–36. Oxford: Oxford UP, 2007.
Zunshine, Lisa. *Why We Read Fiction: Theory of Mind and the Novel.* Columbus: Ohio State UP, 2006.

Index

Abrams, M. H., 34
Across Five Aprils, 58
Adaptationism. *See* Adaptations; Evolutionary biology; Natural Selection
Adaptations, 17, 22–23, 26, 50–51. *See also under* Arts, the; Human nature; *under* Literature; Life history theory
Adapted Mind, The, 7, 276
Adelman, Janet, 124
Adventures of Stanley Kane, The, 58–59
Agassiz, Louis, 209, 210
Agonistic structure: ambiguous characters in, 165–66; definition of, 35, 152; and differences of gender, 167–73; and human nature, 165, 167; is falsifiable, 164–65; is more important than gender, 167–70; in *Mayor of Casterbridge*, 177–93; shaping force of, 155, 166; social function of, 152–53, 170, 173–75; the Tiv understand, 124; in Victorian novels, 151–75. *See also* Antagonists; Protagonists
Alcott, Louisa May: author of: *Jo's Boys*, 59; *Little Men*, 58, 59
Alice in Wonderland, 131–32
Allott, Miriam, 109
Althusser, Louis, 18, 29, 277
Altruism, 40, 41–42, 48, 66. *See also* Social life
American Scholar, 61

Andrews, Alice, 10
Anna Karenina, 81
Annaud, Jean-Jacques: director of *Quest for Fire*, 58
Antagonists: in definition of agonistic structure, 35, 152; in *Dracula*, 164; in *Emma*, 160; emotional responses to, 155, 163–64, 165, 166; exemplify dominance behavior, 152, 155, 158, 165, 168, 169, 170; in *Jane Eyre*, 161–62; in *Mansfield Park*, 159; mate preferences in, 160–61, 187; in *Mayor of Casterbridge*, 180, 184–85, 191; motives of, 158–59; personality of, 161–62, 165; in *Wuthering Heights*, 113, 115. *See also* Agonistic structure; Protagonists
Aristotle, 126
Arnold, Matthew: cites Darwin, 264–65; cultural teleology of, 221, 264–65; demonstrates potential of cultural criticism, 58; echoed by Leavis, 267; exchange between Huxley and, xiv, 260–61, 264–66; on *Hamlet*, 145–46; influenced humanistic ethos, 72, 261, 266, 267, 268–69, 275–76; influenced Trilling, Leavis, and Frye, 266; sadness of, 263–64, 275; on Sophocles, 215; author of: 1853 Preface, 145–46; "Dover Beach," 263–64; "Empedocles on Etna," 145–46; "Literature and Science," 264–65, 266

Index

Arts, the: adaptive functions of, xii 4, 20–29, 49–53, 62–70 *passim;* are universal, 28, 52; definition of, 23, 27; formal properties of, 26, 27, 70; and sexual display, 28. *See also* Literature

Atlantic Monthly, 61

Attachment theory: applies to all mammals, 14; Easterlin invokes, 31; and *Hamlet,* xiii, 138–39, 141

Audience: adaptations for responses in an, 26; authors share understanding with, 18, 20; catching the attention of an, 126; Dissanayake's, 63; Emily Brontë's, 110, 111, 113, 115; for evolutionary social scientists, 5, 152; for *Hamlet* in a Nigerian tribe, 31; Hamlet's, 144, 145; for literary studies, 152, 157; for *Madame Bovary's Ovaries,* 12; Mark Ridley's, 198; Spolsky's, 6. *See also* Readers

Austen, Jane, 152; author of: *Emma,* 160; *Mansfield Park,* 159; *Persuasion,* 164; *Pride and Prejudice,* 32, 156, 160, 161; *Sense and Sensibility,* 156

Austin, Michael, 12

Authors: affirm human nature, 18; control emotional responses of readers, 156; forces shaping perspective of, 48–49; have shared understanding with readers, 18, 20, 30; individual identity in, 10, 30, 61, 81, 93; intentional meaning in, x, 81–82, 83, 100, 107, 116, 131, 135, 156, 183, 187, 188, 189–90, 193; negotiate with cultural traditions, 81; stipulate features of characters, 156. *See also* Literature; Meaning in literature; Point of View; Psychodrama

Autobiography: chapter two is an, xii; Darwin's, 56, 200, 216, 226, 228, 230, 231, 254; Pater's, x; Spencer's, 223

Baer, Karl Ernst von, 207
Baron-Cohen, Simon, 9

Barrett, Louise: co-editor of *Oxford Handbook of Evolutionary Psychology,* 274

Barthes, Roland, 274–75

Beer, Gillian: author of *Darwin's Plots,* 78

Behavioral and Brain Sciences, 85

Behavioral ecology, 41, 274

Benzon, William, 49

Bergman, Ingmar: director of: *Wild Strawberries,* 58; *The Seventh Seal,* 146

Bible, the, 72, 209, 213, 241, 242. *See also* Christianity; Religion

Biology. *See* Evolutionary biology

Boehm, Christopher, 173. *See also* Egalitarianism

Bohannan, Laura, 124, 125

Bordwell, David, 8

Boston Globe, 61

Bowlby, John, 31, 138. *See also* Attachment theory

Boyd, Brian: on the adaptive function of the arts, 49; assimilates cognitive science, 8–9; Deresiewicz denounces, 68; on *Hamlet,* 123, 124–26; on human nature, 130; valorizes expansiveness, 31, 126; author of *On the Origin of Stories,* 11, 62; co-editor of *Evolution, Literature, and Film,* 9–10

Boyd, Robert, 45

Braddon, Elizabeth: author of *Lady Audley's Secret,* 165

Bradley, A. C.: author of *Shakespearean Tragedy,* 132–37 *passim,* 139, 141, 143

Brain, the: and birth canal, 15; cognitive modules in, 21; constructivist view of, 65; and depression, 140; function of, 21–22, 25; Geoffrey Miller's view of, 21–22, 66; Gould's view of, 66; in historical study, 86; is a precondition for culture, 44; is expanded in humans, 16, 40; in literary study, 85–86

Bristol, Michael, 124

Index

Brontë, Charlotte, 85; author of *Jane Eyre*, 162–63
Brontë, Emily: author of *Wuthering Heights*, xiii, 12, 109–22, 165
Browning, Robert: author of "Bishop Blougram's Apology," 146
Buffon, Georges-Louis Leclerc, Comte de, 207, 213
Burnet, Thomas: author of *Sacred Theory of the Earth*, 213
Buss, David, 62; editor of *Handbook of Evolutionary Psychology*, 274
By-product, 21, 28, 40, 50–52, 62
Byron, George Gordon, Lord, 35, 81, 182

Carlyle, Thomas, 221; author of *Sartor Resartus*, 146
Carroll, Joseph: Deresiewicz denounces, 68; intellectual history of, ix–xi, xiii, 55–60, 152; produced an edition of *Origin of Species*, xi; author of: *Cultural Theory of Matthew Arnold*, ix–x; *Evolution and Literary Theory*, x–xi 10–11, 76, 277; *Literary Darwinism*, xi; *Wallace Stevens' Supreme Fiction*, x, 58; co-editor of *Evolution, Literature, and Film*, 9–10
Catch-22, 59
Cather, Willa, 48
Cecil, Lord David, 183
Chambers, Robert, 39, 220–21; author of *Vestiges of Creation*, 220
Characters: elements composing, 155; empirical study of, 11, 35–36, 151–93 *passim;* forces shaping depiction of, 48–49, 94; individual identity in, 10, 83, 85, 155; minor, 35, 152, 156, 157, 163, 178, 184, 185, 186; as organisms, 10; in psychodrama, xiii, 32, 91–108 *passim*; in relation to authors and readers, 12, 32; as represented subjects, 81; Saunders on Wharton's, 12; as thematic personifications, 19; in Victorian novels, xii, xiii–xiv, 91–122 *passim*, 151–93 *passim*. *See also* Agonistic structure; Antagonists; Drama; Fiction; Novels; Protagonists
Charlesworth, Barbara, 94–95
Chaucer, Geoffrey, 85, 146
Chimpanzees, 14, 16, 173
Christianity: Arnold abandons, 265; in *Dorian Gray*, 32, 91–108 *passim*; Frye believed in, 275–76; Hardy's relation to, 183; Huxley on, 262; in a model of tragedy, 181, 182; the New Critics believed in, 275. *See also* Religion
Chronicle of Higher Education, 77
Churchill, Winston, 218
Clough, Arthur Hugh, 146
Cognitive poetics, 6–8, 14
Cognitive science: Brian Boyd's use of, 8–9; and evolutionary psychology, 7–8
Colbert, Stephen, 61
Coleridge, Samuel Taylor, 135
Comedy: latent in *Hamlet*, 137; Nettle on, 31, 126–27. *See also* Romantic comedy
Coming Up for Air, 59
Comte, Auguste, 221
Conrad, Joseph: author of *Heart of Darkness*, 4, 104, 107
Conscientiousness. *See* Personality
Consilience: in literary Darwinism, 5–6, 29–30, 154, 269–70, 276; Menand rejects, 74; Wilson's vision of, 5, 274. *See also* Wilson, Edward O.
Cooke, Brett, 32, 37–38; author of *Human Nature in Utopia*, 11
Cosmides, Leda: on the arts, 21; on *Hamlet*, 123, 131–32, 134, 135, 145; co-author of "The Psychological Foundations of Culture," 51
Crane, Mary, 7
Creationism: Agassiz believed in, 210; Cuvier believed in, 209; as Darwin's foil, 65, 202–203, 221, 234, 271; is a *deus ex machina*, 202; evidence that could support, 245–46;

Creationism *(continued)*: Lyell tried to assimilate, 211; is no longer worth arguing with, 64–65. *See also* Paley, William
Crews, Frederick, 9, 48; author of *Postmodern Pooh*, 5
Criticism. *See* Humanism; Literary Darwinism; Poststructuralism
Crow, 146
Cultural constructivism: attributed to Renaissance writers, 12; vs. biocultural critique, 6, 20, 38, 43–44, 78; and cultural evolution, 44–45; dominates literary criticism, 32, 74; and dualism, 65; B. H Smith on, 260; ideological motives for, 78–79, 273; origins of, 272; and "post-theory," 68; two chief propositions of, 75–76. *See also* Cultural evolution; Culture; Gene-culture coevolution; Poststructuralism
Cultural evolution, 44–46. *See also* Cultural constructivism; Culture; Gene-culture coevolution
Culture: in animals, 43, 44; in biocultural critique, xiii, 6, 12, 53, 54, 70, 125, 151, 154; biological preconditions for, 44–45; C. P. Snow on, 266–67; definition of, 17, 154; depends on the imagination, 17; evolved late, 93; faulty extremes in concept of, 43–44; F. R. Leavis on, 266–68; and human nature, xi, 10, 18, 36, 43–44, 70, 93, 154, 157, 272, 273–74; inadequate evolutionary concepts of, 43, 53; influences fictive depictions, 49; is a capacity for symbolism, 143; Matthew Arnold on, 261, 263–69 *passim*, 275; natural selection produces, 44–46; next phase in theory of, 42, 277; non-literate, 174; in the past 100,000 years, 26, 41; peculiarly human capacity for, 157; protagonistic dispositions for, 157, 158, 165, 172, 184, 185; in recorded history, 43; regulates behavior and thought, 153, 154; scale of time in, 275; specifically human forms of, 23–24, 25; as a subject for the humanities, 43, 80; as a substitute for religion, 265, 275–76; supposed autonomy of, 272–73; teleological versions of, 221; T. H. Huxley on, 261–63, 265, 269; Trilling on, 265. *See also* Cultural constructivism; Cultural evolution; Gene-culture coevolution; Human Revolution, the; Memes
Cuvier, Georges, 202, 207, 209–11, 212, 213, 220–21

Damasio, Antonio, 126
Dante [Alighieri], 145, 146, 147
Darwin, Charles, 58; adaptationist theory of, 51; Arnold cites, 264–65; creationism as foil for, 65, 202–203, 221, 234, 271; education of, 208, 216, 217–18; explained the formation of coral reefs, 213–14, 225, 253; intellectual character of, 214–20; Lamarck and, 39, 198, 211, 221–23, 256; lost religious faith, 56; Lyell influenced, 208, 212, 213–14; moral theory of, 32, 106; studied barnacles, 225, 228, 239; vision of deep time in, 57; Wallace and, 226, 229–30, 234, 240–45 *passim*, 253
Works of Charles Darwin
—1844 manuscript, 228, 230, 231–36, 238–39
—*Autobiography*, 56, 200, 216, 226, 228, 230, 231, 254
—*Descent of Man*: allusion to *Hamlet* in, 136–37; and literary Darwinism, 276; and Wilson's *Sociobiology*, 274; compared with influence of Nietzsche and Spencer, 271; compared with *Origin of Species*, 247; conclusion to, 136–37, 276; idea of maternal bonding in, 138; ideas on ontogeny in, 203; incor-

porates humans in evolution, 198, 246; influence of, x, 56, 57, 58, 271; publication of, 246; rhetoric of, 133, 272; social theory in, 205, 224–25, 247; theory of leisure class in, 217; utopianism in, 225
—*Origin of Species*: as one long argument, 254; Broadview edition of, xi; changes in title of, 230; compared with 1844 ms., 233–38, 239; compared with *Descent of Man*, 248; conclusion to, 38, 56; ecological vision in, 204–05, 232–33; influence of, x, 56–57, 58, 197–98, 206, 240–41, 247, 249–57, 271–72; influence of Malthus on, 241–46; is one long argument, 65, 204, 254; later editions of, 240–41; literary qualities of, 197, 206, 214, 215, 234, 238; structure of argument in, 50, 199–201, 203, 204, 249–57; supposed delay in publication of, 238–40; mentioned, x, xiv, 38, 50, 56, 57, 197–257 *passim*
—*Voyage of the Beagle*, 207, 208, 213, 218, 232
Darwin, Erasmus, 220, 227; author of *Zoonomia*, 220
Darwinian social science. *See* Evolutionary psychology
Dawkins, Richard, 45–46, 51, 62, 274
Deacon, Terrence, 25
Deconstruction: brevity in reign of, 72–73, 276–77; in the casebooks, 77, 82; characteristic activities of, 73; in cognitive poetics, 8; cognitive sensation produced by, 76; core ideas of, 19, 276; in criticism of *Hamlet*, 123–24; vs. the Enlightenment, ix; and "post-theory," 68. *See also* Derrida, Jacques; Poststructuralism
De Man, Paul, ix
Dennett, Daniel, 11
Depression, 133, 136, 139–41
Deresiewicz, William, 68–69
Derrida, Jacques, ix, 274, 276

Desmond, Adrian, 239, 244
De Waal, Frans, 62, 274
Dewey, John, 272
Dickens, Charles, 242
DiSalvo, David, 55
Discourse. *See* Language; Poststructuralism
Dissanayake, Ellen, 35, 49, 63; author of: *Art and Intimacy*, 10, 63; *Homo Aestheticus*, 10
Dobzhansky, Theodosius, 274
Dollimore, Jonathan, 92
Dominance. *See* Agonistic structure; Antagonists; Egalitarianism; Social life
Donne, John, 9
Drama: Aristotle on, 126–27; canonical ranking in, 135; a closet, 145; Darwinists have focused on, 9; Darwinist studies of, 11–12, 31, 123, 124–32, 134; Dryden on, 13; empirical studies of, 35; as escape, 133; finite body of, 80; Hamlet on, 131; pluralism in study of, 80; produced by inclusive fitness, 16, 46; theory of, 9, 13, 46, 126–27, 131. *See also* Shakespeare, William
Dryden, John, 13
Dunbar, Robin: co-editor of *Oxford Handbook of Evolutionary Psychology*, 274
Durkheim, Émile, 272
Dutton, Denis: author of: afterword to *Literary Animal*, 9; *The Art Instinct*, xii, 61–70

Easterlin, Nancy, 31
Ecocriticism, 11
EEA (Environment of Evolutionary Adaptedness). *See under* Evolutionary social sciences
Egalitarianism, 24, 152, 173–74. *See also* Agonistic structure; Boehm, Christopher; Protagonists; Social life
Ekman, Paul, 163
Elegy, 122, 193

Eliot, George, 58; author of *Middlemarch*, 55, 56, 59, 153, 159, 241
Embryology: Darwinism provides context for, 251; Darwin's contribution to, 228; in *Descent of Man*, 247; as evidence for evolution, 50, 234, 245, 256; history of, 203, 207
Emotional Stability. *See* Emotions; Personality
Emotions: are a main form of literary structure, 188–89; are cultivated by art, 66–67; are influenced by culture, 44; are the basis of tone, 64; authors control readers', 156; basic, 40, 93, 125–26, 163; correlate with agonistic role assignments, 155, 156; factor analysis of, 163–64; fiction provides simulated experience of, 152–53; guide decision-making, 126; in *Hamlet*, 145; and literary meaning, 152–53; in a model of human nature, 151, 155; of readers, 12, 111, 118, 121–22, 141, 152–56 *passim*, 163–70 *passim*, 173, 180, 183, 186–93 *passim;* in religion and poetry, 256; in responses to antagonists, 155, 163–64, 165, 166; in responses to protagonists, 155, 163–69 *passim*, 191; traumatic disturbances in, 134. *See also* Tone
Enlightenment, the, ix
Evans, Dylan, 31
Evolution. *See* Adaptations; Cultural evolution; Evolutionary biology; Evolutionary psychology; Gene-culture coevolution; Inclusive fitness; Life history theory; Natural selection
Evolutionary biology, 5–6, 84, 85, 274. *See also* Darwin, Charles; Embryology; Genetics; Hamilton, William; Inclusive fitness; Maynard Smith, John; Mayr, Ernst; Modern Synthesis, the; Natural selection; *under* Paradigm change; Williams, George C.
Evolutionary psychology: and cognitive science, 7–8; concept of modules in, 21, 27, 40; converges with folk psychology, 110, 124; deprecates individual differences, 145; as a distinct school (EP) in evolutionary social science, 7, 20–21, 41; not used in queer theory, 92; and sociobiology, 7, 41, 274. *See also* Cosmides, Leda; Evolutionary social sciences; Pinker, Steven; Symons, Donald; Tooby, John
Evolutionary Review, 10
Evolutionary social sciences: absence of cultural history in, 42–43; are entering a mature phase, 274; are not yet mature, xi, 4, 53, 151, 248–49; are reciprocally dependent on literary Darwinism, 36, 59–60; concept of human nature in, 4, 20–21, 27, 32, 124; consensus in, 39–40; continue to develop, 71; current problems in, 40–41; and the EEA, 40, 41; history of, xiv, 5, 20–21, 24, 39, 61–62, 246–49, 272, 274; inadequate concept of culture in, 43; might transform literary study, 71; various disciplines in, 274. *See also* Behavioral ecology; Evolutionary psychology; *under* Paradigm change; Sociobiology
Extinction: Cuvier's views on, 210–11, 220; Darwin's observations on, 209, 228, 253, 255; Lamarck denied, 211; Lamarck's lineage has succumbed to, 227, 271; Lyell's views on, 209, 211, 214, 253; most species succumb to, 51; Neanderthals have succumbed to, 227. *See also* Natural selection
Extraversion. *See* Personality

Factor analysis, 157–65 *passim*, 172, 189
Fairy tales, 35, 95, 96, 106
Family systems therapy, 129–30
Feminism: in the casebooks, 77, 82; core ideas of, 19; in criticism of *Hamlet*, 123–24; in criticism of

Index

Victorian novels, 169–70; Easterlin on, 31; redemptive tragic model in, 181. *See also* Gender

Fiction: adaptive functions of, 173–74; analyzing meaning in, 153; cognitive adaptations for, 26; Darwinists have focused on, 9; historical, x; Pinker on, 21; poststructuralists view truth as, 57; reviews of, 10; utopian/dystopian, 11; Victorian, xii. *See also* Fairy tales; Folk tales; Genres; Literature; Narrative; Novels; Science fiction

Fisher, Maryanne, 34–35

Fisher, Ronald, 249

Fitzroy, Robert, 208

Flesch, William, 12

Folk psychology: in audience for *Wuthering Heights*, 110; cognitive psychologists study, 7; cognitive structures for, 86; common themes in, 125; converges with evolutionary psychology, 110, 124; gives insight into human nature, 20, 29, 83, 120–21; is implicit in the common idiom, 124; is the basis of shared understanding, 83; kin relations loom large in, 114; in literary study, 86; literature articulates, 13, 17; vs. poststructuralism, 19–20

Folk tales, 34

Form: cognitive structures for, 86, 248; and conflict, 46; critics say Darwinists neglect, xiii; in Donne's poetry, 9; in film, 8; generating novelty in, 9; of imagined virtual worlds, 24; is a main literary category, xiii, 30, 69, 70, 81–82, 84. *See also* Style

Forster, E. M., 48, 152

Foucault, Michel: concept of discursive practices in, 78; is hostile to the Enlightenment, ix; is the dominant poststructuralist theorist, 18, 73–74, 276–77; problematizes individual identity, x; provides content to poststructuralism, 73–74, 276–77; textualist doctrine of, 274–75. *See also* New Historicism; *Poststructuralism*

Frankenstein, 165

Freud, Sigmund. *See* Psychoanalysis

Fromm, Harold, 109; author of *Nature of Being Human*, 11

Frye, Northrop, x, 34, 112, 133, 266, 275–76

Garber, Marjorie, 124

Gender: and agonistic structure, 158–75 *passim;* authors compared by, 170–73; and changing social roles, 172; constructivist view of, 65, 76; and meaning in Victorian novels, 167–70. *See also* Homosexuality; Mate selection; Queer theory; Reproduction

Gene-culture coevolution, 25–26, 41, 44–46

Genes: are the medium of natural selection, 14, 265; constrain behavior, 20, 274; contrasted with memes, 44–46; fuse two individuals, 119; and kinship, 114; for language, 25, 44; and levels of selection, 41–42; and organisms are interdependent, 46; for producing and consuming art, 26; that make culture possible, 25, 26, 44. *See also* Gene-culture coevolution; Genetics; Inclusive fitness; Natural selection

Genesis. *See* Bible, the

Genetics: behavioral, 274; Darwin's version of, 252; as evidence for evolution, 245; generalizing from particulars in, 151–52; humanists need to understand, 84; in the Modern Synthesis, 39, 197, 245, 251, 252, 256, 271. *See also* Gene-culture coevolution; Genes; Inclusive fitness

Genres: are main categories in literary history, 57, 69, 82, 84; diverge in *Wuthering Heights*, 109, 110; framework for analyzing, 12;

Genres *(continued)*:
and life history theory, 112; Nettle on, 9, 31, 123, 126–28; sentimental, 48; shape plot and character, 49. *See also* Autobiography; Elegy; Fairy tales; Folk tales; Heroic valor; Horror; Myths; Naturalism; Psychodrama; Realism; Romantic comedy; Romanticism; Satire; Science fiction; Sublime, the; Supernatural fantasy; Tragedy; Tragicomedy; Travel writing
Geology. *See* Agassiz, Louis; Burnet, Thomas; Cuvier, Georges; Darwin, Charles; Hutton, James; Lyell, Sir Charles; *under* Paradigm change; Sedgwick, Adam; Smith, William; Wegener, Alfred; Werner, Abraham
Georgiades, Stelios, xiii, 32
Ghiselin, Michael, 198, 201, 203, 213–20 *passim*, 243, 253
Goethe, Johann Wolfgang von: author of: *Faust*, 135, 146; *Wilhelm Meister's Apprenticeship*, 143
Goleman, Daniel, 62, 274
Gollum, 76
Goodfield, June, 212
Goodheart, Eugene, 13–14, 34
Gottschall, Jonathan: on the adaptive function of the arts, 50; advocates narrowing possible space of explanation, 53; Deresiewicz denounces, 68; empirical work by, xii, xiii, 9, 10–11, 12, 32, 34, 35, 151, 152; at a symposium, 49; author of: *Literature, Science, and a New Humanities*, 10–11; *Rape of Troy*, 12, 31; co-editor of: *Evolution, Literature, and Film*, 9–10; *Literary Animal*, 9, 62
Gould, Stephen Jay, 50–51, 67, 68
Gray, Asa, 228, 230, 231
Gross, Paul, 260; co-author of *Higher Superstition*, 259
Guardian, 61, 77
Gulliver's Travels, 76

Haeckel, Ernst, 203
Haldane, John, 249

Half Magic, 58
Hamilton, William, 51
Hardy, Thomas: religious views of, 183; temperament of, 191–93; author of: *Far from the Madding Crowd*, 183; "For Life I Had Never Cared Greatly," 191; *Jude the Obscure*, 193; *Mayor of Casterbridge*, xiii–xiv, 177–93; *Return of the Native*, 183; *Tess of the d'Urbervilles*, 55, 183, 193; *Trumpet Major*, 183; *Woodlanders*, 183
Hart, F. Elizabeth, 7
Hayles, Katherine, 54
Hazlitt, William, 131, 132–34, 143
Headlam Wells, Robin: author of *Shakespeare's Humanism*, 12; co-editor of *Human Nature: Fact and Fiction*, 9
Hegel, Georg Wilhelm Friedrich, 221, 276
Heine, Heinrich, 58
Henslow, John, 208
Heroic valor, 137–38
Hill, Kim, 43–44
History of Mr. Polly, The, 59
Homer, 12
Homosexuality: authors suppressing recognition of, 48; controversy over adaptiveness of, 91–92; in *Dorian Gray*, 32, 91–108 *passim*; hostility to, 74; is a puzzle in evolutionary research, 40; male compared with female, 99; in Oscar Wilde, xiii; statistics on, 166; in Walter Pater, 96; Wilde's trial for, 100
Hooker, Joseph, 216, 228, 229, 233
Horace, 13
Horror: in *Dorian Gray*, 92, 95, 96, 103, 104, 107; in *Hamlet*, 134, 139
Horton Hears a Who, 31
Huckleberry Finn, 58
Humanism: Arnold's influence on, 72, 261, 266, 267, 268–69, 275–76; beliefs shared between poststructuralism and, 65, 68–69, 79–80, 167, 261, 268; collapse of, 268; core ideas of, x, 72; dualism

in, 34, 64–65, 79–80; explanatory reduction in, 18, 19; hostility to general ideas in, 34, 68–69; idealizing sentiments of, 29, 269; in interpretation of *Wuthering Heights*, 109; Leavis as instance of, 259, 268; vs. poststructuralism, ix; reverence for literature in, 75; spiritualism in, 56, 72, 266, 275–76; uses the common idiom, 72, 75, 124, 132. *See also under* Paradigm change; Pluralism

Humanities, the: Arnold's influence on, 275; crisis in, 10–11, 277; dualist metaphysic of, 79–80; have been an intellectual backwater, 275–77; mystical fervor in, 275–76; "post-theory" in, 68–69; put humans at the center, 262, 269; resistance to consilience in, 260; study of cultural history in, 43. *See also* Arts, the; Humanism; Literary Darwinism; Literature; Poststructuralism; Two cultures, the

Human nature: and agonistic structure, 165, 167; animates novels, 175; Arnold invokes, 264; as basis for literary theory, x, 4, 33, 60, 69; Boyd on, 130; Christian concept of, 91; conflict is endemic to, 15–16, 113; and culture, xi, 10, 18, 36, 43–44, 70, 81, 93, 154, 272, 273–74; Darwin's concept of, 98, 224, 246–48, 274; in *Dorian Gray*, 91; Dutton on, 66–67; egalitarianism in, 173; elements of, 13, 24, 27, 30, 66–67, 93, 110–11, 157; Emily Brontë invokes, 112–13; E. O. Wilson on, 274; evolutionary changes in, 21, 23–26, 44, 93, 136–37, 151, 173; folk understanding of, 20, 29, 110, 112, 120–21, 124; growing knowledge of, 75, 274; Hamlet on, 130, 137; humanist view of, 19; human universals reflect, 6; imagination is part of, xi, 25, 35; inclusive fitness shapes, 40, 265; in interpretation of *Wuthering Heights*, 110; is finite, 167; limitation in current conception of, 53; literary criticism presupposes, 34; literature affirms, 13, 18; male homosexual version of, 101; Matthew Arnold on, 264–65; models of, 10, 13–18, 151, 157, 166, 167; in models of tragedy, 181–82; moral aspects of, 13, 32, 97–98; Pater's concept of, 91, 107; personality is part of, 142–43; poststructuralism rejects, 13, 19, 32, 33, 269; pressure points in, 30; psychoanalysis distorts, 128–29; Renaissance conception of, 12; social character of, 76, 113, 142, 224–25; specifically human aspects of, xi, 15, 16–17, 23–28, 58, 157, 269; systemic character of, xiii, 39–40, 83; utopianism neglects, 78–79, 273–74; variability in, 83; Wilde penetrates to, 93. *See also* evolutionary social sciences; Human universals; Life history theory; Social life

Human Revolution, the, 26, 41

Human sciences. *See* Evolutionary social sciences

Human universals: and culture, 125; and historical scholarship, 36; include incest avoidance, 128; include the arts, 28, 52; in literature, 10; produced by human nature, 6; should be integrated with individual differences, 145. *See also* Human nature; Life history theory

Humboldt, Alexander von, 207

Hunter-gatherers, 152, 173

Hutton, James, 210, 212–13, 214, 257; author of *Theory of the Earth*, 208

Hutton, R. H., 181

Huxley, Thomas Henry: associated with Weinberg and Wilson, 261, 269; on Darwin's intellectual character, 215–16, 220; exchange between Arnold and, xiv, 260–65; intellectual character of, 199–200, 216; materialist metaphysic of, 262, 268;

Huxley, Thomas Henry *(continued)*: on nature, science, and civilization, 261–62; quoted by Weinberg, 261, 263; on the reception of the *Origin*, 250–51; response to *Origin* by, 199, 254; was H. G. Wells' student, 57; was Darwin's confidant, 228; was hostile to Chambers, 220–21; author of: "On a Piece of Chalk," 263; "Science and Culture," 261

Imagination, the: of Charles Darwin, 200–201, 215–16, 245; co-evolved with intelligence, 24; in current intellectual life, xiv; Darwin's impact on, x, 56–57, 200, 201, 274; of deep time, 57; distinctively human character of, xi, 58; dualist conception of, 64–65, 79; Emily Brontë's, 109, 110, 119, 121, 122; of European civilization, 265; evolutionary context for, 6; evolution of, 26; expanding, 55, 200; extended by culture, 45; *Hamlet* as a touchstone for, 147; Hamlet's, 131, 141, 144, 146–47; as the highest level of phenomenal organization, 5; Homer's, 31; is adaptively functional, 20, 24, 25–26, 52–53, 174; is not restricted to ordinary reality, 64; is part of human nature, x, xiii, 20, 24, 25–26, 28, 35, 52, 57; is the medium for cultural norms, 17, 27, 80, 151, 153, 174; is universal, 52; in late Romanticism, 55–56, 72, 275–76; and life history theory, 30; literary Darwinism focuses on, 6, 30, 80; Lord Henry Wotton on, 97; Lyell's impact on, 213; in Marxism, 272; Marxist conception of, 19; in modern life, 56, 144, 146, 147; Nettle leaves out, 128; novels activate, 188; and Openness to Experience, 142, 143; organizes experience, 57, 64, 132; in *Origin of Species*, 197, 201, 234, 235; Oscar Wilde's, 95, 98; Pinker on, 21; in poststructuralism, ix, 20, 80, 274–75; religion and, 265; repression in, 48–49; science has influenced, 57; Thomas Hardy's, 192; tragic forms of, 180, 181–82; Trilling on, 266; vanities of, 274; varies in quality, 161; in various cultures, 147; in Victorian novels, 175; we live in, 28, 58, 270, 272–73. *See also* Arts, the; Culture; Literature; Symbolism

Inclusive fitness: among hunter-gatherers, 21; binds cooperation and conflict, 46–47, 137, 265; contrasted with meme theory, 46; in families, 16; is the ultimate regulative principle, 69; shaping force of, 14, 121, 265

Individual identity: adaptive significance of, 40; in authors, 10; Boyd on, 126; in characters, 10; continuity of, 97–98, 138; deprecated in evolutionary psychology, xi, 145; elements of, 155; as humanist idea, x; in literary Darwinism, xiii, 10, 81; Oscar Wilde's, 91–108 *passim*; Pater's concept of, 97–98; in personality, xiii, 10; as point of view, 64; poststructuralism problematizes, x, 8, 33, 92–93, 94, 95; in readers, 10; social nature of, 98; and Theory of Mind, 16–17; in three levels of analysis, 10, 32, 81. *See also* Point of View

Intelligence, xi, 7, 23–24, 40
Intelligent design. *See* Creationism
Irony, 48, 56, 182. *See also* Satire

Jackson, Tony, 33-34, 33
James, Henry, 215
James, William, 272
Jameson, Fredric, 18
Jancsó, Miklós: director of *The Peach Thief*, 58
Jenkin, Fleeming, 222, 249
Jobling, Ian, 32, 34–35
Johnson, John, xii, xiii, 12, 32, 35, 152
Johnson, Samuel, 143–44
Jones, Ernest, 124

Jument Verte, La, 59
Jungian psychology, 18

Keats, John, 55, 57, 182
Kelvin, William Thomson, Lord, 222, 245
Knapp, John, 123, 129–31
Kroeber, Alfred, 272
Kronenberg, David: director of *The Fly*, 58
Kruger, Daniel, xii, xiii, 12, 32, 34–35, 152
Kuhn, Thomas, xiv, 199, 250–53, 257. *See also under* Paradigm change; *under* Science

Lacan, Jacques, 18, 29, 31, 124, 277
Lactose tolerance, 25, 44
Lamarck, Jean-Baptiste: and Cuvier, 200–201, 210–11, 220; Darwin's relation to, 39, 198, 211, 221–23, 256; evolutionary ideas of, 209, 210, 211, 220–21; influenced Spencer, 223–24, 227; Lyell's response to, 211, 221, 223; teleology in, 221; was preoccupied with hunger, 241–42; author of *Philosophie Zoologique*, 208
La Nausée, 146
Language: evolutionary conception of, 21; evolved late, 93; exemplifies gene-culture coevolution, 25; function of, 17; and the Human Revolution, 41; is a precondition for culture, 44; is specifically human, 16–17, 23, 269; life precedes, 78; poststructuralist account of, 7, 78, 269
Lawrence, D. H., 267
Leavis, F. R.: author of "Two Cultures? The Significance of Lord Snow," xiv, 267–68. *See also* Two cultures, the
Leibnitz, Gottfried Wilhelm, 222
Levine, George: author of: *Darwin and the Novelists*, 78; *The Realistic Imagination*, 181

Levitt, Norman, 260; co-author of *Higher Superstition*, 259
Lewontin, Richard, 67
Life history theory: and core literary Genres, 112; Darwin uses, 203; elements of, 14, 110; human, xiii, 30, 32, 80, 83, 110–11, 157, 249; and individual point of view, 32; is a comprehensive explanatory model, 14, 40, 110–11, 157, 249; is a reproductive cycle, 14, 111; is implicit in *Wuthering Heights*, 110–11, 113, 122
Linnaeus (Car von Linné), 201, 202, 207
Literary Darwinism: and cognitive poetics, 6–8, 14; consilient perspective of, 5–6, 29–30, 154, 269–70, 276; critics of, xiii, 3, 81, 145; development of, 3, 5, 10, 71–87 *passim*, x, 277; does not mean, 82–83; and evolutionary social science, 36, 39, 59–60; examples of, 9–12, 30–33; explanatory reduction in, 20, 29–33, 126–28; focuses on the imagination, 6, 30; global ambitions of, 5, 37–39, 82; growing sophistication of, 9–10; historical position of, 38–39; idea of repression in, 30; institutional obstructions to, 4–5, 36, 74, 77, 84; vs. poststructuralism, xii, 33; public response to, 77; scope of criticism in, 27, 29–30, 69–70, 80, 81–82, 84, 85–86; three levels of analysis in, 10, 32, 81; vulgar form of, 80–81, 82. *See also under* Paradigm change
Literary theory. *See* Genres; Humanism; Literary Darwinism; Literature; Meaning in literature; *under* Paradigm change; Pluralism; Poststructuralism
Literature: adaptive functions of, 5, 20–29, 38, 40, 41, 42, 49–53, 82, 151, 152; affirms human nature, 18; articulates folk psychology, 13, 17; definition of, 10, 23;

Literature *(continued)*:
elements of, 10; empirical study of, xi, 11, 32–36, 53, 85, 86, 151–75, 177–93; existential function of, 72, 183, 267; generating new knowledge about, 29–33; major themes of, 66–67; psychosexual symbolism in, xiii, 91–108 *passim*. *See also* Arts, the; Audience; Authors; Characters; Drama; Fiction; Form; Genres; Imagination, the; Meaning in literature; Motifs; Narrative; Novels; Plot; Poetry; Point of View; Readers; Style; Symbolism; Themes; Tone

Locke, John: author of *Essay Concerning Human Understanding*, 248

Long Day's Journey into Night, 146

Longinus, 200

Lowie, Robert, 272

Lyell, Sir Charles: founded modern geology, 39, 199, 208–209, 213; influenced Darwin, 208, 212, 213–14; as a species theorist, 208, 209, 211, 223; was a synthesizer, 208–209, 214; was influenced by Hutton, 208–09, 212, 213, 214; author of *Principles of Geology*, 208, 209, 212, 214, 221, 257

Madame Bovary, 81

Malthus, Thomas: author of *Essay on Population*, 207, 228, 229, 241–45, 253

Marx, Carl, 276. *See also* Marxism

Marxism: in the casebooks, 77, 82; core ideas of, 19, 170, 273; in criticism of *Hamlet*, 123–24; historical teleology in, 273, 275; interpretation of evolutionary biology in, 244–45; is not a science, 273; is obsolete, 277; and the old Scholastics, 65; in poststructuralism, 13, 56, 78, 277; view of science in, 243, 244; vulgar form of, 80

Mate selection: in antagonists, 160–61, 187; correlates with agonistic role assignments, 165; factor analysis of, 160; gender differences in, 157, 160, 168–70, 171, 172; in *Mayor of Casterbridge*, 184, 187–88; in a model of human nature, 155, 160–61, 167; Nettle on, 31; in protagonists, 160–61, 168, 169, 170, 172, 187. *See also* Reproduction

Maugham, Somerset, 48

Maupassant, Guy de, 58

Maynard Smith, John, 244, 254

Mayr, Ernst, 197–98, 220, 250–55 *passim*

McEwan, Ian, 9

McFadden, Johnjoe, 9

Meaning in literature: authors control determinate, 156; casebook essays distort, 82; comprehensive framework for analysis of, 30, 81–82, 153–54; *Dorian Gray* as example of, 91–93, 95, 104–05; Dutton on, 64; empirical study of, 33; and explanatory reduction, 29; and gender in Victorian novels, 167–70; *Hamlet* as example of, 124, 125, 127, 128, 131, 132, 135, 138; has finite elements, 167; implications of quantifying, 154, 167; is not reducible to depicted action, 81; *Mayor of Casterbridge* as example of, 186, 192; objective understanding of, 154; poststructuralist view of, 75; and psychological function, 153–54, 156; purported indeterminacy of, 75, 154; and purpose in life, 72, 183, 267; reader responses are integral to, xiii, 81, 153–54, 156; and repression, 30, 47–48; three levels of analysis for, 10; total structure of, xiii, 110, 192; *Wuthering Heights* as example of, 110, 115–16. *See also* Authors; Genres; Literature; Point of View; Readers; Symbolism; Themes; Tone

Mellard, James, 5

Memes, 44–46

Menand, Louis, 74
Mendel, Gregor, 152, 249
Miall, David, 35
Mill, John Stuart, 221
Miller, Geoffrey, 21–22
Milton, John: author of *Paradise Lost*, 218, 272
Mithen, Steven, 26
Modern Language Association (MLA), 4–5, 74, 77, 260
Modern Synthesis, the, 39, 197, 199, 222, 250–57 *passim*, 271
Modules, 7, 21, 27, 40
Moore, James, 239, 244
Moore, John, 203–04
Morgan, Thomas Hunt, 152
Motifs, 34, 139. *See also* Symbolism
Motives: academic ideological, 78–79; in ancestral environments, 17; of antagonists, 158–59; are basic life goals, 157; are chief constituents of human nature, 83; are correlated with emotions, 40, 83; authors stipulate, 156; basic, 12, 27, 31, 157; Boyd on, 31; correlate with agonistic role assignments, 12, 165; Darwin on, 248; differ in authors and characters, 127; disciplinary, 78; and ego psychology, 13; factor analysis of, 157–59; fade in depression, 140; in folk psychology, 18, 83; gender differences in, 170–72; in *Hamlet*, 138, 144; and life history theory, 83, 122; Malthus', 244; in *Mayor of Casterbridge*, 177, 184–85, 187; in models of human nature, x, 13, 127–28, 151, 155, 157–59; and moral consciousness, 98; natural selection shapes, 14, 21; in natural theology, 222; Nettle on, 31, 127–28; Pater on, 91; people know little of their own, 153; of protagonists, 152, 157, 158–59, 165, 168, 169, 170, 172, 185; scientific, 222; in sociobiology, 248; in theory formation, 272; Theory of Mind evolved to read, 126; Wilde's artistic, 92; in *Wuthering Heights*, 12, 118, 119, 122. *See also* Human nature; Life history theory
Mullarkey, Maureen, 64–65
Myths: are part of culture, 17, 25; geological, 213; influence identity, 28; Jungian view of, 19; Kuhn's, 252; in *Wuthering Heights*, 112

Narrative: adaptive functions of, 21, 52, 66–67, 173–74; basic elements of, 125; cognitive adaptations for, 23, 26, 27, 52, 86; *Dorian Gray* is a slender, 94; Dutton on, 66–67; as elementary conceptual schema, 7; graphic, 9; history of, 34; meta-, 68; in *Origin of Species*, 204; theory of, 9, 64, 85–86, 152, 153; in *Voyage of the Beagle*, 228; in *Wuthering Heights*, 115, 116. *See also* Fiction; Novels
Naturalism: Darwin's, 272; Harold Fromm's, 11; Matthew Arnold's, 264; new Darwinian, 5; vs. post-structuralism, 10; quasi-Darwinian character of, 91; in *Wuthering Heights*, 110, 112, 114. *See also* Realism
Natural selection: and artificial selection, 232, 235; constraining force of, 83; contrasted with sexual selection, 66–67; contrasted with meme theory, 45–46; core concepts of, 203, 209, 214, 226, 231, 237, 242; culture is part of, 44–46; delay in establishing idea of, 249, 250–55, 271; and ecological diversification, 233; eliminates teleology, 221, 265; Emily Brontë predates discovery of, 112; evidence that could falsify, 245–46; explains taxonomy, 203–204; H. G. Wells understood, 57; historical context for theory of, 39, 214; is a central organizing idea, 248, 254–55; Malthus influenced idea of, 243; in the Modern Synthesis, 197, 249–50, 256–57; vs. natural theology, 202–203, 205, 255–56;

Natural selection *(continued)*:
necessarily involves conflict, 46–47; one long argument for, 65, 204, 254; operates through inclusive fitness, 14, 21; Spencer's notion of, 223, 226; stimulus for discovery of, 227–28, 243; and tree of life metaphor, 206, 236; Wallace's discovery of, 229, 240. *See also* Adaptations; Darwin, Charles; *under* Works of Charles Darwin; Evolutionary biology; Extinction; Inclusive fitness
Nature (journal), 77
Neill, Michael, 124
Nettle, Daniel, 9, 31, 123, 126–28, 131, 274
New Criticism, the, 266, 275
New Historicism, 6, 18, 77
New Scientist, 61
Newsweek, 61
Newton, Sir Isaac, 263
New Yorker, 61
New York Times, 61, 77
Nietzsche, Friedrich, 271, 272
Nordlund, Marcus, 11
Novels: Darwin enjoyed, 200; Dickens', 242; Edith Wharton's, 12, 32; empirical study of Victorian, xiii–xiv, 12, 32–33, 35–36, 151–93; finite body of, 80; Ian Watt on, 34; Sir Walter Scott's, 32; three levels of analysis in, 32; and the two cultures, 266. *See also* Brontë, Emily; Eliot, George; Fiction; Hardy, Thomas; Literature; Narrative; Wilde, Oscar

Oedipal theory. *See* Psychoanalysis
Oedipus Rex. *See* Sophocles
Olivier, Laurence, 128
Openness to Experience. *See* Personality

Paley, William, 205, 207, 222, 241, 242
Panksepp, Jaak, 24–25
Panksepp, Jules, 24–25

Paradigm change: in evolutionary biology, xiv, 39, 203–204, 249–57, 271–72; in geology, 39, 199, 208–209, 211–13; from humanism to poststructuralism, ix–x, 73–74, 259, 261, 268, 274–76; Kuhn on, xiv, 199, 250–57 *passim*; from poststructuralism to literary Darwinism, x–xi, xiv, 5, 18–20, 54, 71–87 *passim*, 269–70, 276–77; in the social sciences, 5, 71–72, 85, 247–48, 268, 272–74
Pater, Walter: aestheticism of, 32; concept of morality in, 107; homosexuality of, 48; influenced Wilde, 91, 96–98, 100–101, 107; author of: *Marius the Epicurean*, x, 146; *The Renaissance*, 97, 101
Paternal investment, 15
Paterson, John, 180
Personality: of antagonists, 161–62, 165; correlated with agonistic role assignments, 165; five-factor model of, 141–45, 161–63, 167; and gender differences, 170, 173; Henry James' intellectual, 215; is part of human nature, 142–43; is stipulated by authors, 156; in *Mayor of Casterbridge*, 177, 180, 183, 184–87; in a model of human nature, 151; modern conception of, 10, 59; of protagonists, 143–44, 161–63, 165, 168, 186. *See also* Individual identity
Phantom Tollbooth, The, 58
Phenomenology, 13, 18, 19
Philadelphia Inquirer, 61
Philosophy and Literature, 63
Pinker, Steven: on the arts, 9, 21, 22, 49–52, 62; is not a connoisseur of the arts, 63; Tooby and Cosmides influenced, 51; author of: *How the Mind Works*, 62; *The Language Instinct*, 62
Plato, 145, 264, 267
Plays. *See* Drama
Plot: agonistic structure shapes, 170; conforms to an author's emotional

needs, 48–49; defined in biological terms, 10; in *Dorian Gray*, 92, 99, 103, 104, 107; in *Hamlet*, 139; literary meaning consists not only in, 81; in *Mayor of Casterbridge*, 179, 190; in Wharton's novels, 12; in *Wuthering Heights*, 111, 114, 116, 121, 122

Pluralism, 79, 82, 84. *See also* Humanism; Poststructuralism

Poetry: Aristotle valorizes, 133; Arnold's hopes for, 72, 265, 275; cognitive adaptations for, 86; Darwin lost interest in, 200; finite body of, 80; has been neglected by Darwinists, 9; Leavis' ideal of, 267; sensual richness in, 192; of Wallace Stevens, 55, 272; Wordsworth's autobiographical, 31

Point of view: central importance of, 64; in cognitive science, 7, 86; C. P. Snow's, 267; Darwin's, 56, 198; Dissanayake's, 63; in *Dorian Gray*, 105, 106, 107; Eugene Goodheart's, 13; an evolutionary, 11, 12, 17, 20, 30, 32, 33, 38, 43, 47, 69, 72, 77, 85, 91, 110, 121, 123, 137–39, 154, 170; Hamlet's, 130, 135; in Hardy's novels, 193; of Huxley, Weinberg, and Wilson, 269; Huxley's, 261, 269; as a major analytic category, 84, 85; Marxist, 170; in *Mayor of Casterbridge*, 180, 183–84, 185–86, 187, 191, 193; Michael Austin on, 12; in *Middlemarch*, 56; naturalistic, 46; Nettle leaves out, 127; of a Nigerian tribe, 31; pluralist view of, 80; poststructuralist, 19, 29, 36, 47, 79, 269; in readers, authors, and characters, 12, 30, 32, 81, 188; scientific, 244; and tone, 12; of a tragic protagonist, 183; in *Wuthering Heights*, 115, 116, 117. *See also* Individual identity

Polanski, Roman: director of *Tess*, 58

Pornography, 35

Postmodernism. *See* Poststructuralism

Poststructuralism: beliefs shared between humanism and, 65, 68–69, 79–80, 167, 261, 268; chief doctrines of, 68, 72–74, 78; and cognitive poetics, 6–8; concept of language in, 7; in criticism of *Hamlet*, 123–24, 136; cultural critique in, 80; currently dominates the humanities, ix–x, 74–75; dualism in, 64–65; vs. the Enlightenment, ix; explanatory reduction in, 18–19, 20; vs. folk psychology, 19–20; fundamental axiom of, 75; hermeneutics of suspicion in, 47; vs. humanism, ix; ideological motives of, 78–79, 268–69; inner inanition of, 75, 84–85; in interpretation of *Wuthering Heights*, 109; is incompatible with Darwinism, 78, 79; is presupposed in "post-theory," 68; vs. literary Darwinism, xii, 6, 12, 33; vs. nature, 10, 124; negativity of, 268, 276; political stance of, 73–74; primacy of language in, 78, 268–69, 274–75; problematizes individual identity, 94, 268; rejects human nature, 13, 19, 32, 33, 269; relativism in, 29; rescuing literary theory from, xi, 57; revolutionary upsurge of, ix, 73, 259, 268; sophistry in, 59, 75–76, 274–75; theoretical components of, 5, 13, 18–19, 29, 78, 181; "Theory" as synonym for, ix, xi, 59, 65, 68, 75; utopianism of, 47, 57, 78–79; view of science in, 36, 181, 259–60, 261, 269. *See also* Cultural constructivism; Deconstruction; Derrida, Jacques; Foucault, Michel; Lacan, Jacques; *under* Paradigm change

Protagonists: Austen's male, 161; in definition of agonistic structure, 35, 152; in *Dorian Gray*, 92, 165–66; emotional responses to, 155, 163–64, 165, 166, 168–69, 191; ethos of, 170; gender differences in, 158, 168–69, 172; in *Hamlet*, 125, 127, 144; in *Jane Eyre*, 162;

Protagonists *(continued)*:
by male and female authors, 170–73; mate preferences in, 160–61, 168, 169, 170, 172, 187; in *Mayor of Casterbridge*, 180–93 *passim;* in *Middlemarch*, 59, 159; in models of tragedy, 180–83; motives of, 152, 157, 158–59, 165, 168, 169, 170, 172, 185; Nettle on, 131, 132; in novels by Charlotte and Emily Brontë, 162; passional and perspectival, 183–84; personality of, 143–44, 161–63, 165, 168, 186; in *Persuasion*, 164; in *Pride and Prejudice*, 156, 160–61; Tiv understand, 125; in *Wilhelm Meister's Apprenticeship*, 143; in *Wuthering Heights*, 109, 111, 116, 117. *See also* Agonistic structure; Antagonists
Proust, Marcel, 48
Psychoanalysis: in the casebooks, 77, 82; concept of repression in, 47; core ideas of, 19, 128–29, 138–39; in criticism of *Hamlet*, 123–24; Easterlin on, 31; emphasizes continuity of identity, 138; idea of symbolism in, xiii; is obsolete, 277; Oedipal theory in, 31, 128–30, 138. *See also* Psychodrama
Psychodrama, xiii, 32, 91–108 *passim*
Queer theory, 18–19, 32, 92–93. *See also* Gender

Ray, John, 207
Readers: affirm egalitarianism, 174; agonistic structure and, 151–93 *passim;* are part of literary meaning, xiii, 12, 70, 81–82; authors interact with, 18, 30, 153, 156, 174, 188; authors largely control responses of, 156; community of, 174; Darwin's, 197, 232, 234, 235, 238; of *Dorian Gray*, 92–105 *passim;* educated general, xii, 63, 197; emotional responses of, 12, 111, 118, 121–22, 141, 152–53, 155, 156, 163–64, 165, 166–67, 170, 173, 180, 183, 186–87, 188–89, 191, 193; of *Hamlet*, 123–47 *passim;* in the humanities, 167; individual identity in, 10, 64; and life history theory, 30, 32; of *Mayor of Casterbridge*, 177–93 *passim;* need not be conscious of effects, 153; pluralism in, 37–38; provided empirical data, xiii, 12, 32, 151–93 *passim;* of *Wuthering Heights*, 111, 115, 116, 118, 121, 296–97n1, 297–98n8, 297n7, 298n9. *See also* Audience
Realism: vs. agonistic structure, 174; Darwin's, 206; Frye on, 133; in *Middlemarch*, 56; vs. minimal ontological violations, 136; naive, 64; quasi-Darwinian character of, 91; in theory of science, 243; in Victorian novels, 175; Watt on, 34; in *Wuthering Heights*, 118. *See also* Naturalism
Reduction: in humanism, 18, 19; in life history theory, 40; in literary Darwinism, 20, 29–33, 126–28; Nettle's, 126–28; in poststructuralism, 18–19, 20; is the ultimate explanatory aim, 29
Religion: Adam Sedgwick's, 210; of culture, 264–65, 275; in *Dorian Gray*, 91–108 *passim;* and dualism, 64–65; evolutionary understanding of, 41, 83, 91; forms imaginative communities, 28; Hamlet on, 130; and the humanities, 275–76; Huxley on, 262; is part of culture, 17, 24, 25; and late Romantic spiritualism, 56; modern loss of faith in, 56, 264, 275; and poetry, x, 55, 72, 265–66, 275; in premodern societies, 63; subordinates truth to value, 273; substitutes for, 56, 72, 265–66, 267, 273, 275–76; and tragedy, 180; Victorians were troubled about, 145–46, 263–64, 275. *See also* Bible, the; Christianity; Creationism; Theodicy

Reproduction: emphasized in sociobiology, 7; and evolved sex differences, 15–16, 99, 160; in life history theory, 14, 111; and sexual display, 22; in *Wuthering Heights*, 120, 121. *See also* Mate selection; Sexual selection
Republic, The, 145
Richardson, Alan, 6–7, 14
Richerson, Peter, 45
Ridley, Mark, 252; author of *Evolution*, 198
Ridley, Matt, 62, 274
Rifles for Watie, 58
Rofé, Yacov, 47, 48
Romantic comedy: characteristic features of, 112, 187–88; latent in *Hamlet*, 137; in *Wuthering Heights*, 112–22 *passim*
Romanticism: identification with nature in, 118; in interpretation of *Hamlet*, 134; lyricism of, 192; and Promethean model of tragedy, 180, 181, 182; vs. science, 57; spiritualism of, 55, 56. *See also* Byron, George Gordon, Lord; Coleridge, Samuel Taylor; Keats, John; Shelley, Mary; Shelley, Percy Bysshe; Wordsworth, William
Ruse, Michael: author of *The Darwinian Revolution*, 239

Salammbô, 59
Salmon, Catherine, 35
Satire, 48, 97, 104, 144. *See also* Irony
Saunders, Judith, 12, 32
Scalise-Sugiyama, Michelle, 9, 12, 31, 123, 124–25
Schiller, Friedrich, 58
Science: cherry-picking, 42; creativity in, 216; Darwin's inclination toward, 217, 218; disciplinary linkages in, 5–6; discursive study of, 36, 68; dualist view of, 34, 79, 167; ethos of, 36–37, 217, 219, 222, 262; Failed efforts to integrate literature and, 38–39, 42; vs. fanciful speculation, 13, 39, 45, 53, 54, 59, 64, 72, 74, 166, 213–23 *passim;* generalizing from particulars in, 151–52; gets things done, 73; has altered imaginative experience, 57, 198, 261–70 *passim;* historians of, 238; Huxley on, 261–62; ideological approach to, 36, 68, 259–60, 269; Importance of individual researchers in, 234, 253–54; is a collective effort, 217, 234, 253–54, 270; is impersonal, 37, 53; is part of culture, 17; vs. journalism, 220; Kuhn on revolutions in, xiv, 199, 250–53, 257; Lydgate's ambition in, 59; Marxism is not a, 273; methods of, 11, 36–37, 50, 53, 86, 213, 222–23, 262; objectivist view of, 242–46; and positivism, 221; simplicity as criterion in, 200; Snow valorizes, 266–67; social constructionist view of, 242–46; standing of *Origin of Species* in, 197, 200, 214, 215, 234, 240, 271–72; vs. theology, 222; tone and manner in, 216; unity of, 5–6, 261–62, 263; Weinberg on, 263. *See also* Cognitive science; Consilience; Embryology; Evolutionary biology; Evolutionary psychology; Evolutionary social sciences; Genetics; Literature, empirical study of; *under* Paradigm change; Two cultures, the
Science (journal), 77
Science fiction, 57
Science wars, the, 259–70. *See also* Gross, Paul; Levitt, Norman; *under* Poststructuralism; Science; Sokal, Alan; Two cultures, the
Scott, Sir Walter, 35; author of *Ivanhoe*, 32
Sedgwick, Adam, 210
Seventh Seal, The, 146
Sexual selection, 21–22, 65–67, 66, 66–67. *See also* Dutton, Denis; Mate selection; Miller, Geoffrey; Reproduction

Shakespeare, William: Darwin's response to, 218; lyricism of, 192; methods in study of, 85, 123–24, 126–29, 131, 141–43; Nordlund on, 11–12; author of: *Hamlet*, xiii, 2, 4, 12, 123–47; *King Lear*, 59
Shelley, Mary: author of *Frankenstein*, 165
Shelley, Percy Bysshe, 182
Showalter, Elaine, 101, 181
Siècle de Louis Quatorze, Le, 59
Simpson, George Gaylord, 220
Smee, Sebastian, 67–68, 69
Smith, Barbara Herrnstein, 260
Smith, William, 210
Snow, C. P., xiv, 238, 259, 266–67. See also Two Cultures, the
Social constructivism. See Cultural constructivism
Social life: changing ideas about, 41–42; Darwin's views on, 217, 223–25; elements of, 13, 16, 152, 182; evolutionary conception of, 21; human nature is adapted to, 83, 110, 113, 273; idealist experiments in, 78–79, 273–74; importance of, 17; levels of selection in, 41–42; moral aspects of, 41–42, 66, 127–28, 224–25; necessarily involves conflict, 47, 265; scholarship as form of, x; Spencer's views on, 223–24; tribal instincts in, 40, 66, 173–74. See also Culture; Egalitarianism; Utopianism
Social sciences. See Evolutionary social sciences
Social Text, 259
Society. See Social life
Sociobiology, 7–8, 41, 59, 248
Socrates, 145
Sokal, Alan, 259, 260
Sophocles, 146, 147, 215; author of *Oedipus Rex*, 31, 145
Spandrel. See By-product
Spencer, Herbert: Darwin's relation to, 39, 198, 220, 223–27; is obsolete, 227, 271; was preoccupied with hunger, 241; author of: *Autobiography*, 223; *First Principles*, 225; *Social Statics*, 223–24; "The Development Hypothesis," 225
Spiegelman, Art, 31
Spolsky, Ellen, 6, 7, 78
Stevens, Wallace, 55, 56, 58, 272
Stiller, James, 35
Stoker, Bram: author of *Dracula*, 164
Storey, Robert, 10, 12, 31, 123, 124–25
Structuralism, 72–73
St. Simon, Henri de, 221
Style: cognitive structures for, 86; Darwin's, 235–38; in *Dorian Gray*, 92, 99–102; is a main literary category, 30, 70, 84, 85; sensitivity to, 36; vs. substance, 73. See also Form
Style (journal), 3, 5, 37
Sublime, the, 96, 183, 193, 197, 200
Supernatural fantasy: in *Dorian Gray*, 91–108 *passim*; in *Hamlet*, 126; in *Wuthering Heights*, 110, 112, 118, 121
Symbolism: of the Berlin Wall, 85; culture is a capacity for, 143; in *Dorian Gray*, 92, 93, 103, 107; EP explains away, 27; in *The Fly*, 58; in *Hamlet*, 131–39 *passim*, 145; in *Marius the Epicurean*, x; in *The Mirror and the Lamp*, 34; peculiarly human capacity for, xi, 24, 25, 26–27, 41, 143; poststructuralist conception of, 19; psychological, xiii, 19, 93, 137–39; in *Wuthering Heights*, 109, 110, 117, 118. See also Themes
Symons, Donald, 32, 35; author of *Evolution of Human Sexuality*, 91, 98–99

Tacitus, 59
Teleology, 221, 223, 225, 264–65, 273. See also under Arnold, Matthew; Lamarck, Jean-Baptiste; Marxism; Utopianism; Veblen, Thorstein
Tennyson, Alfred, Lord: author of *In Memoriam*, 146

Index

Thackeray, William Makepeace: author of *Vanity Fair*, 165
Theater. *See* Drama
Themes: and agonistic structure, 170; are a main literary category, 57, 70, 126; are what a work is about, 132; Darwinian, 78, 81; in *Dorian Gray*, 32, 91, 92, 94, 95–96, 100, 102; Dutton lists the major literary, 66–67; in the essay on Darwin's *Origin*, 199; in the final essay, xiv; in *Hamlet*, 124, 130, 132, 142; in humanist criticism, 19; in literary scholarship, 69; in *Mayor of Casterbridge*, 192; Nettle on, 127; personality factors as, 189; in poststructuralist criticism, 19; are reduced by Nettle to two, 126–28; in utopian/dystopian fiction, 32; in Wilde's fairy tales, 95–96; in *Wuthering Heights*, 109, 110, 114, 117. *See also* Meaning in Literature; Symbolism
Theodicy, 222
"Theory". *See* Poststructuralism
Theory of Mind, 7, 8, 16–17, 40, 86, 126
Time, 73
Tithonus, 76
Tiv, the, 31, 125, 126, 128, 147
TLS, 61, 77
Tom Jones, 59, 81
Tone: and authorial stance on life history, 30; and basic emotions, 12; in Darwin's writing, 206, 216; in *Dorian Gray*, 92, 107; as emotional continuum in art, 64; of exchange between Arnold and Huxley, 265; in *Hamlet*, 124, 142; humanistic sensitivity to, 36, 192; is a major analytic category, 70, 84, 85, 92, 126; in *Mayor of Casterbridge*, xiii, 177, 180, 182, 184, 186, 192; neuroscientific access to, 85–86; and point of view, 12; quantitative analysis of, 33, 177, 178, 188, 192; is complex in *Wuthering Heights*, 12. *See also* Emotions; Genres; Point of view

Tooby, John: on the arts, 21; on *Hamlet*, 123, 131–32, 134, 145; co-author of "The Psychological Foundations of Culture," 51
Toulmin, Stephen, 212
Tragedy: Bradley's book on, 131; characteristic features of, 112; *Hamlet* as a, 12, 130, 135, 144, 147; Hardy's concept of, 182; in Homer, 31; Lord Henry Wotton on, 102; models of, 180–83; Nettle on, 31, 126–27, 131; Sybil Vane's death as a, 106
Tragicomedy, 144
Travel writing, 207
Trilling, Lionel, 266
Trollope, Anthony, 59
Turner, Frederick, 9
Twain, Mark, 275
Two cultures, the: different methods of, 4, 35, 36–37; history of debate over, 259–70; linked by evolutionary biology, 5–6; and metaphysical dualism, 69; obstacles to integrating, 53–54; *Origin of Species* falls between, 238; poststructuralist view of, 73; recent works bridging, 10; removing barriers between, 84–85. *See also* Humanities, the; Leavis, F. R.; Science; Science wars, the; Snow, C. P.

Universals. *See* Human universals
Utopianism: Brett Cooke on, 11, 32; in *Descent of Man*, 225; is incompatible with Darwinism, 79; is not active in the resolution of *Wuthering Heights*, 122; Marxist, 221; in poststructuralism, 47, 57, 79; Spencer's, 223; utilitarian, 221; Victorian, 72, 221, 225

Van Ghent, Dorothy, 119
Veblen, Thorstein, 272, 273

Wade, Nicholas, 62, 274
Waiting for Godot, 146
Wallace, Alfred Russel: also formulated the theory of natural selection, 229–30, 234, 240, 243, 253; on survival of the fittest, 226; was influenced by Malthus, 241–42, 243, 245
Washington Post, 61
Waste Land, The, 146
Watt, Ian, 34
Wegener, Alfred, 209
Weinberg, Steven: 261; author of: *Dreams of a Final Theory,* 263; *Facing Up,* 263
Weismann, August, 222
Wellington, Arthur Wellesley, Duke of, 218
Wells, H. G.: was Huxley's student, 57; author of: *Outline of History,* 56–57; *The Island of Dr. Moreau,* 57; *The Time Machine,* 57
Werner, Abraham, 210, 212, 213, 257
Wharton, Edith, 32
White Panther, The, 58
Wilde, Oscar: divided identity of, 91–108 *passim;* homosexuality of, xiii, 48, 91–108 *passim;* author of: fairy tales, 95–96, 106; *Letters,* 105; *Picture of Dorian Gray,* xiii, 32, 91–108, 165
Williams, George C., 51
Wilson, David Sloan, 274; co-editor of *Literary Animal,* 9, 62
Wilson, Edward O.: on the arts, 23–24, 62; cites Huxley, 261; influenced literary Darwinism, 5–6, 276; is not a connoisseur of the arts, 63; set off the sociobiological revolution, 274; author of: *Consilience,* 5, 23, 62, 74, 260, 276; foreword to *Literary Animal,* 9; *On Human Nature,* 62; *Sociobiology,* 61–62, 260, 274
Wilson, J. Dover, 126
Wilson, Timothy, 48
Wilson Quarterly, 61
Winkelman, Michael, 9
Woolf, Virginia, 85
Wordsworth, William, 31, 182, 266
Wright, Sewall, 249
Wrinkle in Time, A, 58

Zahavi, Amotz, 9
Zamyatin, Yevgeny, 11
Zauberberg, Der, 59
Zunshine, Lisa, 8

www.ingramcontent.com/pod-product-compliance
Ingram Content Group UK Ltd.
Pitfield, Milton Keynes, MK11 3LW, UK
UKHW041915140426
5217IPUK00013B/163